# Optical Forces on Atoms

**Online at: https://doi.org/10.1088/978-0-7503-2308-6**

# Optical Forces on Atoms

**Farhan Saif**
*Department of Electronics, Quaid-i-Azam University, Islamabad, Pakistan*
*and*
*Physics department, The University of Jordan, Amman, Jordan*

**Shinichi Watanabe**
*Director of The International Education Center,*
*University of Electro-Communications, Tokyo, Japan*

**IOP** Publishing, Bristol, UK

ISBN 978-0-7503-2308-6 (ebook)
ISBN 978-0-7503-2306-2 (print)
ISBN 978-0-7503-2309-3 (myPrint)
ISBN 978-0-7503-2307-9 (mobi)

DOI 10.1088/978-0-7503-2308-6

Version: 20231201

IOP ebooks

British Library Cataloguing-in-Publication Data: A catalogue record for this book is available from the British Library.

Published by IOP Publishing, wholly owned by The Institute of Physics, London

IOP Publishing, No.2 The Distillery, Glassfields, Avon Street, Bristol, BS2 0GR, UK

US Office: IOP Publishing, Inc., 190 North Independence Mall West, Suite 601, Philadelphia, PA 19106, USA

# Contents

# Prologue

The study of light and the atom is by now itself a relatively old subject in modern physics. At around the turn of the 20th century, it helped the discovery of quantum mechanics considerably by offering a clear-cut example of quantized energy levels, namely Fraunhoffer's dark lines in the Sun's optical spectra. Bohr's celebrated theory of atomic structure proved to be a key concept that connects quanta of light, namely photons, with the quantized atomic energy levels. Half a century later, photon sources based on stimulated emission from atoms came into being, i.e. the laser; incidentally, its predecessor, the 'maser', should not be forgotten. In the process of seeking ever higher precision in frequency measurements, techniques to access atomic levels of interest via intermediate states found their way into new and innovative methods to control the center-of-mass motion of the atoms. Thus, the study of photons and atoms culminated in the production of ultra-cold atoms. This book deals with certain aspects of coherent photons and coherent ultra-cold atoms.

Atoms at room temperature move at ultra-sonic velocities, thus their de Broglie wave length is short. They also manifest decoherence due to numerous mutual collisions. In the last three decades, the control of coherence in atomic dynamics was made possible by cooling them to ultra-low temperature scales using laser beams, namely laser cooling, and to even lower temperatures of nanokelvin order by Sisyphus cooling and evaporative cooling. These ultra-cold atoms have de Broglie wavelengths comparable to the optical field. In this book, we use the term 'ultra-cold atoms' for the atoms at these ultra-low temperatures. Ultra-cold atoms thus bequeathed the world of innovation a toolbox with which to explore the laws of quantum mechanics, and opened a gateway to fascinating applications.

Coherent control of ultra-cold atoms is performed by the mechanical action of optical fields. These optical fields manipulate the matter waves by deflecting, focusing, cooling, and trapping them. As per demand, the shape of the optical potential, governed by its depth, spacing between potential minima, and moreover by its interactions with the atoms, can be adjusted at will by controlling physical experimental parameters. For example, a periodic intensity pattern of light is formed when two laser beams interfere. The varying light intensity creates a periodic potential landscape—an optical lattice—for the atoms via the AC-Stark effect. The optical lattices are defect-free for the atoms. Hence, we can simulate electron dynamics in ideal solid-state materials using ultra-cold atoms in an optical lattice. Ultra-cold atoms therefore can serve as a testing ground for modern condensed matter theories with unique manipulation and control possibilities. The complete control over the periodic potential and its interaction with ultra-cold atoms lead to the use of artificial lattices of light to investigate the wondrous quantum characteristics of many-body bosonic systems, Fermi degeneracy in ultra-cold atoms, quantum simulators, quantum computers, and quantum matter manipulated by optical fields.

Ultra-cold atoms in evanescent optical fields provide atomic mirrors, which are basic elements of atomic devices. An evanescent field is able to create a strong optical dipole potential for the ultra-cold atoms which interact with it. It can be created, for example, by the total internal reflection of laser radiation on a dielectric surface. This makes it convenient to build atomic mirrors and surface traps using dielectric slabs and nano-fibers, which serve as tools for trapping, guiding, and manipulating atoms. Ultra-cold atoms are charge neutral. Their interaction with optical fields makes it possible to generate artificial gauge fields under specific conditions such that the motion of neutral particles simulates the dynamics of charged particles in an effective magnetic field.

Controlling ultra-cold atomic systems with time-wise periodic external fields has attracted much research attention, particularly since the experimental realization of atomic Bose–Einstein condensation. An enormous volume of ingenuous applications have been made possible in recent years. Realization of tunable gauge fields can be regarded as a striking example of wave packet manipulation. Amplitude- and phase-modulated optical fields serve as tools to generate quantum chaos. Coherent control of ultra-cold atoms is at the heart of many theoretical and experimental interests. Time-periodic modulations have given birth both to hybrid nano-opto-mechanical systems and driven quantum billiards. Hence, the theoretical and experimental advancements have led to the exploration of coherent transport, controlling ratchet effects, coherent destruction of tunneling, entanglement, and precision measurement of gravitational acceleration. The presence of modulation can simulate impurities in solid-state materials, thus demonstrating newer effects such as quantum recurrences, tunneling, and dynamical localization, which is an analog of the Anderson localization in solid-state materials.

Dr Farhan Saif
Dr Shinichi Watanabe

# Acknowledgments

Many people have contributed directly or indirectly to the development of the manuscript. We thank them for their deep interest in the book and consistent support. Over the past years we have benefited from the views and suggestions of Gernot Alber, Abderrahim Al Allati, Iwo Bialynicki-Birula, Reinhold Blümel, John Bohn, Andreas Buchleitner, Guilio Casati, Hilda Cerdeira, Kwek Leong Chuan, Charles W Clark, Alex Cronin, John B Delos, Michael Fleischhauer, Mauro Fortunato, Jean-Pierre Gazeau, Chris Greene, Martin Holthaus, Shocket Hameed, Pervez Hoodbhoy, Zafar Iqbal, Markku Jaaskelainen, Hui Jing, Hans Jürgen Korsch, Edson Denis Leonel, Margarita Manko, Irene Marzoli, Michio Matsuzawa, Pierre Meystre, Yoko Miyamoto, Gerard Milburn, Toru Morishita, Kali P Nayak, Asghar Qadir, Hiroki Saito, Barry C Sanders, Wolfgang P Schleich, Katepalli R Sreenivasan, Dieter Suter, Tomotake Yamakoshi, and M Suhail Zubairy. We thank John B Delos and Asghar Qadir in particular for carefully reading the manuscript. In addition, we thank our editors for their patience and encouragement that led us to complete the work.

One of us, FS is greatly indebted to Hiroki Saito for hosting him at the Department of Engineering Sciences, University of Electro-Communications as a Japan Society for the Promotion of Science Fellow while we were completing this book. We thank the UEC, JSPS, Quaid-i-Azam university and the Higher Education Commission, Pakistan for providing financial support. We are grateful to our family members, Rabia, Manaal, AbdurRafey, Eshaal, and Shuko and Kai for their support and understanding when this book was contemplated, planned, written, and published.

# Author biographies

## Dr Farhan Saif

**Dr Farhan Saif** has served as the chairman, department of Electronics, Quaid-i-Azam University (QAU) Islamabad and as the Founding Head of Department of Physics, National University of Science and Technology. He obtained teaching and research experience at the University of Ulm, University of Arizona, University of Sao Paulo (UNESP), and University of Electro Communication, Tokyo. Dr Saif obtained his PhD degree at the University of Ulm, Germany in 1998. He has been an Associate Member of the Abdus Salam International Center for Theoretical Physics, Trieste, Italy. He served as a professor of department of Physics QAU from 2016 till 2018, and is presently serving as a tenured professor of Electronics, QAU, Islamabad and honorary professor at Physics Department, The University of Jordan, Amman.

Dr Saif's research areas include ultracold atoms and Bose–Einstein condensates in optical fields, quantum computation, quantum information, nano optomechanics, nano-devices, quantum optics and dynamical systems. He has also contributed in health sciences. In recognition of his many pioneering contributions as a researcher and educator he has received numerous national and international awards. Dr Saif is the principal researcher at Quantum Computing and nano-Devices Laboratory, QAU. He has supervised 12 PhDs and 55 MPhil/MS theses.

Dr Saif has collaborated with theorists and experimentalists alike in research institutions including University College London, Hunan Normal University, Turin Polytechnic university, College of William and Mary, Virginia, Abdelmalek Essaadi University Morocco, Tokyo University and the University of Electro-Communications.

## Dr Shinichi Watanabe

**Dr Shinichi Watanabe** worked as a Tenured Professor at the Department of Engineering Science, University of Electro-Communications (UEC Tokyo), Tokyo, Japan. Previously, he worked as a researcher of the CNRS at the Observatory of Paris at Meudon, France from 1981 till 1987. Dr Watanabe obtained his PhD degree in 1982 at the University of Chicago, USA. Earlier, he completed his BSc in 1977 at Brown University, USA. He served as the Chairman, Department of Applied Physics and Chemistry at UEC Tokyo, and currently he is serving as Director of the International Education Center at the same university.

Dr Watanabe's research interests primarily concerns the dynamics of atoms and their interaction with light. He worked on the double ionization threshold law of the two-electron atom/ion, high-resolution continuum spectra of diamagnetic hydrogen-like atoms, hyperspherical adiabatic theory, Bose–Einstein condensation, ultra-cold molecular formation, and the dynamics of ultra-cold atoms in optical lattices. He led a group of scientists working on Atomic, Molecular, and Optical (AMO) Physics, at the Department of Engineering Science at UEC Tokyo, and has supervised five PhD and 33 MSc theses.

Dr Watanabe has collaborated worldwide with researchers at Kansas State University, USA; the Observatory of Paris at Meudon, France; the University of Tennessee, Knoxville, USA; JILA at the University of Colorado at Boulder, USA; and Quaid-i-Azam University, Pakistan, among others.

# Chapter 1

# Introduction and overview

*'The great revelation of the quantum theory was that features of discreteness were discovered in the Book of Nature, in a context in which anything other than continuity seemed to be absurd according to the views held until then.'*

—Erwin Schrödinger

## 1.1 Historical background

The discussion of light has gone on for eons. It may rightly be traced back to the words of the good Lord as He said, 'Let there be light', as they appear in the translation of Genesis in the Old Testament. Critical analysis of light started with the work of Danish astronomer Ole Romer. In 1676, Romer estimated the time taken by light to reach Earth as he was studying the eclipse of a Jupiter moon. A seminal 19th-century work by Thomas Young and Augustin Fresnel in optics accumulated the existing knowledge of light and explained its corresponding properties. The groundbreaking achievements of Faraday and Maxwell developed an equivalence of electromagnetics, and thus paved the way to our theoretical understanding of light as electromagnetic waves. In a series of experiments, Heinrich Hertz showed the actual existence of electromagnetic waves. He proved that transverse free-space electromagnetic waves can travel over some distance, as predicted by Faraday and Maxwell. In general, the theoretical and experimental advances in electromagnetism made by Volta, Ohm, Ampere, Hertz, Weber, Helmholtz, and Clifford, as well as numerous other excellent scientists and engineers had led, by the end of the 19th century, to a communications revolution in the form of radio, television, mobile phone, laser, optical fiber, internet, wifi, smart phones, etc.

At the dawn of the 20th century, Max Planck solved the theretofore unsolved puzzle of black-body radiation. In 1886, Wien had been the first person to provide a

doi:10.1088/978-0-7503-2308-6ch1  

theoretical underpinning to black-body radiation. His work, which led to the Maxwell–Boltzmann distribution, was based on the frequency of emitted radiation corresponding to the thermal energy of emitting particles. Shortly after Wien, Lord Rayleigh presented his law of radiation, which showed an increase in energy density with increasing frequency, that is

$$I(\nu) = \frac{8\pi\nu^2}{c^3}\frac{\int_0^\infty E\exp(-E/k_BT)dE}{\int_0^\infty \exp(-E/k_BT)dE} = \frac{8\pi\nu^2}{c^3}k_BT, \tag{1.1}$$

where $\nu$ is the frequency, $T$ is the temperature, and $k_B$ is the Boltzmann constant. Rayleigh's distribution law worked well at low frequencies. However, it led to so-called 'ultraviolet catastrophe' at high frequencies, predicting unlimited growth of energy density. Max Planck's distribution of radiation at low frequencies was in agreement with Rayleigh's distribution, whereas it approached a Maxwell–Boltzmann distribution of the emitters at high frequencies. His work showed that electromagnetic radiation can be absorbed or emitted by the oscillators situated in the walls (atoms) of the resonator only in discrete amounts. The smallest energy of the resonator with frequency $\nu$ is suggested to be $h\nu$, where $h$ is the Planck constant. The energy density of emitted radiation has the form

$$I(\nu) = \frac{8\pi\nu^2}{c^3}\frac{\sum_{n=0}^{\infty} nh\nu\exp(-nh\nu/k_BT)}{\sum_{n=0}^{\infty}\exp(-nh\nu/k_BT)} = \frac{8\pi\nu^2}{c^3}\frac{h\nu}{e^{\frac{h\nu}{k_BT}} - 1}. \tag{1.2}$$

Planck's hypothesis is non-trivial as it led to the discrete nature or quantization of radiation. It was soon understood that it is the energy of the field that appears in the discrete amount, thus setting the stage for the quantum theory of radiation. Planck's hypothesis indicates that a single-mode field of frequency $\nu$ has an energy that is an integer multiple of the quantum of energy $h\nu$; however, Planck himself never agreed to this. Planck's expression reduces to Wien's law for high frequencies as

$$I(\nu) = \frac{8\pi\nu^2}{c^3}\frac{h\nu}{e^{\frac{h\nu}{k_BT}}}, \tag{1.3}$$

as shown in figure 1.1.

In 1900, Max Planck reported his seminal work at a meeting of the Deutschen Physikalischen Gesellschaft (German Physical Society). In later years, the work was greatly supported by Albert Einstein as he explained the photoelectric effect based on Planck's hypothesis as

$$E = h\nu - \Phi, \tag{1.4}$$

a phenomenon already reported by Heinrich Hertz in 1887. Here, $E$ is the energy of the emitted electron from metal due to the photoelectric effect and $\Phi$ is the work function. In 1916 Robert Millikan, based on the photoelectric emission of electrons, determined the value of Planck's constant. Experimental advances made in a series of sophisticated experiments by Millikan led to the calculation of the Planck

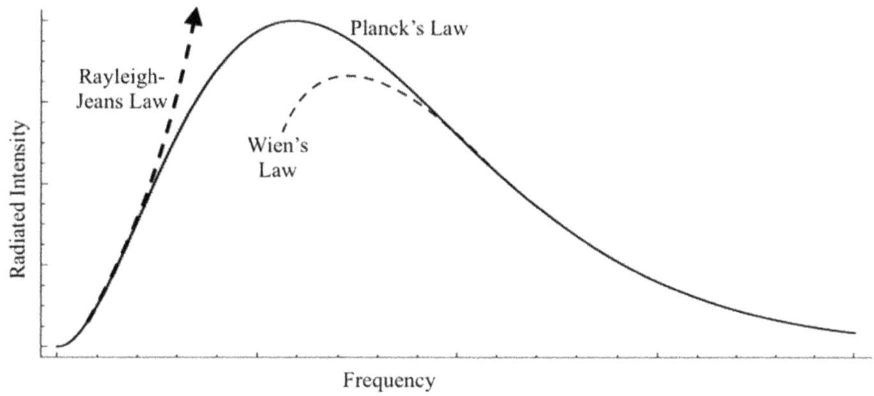


**Figure 1.1.** The range of frequencies obeying Wien's law, the Rayleigh–Jeans law, and Planck's law. The arrow indicates ultraviolet catastrophe for higher frequencies.

constant with high accuracy and a value better adduced than any known previously. In 1918, Max Planck received the Nobel Prize, 'In recognition of the services he rendered to the advancement of Physics by his discovery of energy quanta'. [1, 2] In 1921, Albert Einstein received his Nobel Prize, 'For his services to Theoretical Physics, and especially for his discovery of the law of the photoelectric effect' [3].

Einstein correctly recognized that light has a particle nature, carrying an indivisible form of energy $h\nu$. He called this the 'light quanta'. The light quanta was later named the 'photon', a name coined by chemist Gilbert Lewis in 1926[1] [4]. However, in addition to quantization, the other non-trivial discovery of Planck was a change in the statistics of photons. A complete transition from a classical distribution, as it was in Maxwell–Boltzmann statistics, came with the work of Satyendra Nath Bose [5]. His work was complemented by Einstein [6] and introduced the famous Bose–Einstein distribution based on the integer spin of the particles. A photon with zero spin is a popular example of a boson.

Subsequent years saw huge activity in explaining the nature of light and matter, and their interaction. The so-called Copenhagen group, led by Niels Bohr and Werner Heisenberg, in addition to Louis de Broglie, Erwin Schrödinger and Paul Dirac, developed the tools necessary to understand these foundational concepts. The mechanics that evolved from these efforts was first called quantum mechanics by Max Born[2] [7]. Arthur Compton proved the particle nature of light [8] in 1923 (see figure 1.2). De Broglie presented his hypothesis of symmetry in the nature of light

[1] 'It would seem inappropriate to speak of one of these hypothetical entities as a particle of light, a corpuscle of light, a light quantum, or light quanta, if we are to assume that it spends only a minute fraction of its existence as a carrier of radiation energy, while the rest of the time it remains as an important structural element within the atom … I therefore take the liberty of proposing for this hypothetical new atom, which is not light but plays an essential part in every process of radiation, the name photon.'—Gilbert Lewis.

[2] 'We became more and more convinced that a radical change of the foundations of physics was necessary, i.e., a new kind of mechanics for which we used the term quantum mechanics. This word appears for the first time in physical literature in a paper of mine.'—Max Born.

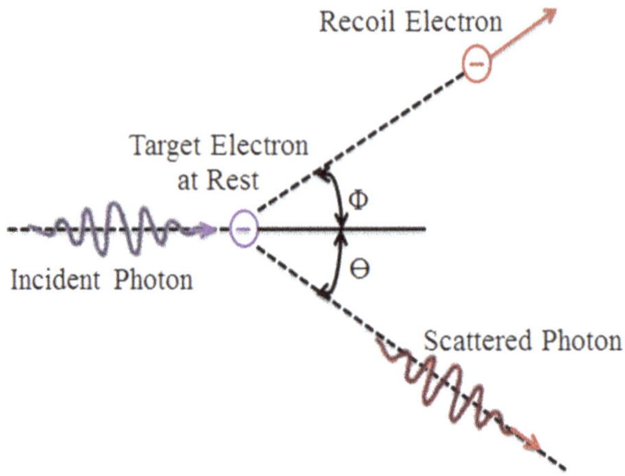


**Figure 1.2.** Compton scattering, i.e. the scattering of a photon following collision with a particle, here an electron. The formula gives the final photon wavelength as a function of its initial wavelength and the scattering angle theta.

and matter, thus postulating the wave nature of particles. He presented his famous de Broglie wavelength relation in his PhD thesis, published in 1924 [9]. The wave nature of the particle was verified in the work of George Paget Thomson [10] as he passed a beam of electrons through a thin metal film and observed the predicted interference patterns, and in a separate experiment by Clinton Joseph Davisson and Lester Halbert Germer as they guided the electron beam through a crystalline grid [11].

In 1916, Einstein derived Planck's distribution formula by discussing the rate at which atoms radiate. The work introduced Einstein's famous $A$ and $B$ coefficients, respectively, corresponding to stimulated emission and spontaneous emission of radiation due to electronic transition within the internal states of atoms [12]. The coherent control of these internal quantum states of atoms was a major achievement. In 1938, Rabi showed the change of internal states using rf resonance [13]. Further, in 1949, Ramsey developed his famous interferometry techniques to engineer long-lived coherent quantum superposition of the internal states of atoms.

Soon after the discovery of optical pumping by Alfred Kastler in the early 1950s, the work of Einstein eventually helped to develop micro-masers and lasers in the 1950s and 60's. Kastler was awarded the Nobel Prize for Physics in 1966 for the discovery and development of optical methods for studying resonances in atoms (figure 1.3). Rapid advances in the conceptual and technological understanding of the atom–field interaction led to the demonstration of the first working lasers in 1960 [14–16], ushering in another great technological and industrial boom.

The first formulation of a quantum theory describing the interaction of radiation and matter is attributed to Paul Dirac, who computed the coefficient of spontaneous emission of an atom. He described the quantization of the electromagnetic fields, understood as an ensemble of harmonic oscillators, by introducing creation and annihilation operators. In the next two decades contributions from Wolfgang Pauli,

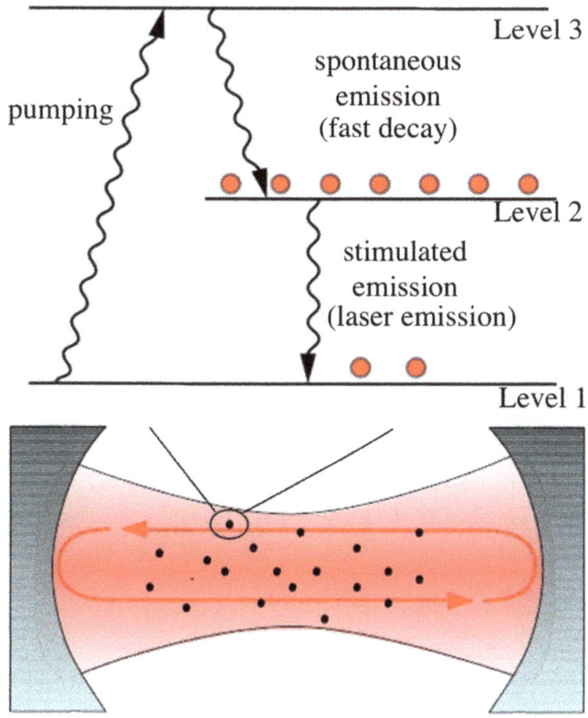


**Figure 1.3.** Optical pumping using three-level atoms for population inversion. Level 2 is the metastable state of a three-level atom. The stimulated emission phenomenon from level 2 to level 1 exhibits the amplification of light.

Eugene Wigner, Pascual Jordan, Erwin Schrödinger, Werner Heisenberg, and Enrico Fermi, and a list of other physicists led to the formulation of quantum electrodynamics. In 1977, Kimble *et al* demonstrated a single atom emitting one photon at a time, providing further compelling evidence that light consists of photons.

Experimental advances in the study of the matter–field interaction drastically improved quantum engineering, that is, the preparation, manipulation, control, and detection of quantum systems. The early experimental measurement of optical force on metallic plates performed by Pyotr Nikolayevich Lebedev [17, 18] and independently by Ernest Fox Nichols and Gordon Ferrie Hull [19, 20] was extended, in 1970, by Arthur Ashkin as he developed control of atoms using optical forcing. This work is famously known as 'atomic tweezers' [21, 22]. In 1973, Saul Altshuler and L M Frantz presented their idea of directly observing the de Broglie wave interference by propagating neutral atoms using optical fields, thus directly testing the laws of quantum mechanics by atoms [23]. Alexander Kazantsev showed the coherent acceleration of atoms due to optical potential in 1974 [24, 25]. In addition, in 1975 the possibility of cooling free atoms and ions captured in electromagnetic traps by weak resonance fields was explained [26, 27]; in 1978, Ashkin described how one might slow down an atomic beam of sodium using the radiation pressure of a laser beam tuned to an atomic resonance [28]. The coherent interaction of a laser–matter introduced new subjects such as atom optics, quantum optics, Bose–Einstein condensation, quantum metrology, quantum communication and computation, and ultra-high-precision spectroscopy.

### Optical lattices

Similar to the potential seen by an electron in a metallic crystal, the standing light field exhibits a periodic potential to cold atoms. Thus, the periodic structures of light, also called optical lattices, can act as refractive, reflective, and absorptive structures for matter waves. The first generation of optical lattices was introduced in the 1990s, as the lower limit on the recoil energy scale of atoms was achieved. In an optical lattice the hopping amplitude is much smaller than the recoil energy, typically in the kHz regime. Thus, an optical lattice becomes a periodic array of micro-traps for cold atoms, which are a lattice spacing ($\lambda/2$) apart. A one-dimensional optical lattice is obtained by the interference of two counterpropagating identical fields, such that

$$V(x) = \frac{V_0}{2} \cos 2k_{\mathrm{L}}x, \tag{1.5}$$

where $V_0$ is directly proportional to the intensity of the optical field, with wave number $k_{\mathrm{L}} = 2\pi/\lambda$.

In the first experiments on atoms, the reflection of electrons from standing light waves as reported by Pyotr Kapitza and Paul Dirac [29], was extended to the reflection of atoms from standing light waves by Saul Altshuler, L M Frantz, and Rubin Braunstein [30]. In optical lattices the phenomenon of Bloch oscillations of atoms within the lattice was reported by Dahan *et al* in [31], who studied the motion of thermal atoms in an accelerated lattice. Since optical lattices can be controlled precisely, Bloch oscillations have been used for precision measurements of quantities related to acceleration such as $g$ [32].

When the lattice potential height, $V_0$, becomes higher than the recoil energy, $\hbar\omega_{\mathrm{rec}}$, the atoms can be localized in the standing light wave. Atoms impinging on such a strong optical lattice are then guided in the troughs that can interfere afterwards [33]. Such guiding is called channeling. For a weak optical potential, $V_0$ is smaller than the recoil energy when we observe the Bragg scattering.

Presently, the ability to make extremely sensitive measurements on individual atoms at lattice sites can be used in quantum metrology, few- and many-body physics, quantum magnetism, and unconventional superconductivity. Optical lattices as arrays of micro-traps of atoms are useful in quantum computation, cooling, and trapping of atoms. To date, single trapped ions [34–36] and ultra-cold neutral atoms in free fall [37, 38] have shown record high performance that is approaching that of the best cesium fountain clocks [39]. Cold atoms trapped in an optical lattice constitute a high-precision quantum clock. Primary cesium fountain standards currently reach an uncertainty of around $2 \times 10^{-16}$, and the improved accuracy and stability of optical clocks motivates a future redefinition of the SI second [40].

### Radiation pressure force

The radiation pressure force is a result of the interaction of laser light with atoms, allowing it to modify the motion of the atoms. This optical force on atoms leads to the atomic tweezer effect, coherent acceleration of atoms, and subsequently the development of tools to control atomic dynamics. It is also the basis for laser cooling and trapping, a technique which enables the production of sufficiently dense cold atomic samples.

In an electromagnetic field a photon is the basic building block. A photon of angular frequency $\omega$ and wavelength $\lambda$ has an energy $\hbar\omega$ and momentum $\hbar k_{\mathrm{L}}$. Here, $\hbar$ is the reduced Planck's constant and $k_{\mathrm{L}}$ is the wave number associated with the wavelength as $k_{\mathrm{L}} = 2\pi/\lambda$. (For simplicity we write $\hbar$ as Plank's constant and $\omega$ as frequency.) If the photon impinges on an atom of mass $m$, it imparts a momentum $2\hbar k_{\mathrm{L}}$ that modifies the kinetic energy of the atom. The force experienced by a two-level atom due to absorption of light followed by spontaneous emission is

$$F_{\mathrm{sp}} = \hbar k_{\mathrm{L}} \Gamma \rho_{\mathrm{ee}}. \tag{1.6}$$

Here, $\rho_{\mathrm{ee}}$ is the probability of the atom being in the excited state and $\Gamma$ is the rate of decay from the excited state.

In the past decade, there has been renewed interest in radiation pressure due to the development of atomic tweezers, Bose–Einstein condensates (BECs) in outer-space, nano-opto-mechanical devices, and the practical development of space propulsion techniques using solar- or laser-based sails (see figure 1.4). Radiation pressure has also been considered for use as a calibration method for the cantilever spring constant in atomic force microscopy and high-power measurements. Endeavors have also been made to explore the effect of radiation pressure in novel systems involving plasmonic materials and negative-index metamaterials.

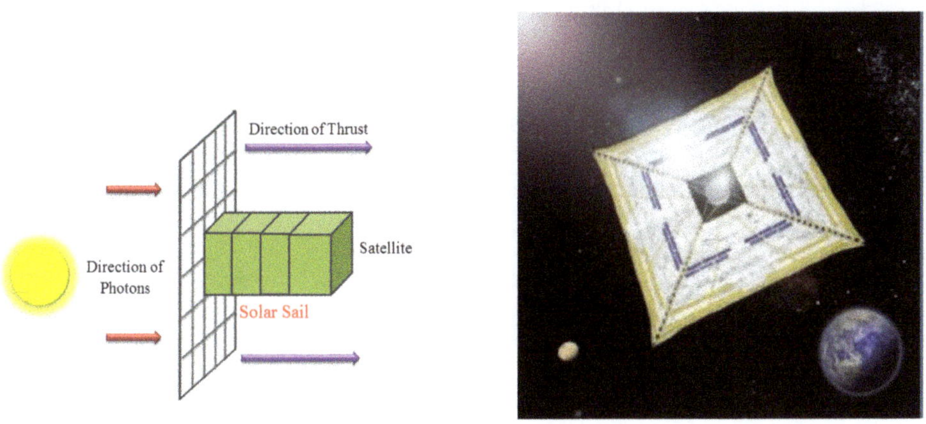


**Figure 1.4.** Left: A sketch showing the Interplanetary Kite-craft Accelerated by Radiation Of the Sun (IKAROS). Right: Radiation pressure force is essential to propel the solar kite satellite, as developed first by the Japan Aerospace Exploration Agency. Reprinted, with permission, from the JAXA Digital Archives.

### Cooling and trapping

The coherent interaction of electromagnetic fields with the internal and external degrees of freedom of an atom has led to the development of methods to cool it to sub-microkelvin temperatures. Details are given in chapter 4. The laser cooling is performed by combining the optical potential with the magnetic potential, which leads to atoms being trapped and cooled to tens of millikelvin temperature. Sisyphus cooling lowers the temperature to microkelvin scales. By contrast, in the presence of evaporative cooling the temperature of the atoms is reduced to sub-microkelvin, where the neutral atoms acting as bosons are found in the same state (see figure 1.5).

Cooling techniques have evolved to develop supersonic beams [41, 42], optical molasses [43, 44], optical traps [45–47], magneto-optic traps [48], magnetic traps [49, 50], atomic fountains [51], velocity-selective coherent population trapping [52], sideband cooling [53–55], cooling to the ground state of a trap [56, 57], and Bose–Einstein condensation [58].

### Ultra-cold atoms and Bose–Einstein condensation

The particle density at the center of an ultra-cold atomic distribution in a magneto-optic trap is of the order of $10^{12}$ cm$^{-3}$ to $10^{14}$ cm$^{-3}$, much lower than the density of room-temperature molecules in air ($10^{19}$ cm$^{-3}$), the density of atoms in liquids ($10^{22}$ cm$^{-3}$), and the density of nucleons in atomic nuclei ($10^{38}$ cm$^{-3}$). Since the temperature of the atoms is directly proportional to the kinetic energy and average of the square of the velocity, $\bar{v}^2$, that is

$$k_B T = \frac{1}{2} M \bar{v}^2 = \frac{\bar{p}^2}{2M}, \tag{1.7}$$

the de Broglie wavelength is

$$\Lambda_{dB} = \frac{h}{\sqrt{\bar{p}^2}} = \frac{h}{\sqrt{2Mk_B T}}. \tag{1.8}$$

Hence, we note that the de Broglie wavelength is inversely proportional to the square root of the temperature $T$. This implies that, as the temperature lowers, the de Broglie wavelength becomes longer, which is essential for observing quantum effects. For example, at room temperature, that is, around 300 K, the de Broglie

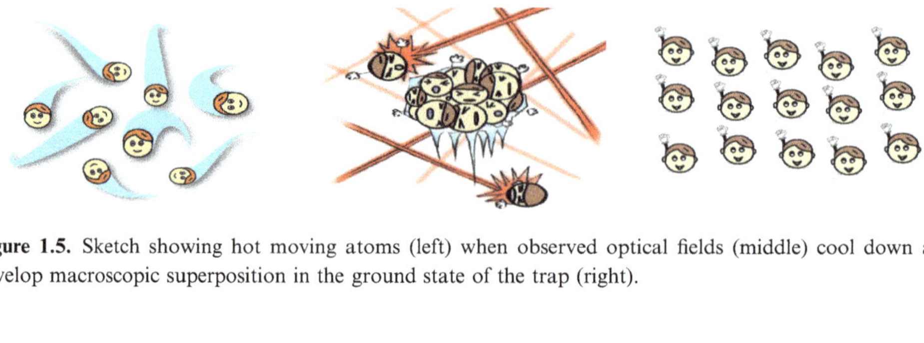

**Figure 1.5.** Sketch showing hot moving atoms (left) when observed optical fields (middle) cool down and develop macroscopic superposition in the ground state of the trap (right).

wavelength of a rubidium atom is 0.2 pm, much smaller than the size of the atom. At 1 K temperature scale the corresponding de Broglie wave associated with the rubidium atom has a wavelength near 0.3 nm, whereas at an eight orders of magnitude further lower temperature it is 3 $\mu$m.

In addition, ultra-cold atoms and BECs can be understood by considering their recoil energy, $E_R = \hbar^2 k_L^2/2M$, in comparison with the ensemble average energy, $k_B T$, where $k_L = 2\pi/\lambda$. For this reason, we find the de Broglie wavelength as

$$\Lambda_{dB} = \frac{h}{\sqrt{2Mk_B T}} = \frac{h}{\sqrt{2ME_R}} = \lambda, \tag{1.9}$$

when $E_R$ is taken to be the same as $k_B T$. This implies that the de Broglie wavelength becomes comparable to optical wavelengths under this condition. However, quantum characteristics only become visible if the de Broglie wavelength is spread over a few optical wavelengths. This requires the recoil energy, $E_R$, to be larger than the thermal energy, $k_B T$, as appears from the above equation. Therefore, we note that for ultra-cold atoms and BECs the de Broglie wavelength is bigger than the wavelength of the optical field, $\Lambda_{dB} > \lambda$.

**Quantum simulators**

According to Richard Feynman, mapping one quantum system onto the other makes a 'quantum simulator'. The wave packet dynamics of cold atoms in optical lattices simulate the electron dynamics in solid-state crystals. In the interaction of matter waves with an optical lattice, we can engineer simple quantum systems with tunable parameters, which provide an understanding of the various characteristics of semi-conductor electronics.

In most cases, the solid-state crystals are composed of complex structures. This makes it difficult to control the electron dynamics in crystals during experiments, for example due to defects and impurities. This lack of control introduces randomness in the electron dynamics. Thus, it becomes a hurdle to obtain predictable results for an ideal crystal, that is, without impurities and defects. Therefore, periodic potentials obtained in an optical lattice have become a substitute for ideal crystals. Atoms at very low temperature occupy lattice sites, in the same way as electrons in a crystal (see figure 1.6), and provide better control and efficiency. The motion and interaction of the particles, whether ultra-cold atoms or electrons, determine the physics of the material.

For a rubidium atom $^{87}$Rb of mass $m = 1.443 \times 10^{-25}$ kg interacting with an optical lattice of wavelength of $\lambda = 842$ nm, the recoil energy becomes $E_R = 1.34 \times 10^{-11}$ eV. This indicates a typical lattice depth of five to ten recoil energy of the order of $10^{-10}$ eV. Hence, a comparison with solid-state electronics implies that various phenomena with ultra-cold atoms are scaled down in energy by no less than ten orders of magnitude.

The dynamics of ultra-cold atoms in optical lattices is in many ways experimentally more accessible than the dynamics of electrons in crystal lattices: (i) the initial

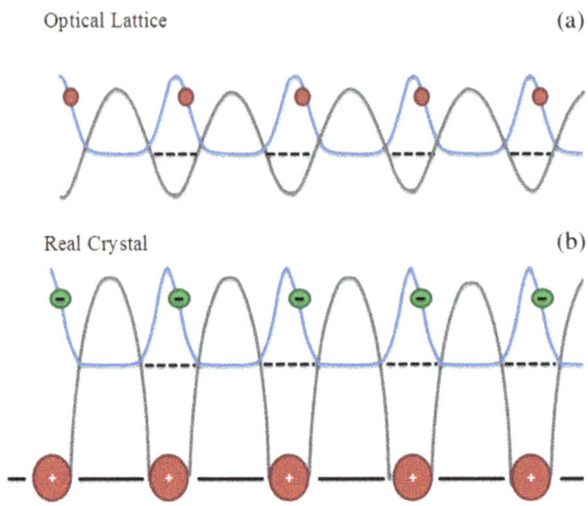


**Figure 1.6.** In an optical lattice (a) atoms are trapped in a sinusoidal potential well (gray) created by a standing-wave laser beam. The atoms' wave functions (blue) correspond to those of valence electrons in a real crystal (b). Here, the periodic potential is caused by the attractive electrostatic force between the electrons and the ions forming the crystal.

momentum distribution of cold atoms can be tailored as per requirements; (ii) in these experiments, the potential depth, lattice period, and acceleration of atoms can be controlled at will with high precision and experimental control; (iii) defects due to impurities and phonon vibrations are absent in optical lattices; and (iv) atomic dynamics in optical lattices occur on a time scale many orders longer than the dynamics of electrons in crystal lattices. These experimental advantages make atom dynamics reliable for demonstrating quantum effects which can only be realized indirectly in a crystal lattice. The ultra-cold atoms loaded in the optical lattice display weak interaction and strong interaction regimes based on the lattice depth. Impressive experiments have been performed in these regimes and have provided strong evidence of band structure, coherent matter wave interferometry, Bloch oscillations, Landau–Zener transitions, and so forth.

### Quantum computers

Ultra-cold dilute gases serve as a reliable tool to understand the many-body physics and characteristic effects that have been verified quantitatively. This understanding has led us to develop quantum bits (qubits or qbits), circuits and logic gates for quantum computers and quantum communication, based on cold atoms, ions, and optical fields.

A qubit may be defined in a two-dimensional Hilbert space spanned by atomic internal states, or the spin of a laser-cooled ion in a penning trap [59], such that

$$|\psi\rangle = a|0\rangle + b|1\rangle, \tag{1.10}$$

where $|0\rangle$ and $|1\rangle$ may describe the ground state and excited state of an atom or the spin-up state and spin-down state of a charged particle. The probability amplitudes $a$ and $b$ satisfy the condition that $|a|^2 + |b|^2 = 1$. In the most general form, we may express the probability amplitudes as complex numbers, i.e. $a = \cos\left(\frac{\theta}{2}\right)\mathrm{e}^{i\varphi_a}$ and $b = \sin\left(\frac{\theta}{2}\right)\mathrm{e}^{i\varphi_b}$. Here, $\varphi_a$ and $\varphi_b$ are the phase terms associated with the probability amplitudes $a$ and $b$. Therefore, we can write a qubit as

$$\begin{aligned} |\psi\rangle &= \cos\left(\frac{\theta}{2}\right)\mathrm{e}^{i\varphi_a}|0\rangle + \sin\left(\frac{\theta}{2}\right)\mathrm{e}^{i\varphi_b}|1\rangle \\ &= \mathrm{e}^{i\varphi_a}\left\{\cos\left(\frac{\theta}{2}\right)|0\rangle + \sin\left(\frac{\theta}{2}\right)\mathrm{e}^{i\varphi}|1\rangle\right\}. \end{aligned} \tag{1.11}$$

The symbol $\varphi$ expresses the relative phase difference, that is, $\varphi = \varphi_b - \varphi_a$. Hence, a qubit is defined as a point $(\theta, \varphi)$ on the surface of a sphere of unit radius as $\left|\cos\left(\frac{\theta}{2}\right)\right|^2 + \left|\sin\left(\frac{\theta}{2}\right)\mathrm{e}^{i\varphi}\right|^2 = 1$. This is known as a Bloch sphere (see figure 1.7). The interacting light or magnetic field provides logic gates with which to manipulate the qubits [60–62].

Cold neutral atoms arranged in an optical lattice display a qubit array [63]. The qubit array obtained in this way can be extended to contain a large number of qubits. These qubits have a long coherence time and allow individual laser pulses to be used to initialize single qubits and to control their state and position. Hence, quantum operations can be performed in parallel on the whole lattice.

Consider a set of $N$ number of qubits. Let us apply a $\pi/2$ pulse to all the spins simultaneously, which leads to $\theta = \pi/2$ and $\varphi = 0$. This turns the initial product state of the $N$ spins $|0\rangle|0\rangle\cdots|0\rangle$ into

$$\left(\frac{1}{\sqrt{2}}\right)^N (|0\rangle + |1\rangle)(|0\rangle + |1\rangle)\cdots(|0\rangle + |1\rangle). \tag{1.12}$$

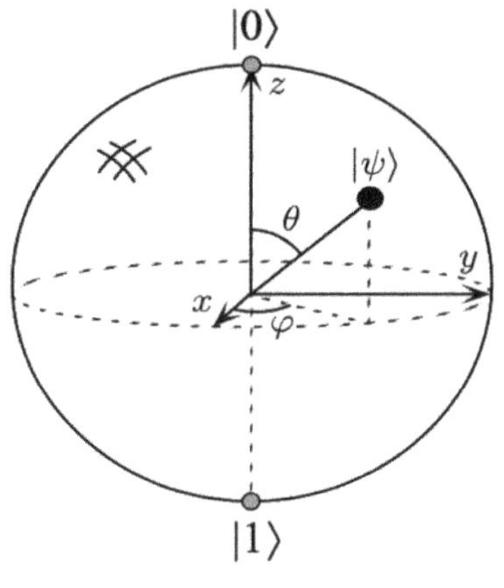


**Figure 1.7.** Description of a qubit on a Bloch sphere.

This state is equivalent to a superposition of $2^N$ product states, $|1\rangle|0\rangle\cdots|0\rangle|0\rangle$, $|0\rangle|1\rangle\cdots|0\rangle|0\rangle$, etc, which may be interpreted as a superposition of binary number states $10\cdots00$, etc.

In this particular example, all the products states may be found with equal probability. Quantum computing applies a sequence of unitary operations, called a quantum algorithm, that singularly enhances a particular collection of binary number states representing the answer to a given mathematical problem. Insofar as the intermediate states are not observed during the operations, the final state may be reversed into the initial states by applying the unitary operations in a reversed order. Like a reversible engine, quantum computing is expected to be energy efficient. At the same time, all the operations are applied to all the spin states in parallel. The parallel nature of its computing scheme is expected to make quantum computing much faster than the classical one if the right algorithm is available. Would it not be wonderful if a quantum computer becomes available for drug development, investment strategies, encryption, route scheduling for large vehicle fleets, etc, one day?

### Modulating optical fields

Ultra-cold bosonic and fermionic atoms in optical lattices in the presence of an external periodic-in-time modulation provide a new area of research both in theoretical and experimental matter wave optics. The increase of one degree of freedom due to the periodic modulation leads to complexity, which generates stable and stochastic atomic dynamics simultaneously, a signature of quantum chaos. The subject of quantum chaos, as the study of the quantum characteristics of classically chaotic systems, has received immense attention following the work of J E Bayfield and P M Koch on microwave ionization of hydrogen [64, 65].

A metal-insulator transition of ultra-cold atoms in quasi-periodic optical lattices is controllable by adjusting the amplitude of the periodic external forcing. Experimental work in this direction has gained momentum with the novel observation of dynamical suppression of tunneling, and even a reversal of the sign of the tunneling matrix element induced by shaken optical lattices. An analog of photon-assisted tunneling with BECs in driven optical crystals has been observed. The same principle reveals that this form of coherent control has recently been explored successfully for frustrated magnetism in driven triangular optical lattices.

We observe various dynamical modes in a system modulated by time periodic forcing. In the corresponding classical systems, the stable non-linear resonances are immersed in a stochastic sea, and the system may display global chaos beyond a critical value of coupling or modulation strength. In the case of spatial periodic potential driven by a periodic force, the classical counterpart of the dynamical system displays dominant regular and stochastic dynamics alternatively as a function of increasing modulation amplitude. In periodically driven systems, Floquet analysis gives quasi-energy eigenstates and quasi-energy eigenvalues.

## 1.2 Structure of the book

The layout of this book is presented in figure 1.8. The book is divided into three parts. Our general strategy in developing the material is to follow experimental developments and technological advances in the field. The step-by-step approach covers the necessary theoretical tools, experimental systems, fundamental phenomena, and the latest applications. The material is structured so as to be accessible to any reader with a basic understanding of quantum mechanics. We have added a brief discussion on Bose–Einstein condensation. However, the focus of the book is on linear matter wave optics, i.e. phenomena where each single atom interferes with itself. Wherever necessary, we have introduced simple examples and exercises to explain the topics. The book is intended to be helpful for students, academics, researchers, technologists, and experts in industry.

Part I provides a general understanding of optical fields. In chapter 2, following the Maxwell description, we discuss vector potential, the continuity equation, and energy and momentum corresponding to optical fields. Based on the Maxwell depiction, quantization of electromagnetic fields is obtained. Quantized electromagnetic fields, such as Fock states, coherent states, squeezed states, and Schrödinger cat states,

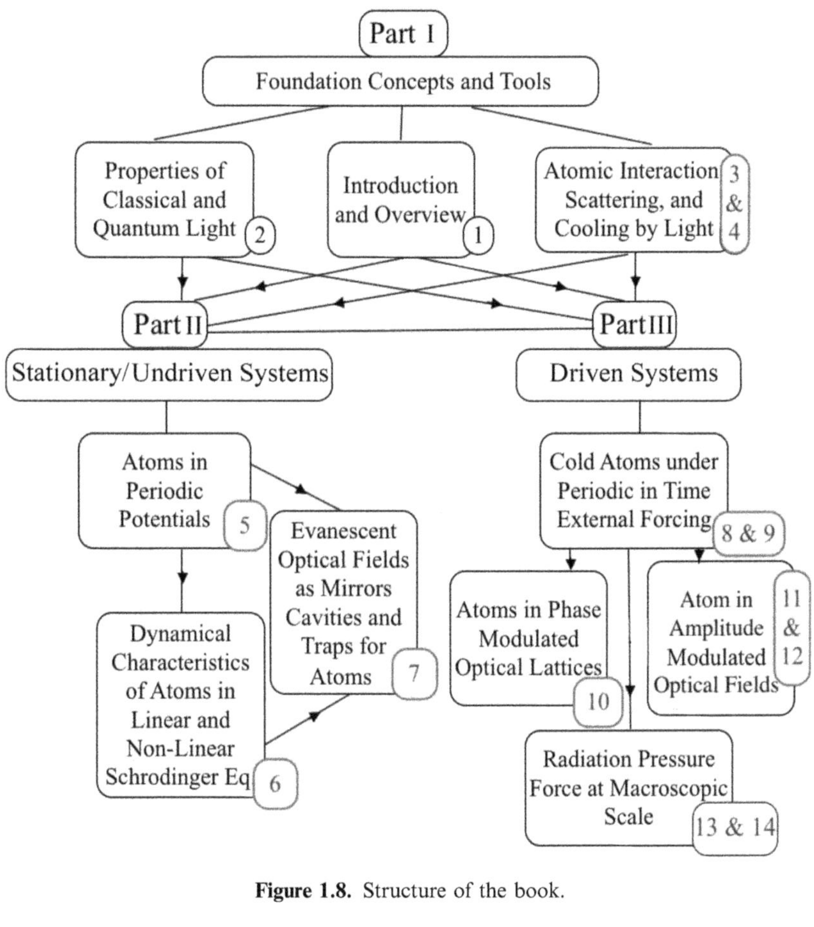


**Figure 1.8.** Structure of the book.

provide a general understanding of the quantum nature of the field. Thereby, we build the necessary foundation-level knowledge required to understand electromagnetic fields in the quantum domain. Chapters 3 and 4 explain the interaction of optical fields with atoms. The radiation pressure force on an atom, which scatters and cools atoms to an ultra-low temperature, is explained.

Part II describes the characteristics of ultra-cold atoms in optical lattices and evanescent optical fields. Chapters 5 and 6 present the basic elements needed to explain ultra-cold atoms in optical lattices in detail. Bloch states and Wannier states, respectively, are discussed in momentum space and position space. Moreover, an understanding of energy band formation is explained. In chapter 6, we explain modified atomic dynamics due to linear confinement. Bloch waves and Landau–Zener transitions are explained. The inter-atomic interaction is presented as the Bose–Hubbard model, which is helpful in understanding advanced topics such as Mott insulators as superfluidity in ultra-cold atoms. In chapter 7, we explain optical forcing due to evanescent optical fields on ultra-cold atoms, which has led to the development of atom interferometers and cavities based on atomic mirrors and nano-optical fibers. We explain atom surface traps, wave packet collapse and revival phenomena.

Part III is dedicated to explaining multi-dimensional coupled optical systems displaying complex atomic dynamics. Chapter 8 provides a basic understanding of complex dynamics. Chapters 8 and 9 explain the effect of time periodic forcing on ultra-cold atoms in optical potentials in a general context. We discuss the Floquet–Bloch solutions and mapping the general three-dimensional coupled Schrödinger equation to the Mathieu equation corresponding to non-linear resonances. In chapter 10, we describe atomic dynamics in phase-modulated optical lattices. The chapter covers the basic tools that provide us with quantitative measures to explain dynamical localization and delocalization. In chapter 11, we explain modified atomic dynamics due to quadratic confinement and amplitude modulation. Chapter 12 describes atomic dynamics in amplitude-modulated evanescent fields making an atomic modulated trampoline under the action of gravitational potential. In the chapter we explain the quantum recurrences in various modulation regimes, acceleration and dynamical localization, leading to the important topic of non-dispersive and accelerated atomic wave packets. Light force in macroscopic systems is useful for the development of quantum devices and sensors at macroscopic scales. We thus focus on this in chapters 13 and 14, and discuss radiation pressure force to develop macroscopic opto-mechanical systems. The chapter explains the mathematical tools needed to handle such systems, leading on to important characteristics like entanglement, bistability, four-wave mixing, and slow and fast light.

Almost every chapter has useful examples and exercises that can be solved easily with a little brainstorming. The examples and exercises are designed to improve the understanding of the relevant topics in the chapters and occasionally to introduce new interesting topics. The dependency diagram (see figure 1.8) will help the reader to determine the order in which to cover the material provided. The book covers the material of two semester courses. It will be useful in the areas of quantum optics, quantum electronics, quantum metrology, photonics, and nano-science and technology. There are three appendices, presenting extended material and related extensive mathematical work.

## Exercises

1. We express the Planck black-body radiation law as

$$P(\nu) = A\exp\left(-\frac{nh\nu}{k_B T}\right).$$

Find the normalization coefficient, $A$.

2. Show that the average energy of a mode is

$$\langle E\rangle = \frac{h\nu}{\exp\left(\frac{h\nu}{k_B T}\right) - 1}.$$

[Hint: $\langle E\rangle = \sum_{n=0}^{\infty} nh\nu P(\nu)$.]

3. Explain equation (1.2) in the limit of (i) $\nu$ approaching zero, and (ii) temperature $T$ approaching infinity.
4. For a cavity volume $V = 1\ \text{cm}^3$ calculate the number of modes that fall within a bandwidth $\Delta\lambda = 10$ nm centered at $\lambda = 600$ nm.

[Hint: The number of modes per unit volume per frequency range is

$$\frac{1}{V}\frac{dN}{d\nu} = \frac{8\pi\nu^2}{c^3}].$$

5. The photoelectric threshold of tungsten is 2300 Å. Determine the energy of the electrons ejected from the surface by ultraviolet light of wavelength 2000 Å.
6. Photoelectrons emitted from a cesium plate illuminated with ultraviolet light of wavelength 1900 Å are stopped by a potential of 3.52 V. What is the work function of cesium?
7. Calculate the average momentum and the de Broglie wavelength of a rubidium atom at temperature (i) 1 K, (ii) $10^{-6}$ K, and (iii) $10^{-9}$ K.

## References

[1] Planck M 1920 Entstehung und bisherige Entwicklung der Quantentheorie: Nobelvortrag (Nobel lecture) *gehalten vor der Königlich Schwedischen Akademie der Wissenschaften zu Stockholm on 2. June 2 1920* (Leipzig: J.A. Barth Verlag)

[2] Planck M 1922 *The Origin and Development of the Quantum Theory* transl. ed H T Clarke and L Silberstein (Oxford: Clarendon)

[3] Einstein A 1905 Über einen die Erzeugung und Verwandlung des Lichtes betreffenden heuristischen Gesichtpunkt *Ann. Phys., Lpz.* **17** 132

[4] Lewis G N 1926 The conservation of photons *Nature* **118** 784

[5] Bose S N 1924 Plancks Gesetz und Lichtquantenhypothese *Z. Phys.* **26** 178–81

[6] Einstein A 1924 Quantentheorie des einatomigen idealen Gases *Sitzungsber. K.P. d. Akad. d. Wiss.* **1** 261–7

[7] Born M 1978 *My Life: Recollections of a Nobel Laureate* (London: Taylor and Francis)

[8] Compton A H 1923 A quantum theory of the scattering of X-rays by light elements *Phys. Rev.* **21** 483

[9] de Broglie L 1924 Recherches sur la theorie des quanta (Researches on the quantum theory) *PhD Thesis* University of Paris (Sorbonne)
de Broglie L 1925 *Ann. Phys.* **10** 22

[10] Thomson G P 1927 Diffraction of cathode rays by a thin film *Nature* **119** 890

[11] Davisson C J and Germer L H 1928 Reflection of electrons by a crystal of nickel *Proc. Natl. Acad. Sci. USA* **14** 317–22

[12] Einstein A 1916 Strahlungs-Emission und Absorption nach der Quantentheorie *Verh. Dtsch. Phys. Ges.* **18** 318–23

[13] Rabi I I, Zacharias J R, Millman S and Kusch P 1938 A new method of measuring nuclear magnetic moment *Phys. Rev.* **53** 318

[14] Maiman T H 1960 Stimulated optical radiation in ruby *Nature* **187** 493–4

[15] Maiman T H 1960 Optical and microwave-optical experiments in ruby *Phys. Rev. Lett.* **4** 564

[16] Javan A, Bennett W R and Herriot D R 1961 Population inversion and continuous optical maser oscillation in a gas discharge containing a He-Ne mixture *Phys. Rev. Lett.* **6** 106

[17] Lebedev P N 1901 Experimental examination of light pressure *Ann. Phys., Lpz.* **6** 433

[18] Lebedev P N 1902 Experimental investigation of the pressure of light *Astrophys. J.* **15** 60

[19] Nichols E F and Hull G F 1901 A preliminary communication on the pressure of heat and light radiation *Phys. Rev. (Ser. I)* **13** 307

[20] Nichols E F and Hull G F 1903 The pressure due to radiation (second paper) *Phys. Rev. (Ser. I)* **17** 26

[21] Ashkin A 1970 Acceleration and trapping of particles by radiation pressure *Phys. Rev. Lett.* **24** 156

[22] Ashkin A 1970 Atomic-beam deflection by resonance-radiation pressure *Phys. Rev. Lett.* **25** 1321

[23] Altshuler S and Frantz L M 1973 Matter wave interferometric apparatus *US Patent* **3761721**

[24] Kazantsev A P 1974 Acceleration of atoms by light *Zh. Eksp. Teor. Fiz.* **66** 1599

[25] Kazantsev A P 1974 The recoil effect in a force resonant field *Zh. Eksp. Teor. Fiz.* **67** 1600

[26] Hänsch T W and Schawlow A L 1975 Cooling of gases by laser radiation *Opt. Commun.* **13** 68

[27] Wineland D J and Dehmelt H G 1975 Laser fluorescence spectroscopy on Tl + mono-ion oscillator *Bull. Am. Phys. Soc.* **20** 637

[28] Ashkin A 1978 Trapping of atoms by resonance radiation pressure *Phys. Rev. Lett.* **40** 729

[29] Kapitza P and Dirac P 1933 The reflection of electrons from standing light waves *Math. Proc. Camb. Phil. Soc.* **29** 297–300

[30] Altshuler S, Frantz L M and Braunstein R 1966 Reflection of atoms from standing light waves *Phys. Rev. Lett.* **17** 231

[31] Dahan M B, Peik E, Reichel J, Castin Y and Salomon C 1996 Bloch oscillations of atoms in an optical potential *Phys. Rev. Lett.* **76** 4508

[32] Robert-de-Saint-Vincent M, Brantut J-P, Bordé Ch J, Aspect A, Bourdel T and Bouyer P 2010 A quantum trampoline for ultra-cold atoms *Europhys. Lett.* **89** 10002

[33] Tolstikhin O I, Morishita T and Watanabe S 2005 Gyroscopic effects in interference of matter waves *Phys. Rev.* A **72** 051603

[34] Diddams S A *et al* 2001 An optical clock based on a single trapped $^{199}Hg^{+}$ ion *Science* **293** 825–8
[35] Peik E, Lipphardt B, Schnatz H, Schneider T, Tamm Chr and Karshenboim S G 2004 Limit on the present temporal variation of the fine structure constant *Phys. Rev. Lett.* **93** 170801
[36] Margolis H S, Barwood G P, Huang G, Klein H A, Lea S N, Szymaniec K and Gill P 2004 Hertz-level measurement of the optical clock frequency in a single $^{88}Sr^{+}$ ion *Science* **306** 1355–8
[37] Wilpers G, Binnewies T, Degenhardt C, Sterr U, Helmcke J and Riehle F 2002 Optical clock with ultracold neutral atoms *Phys. Rev. Lett.* **89** 230801
[38] Oates C W, Curtis E A and Hollberg L 2000 Improved short-term stability of optical frequency standards: approaching 1 Hz in 1s with the Ca standard at 657 nm *Opt. Lett.* **25** 1603–5
[39] Pereira Dos Santos F, Marion H, Bize S, Sortais Y, Clairon A and Salomon C 2002 Controlling the cold collision shift in high precision atomic interferometry *Phys. Rev. Lett.* **89** 233004
[40] Takamoto M, Hong F-L, Higashi R and Katori H 2005 An optical lattice clock *Nature* **435** 321–4
[41] Beijerinck H C W and Verster N F 1981 Absolute intensities and perpendicular temperatures of supersonic beams of polyatomic gases *Physica* B **111** 327
[42] Campargue R 1984 Progress in overexpanded supersonic jets and skimmed molecular beams in free-jet zones of silence *J. Phys. Chem.* **88** 4466
[43] Aspect A, Dalibard J, Heidmann A, Salomon C and Cohen-Tannoudji C 1986 Cooling atoms with stimulated emission *Phys. Rev. Lett.* **57** 1688
[44] Chu S, Hollberg L, Bjorkholm J E, Cable A and Ashkin A 1985 3-dimensional viscous confinement and cooling of atoms by resonance radiation pressure *Phys. Rev. Lett.* **55** 48
[45] Ashkin A 1970 Acceleration and trapping of particles by radiation pressure *Phys. Rev. Lett.* **24** 156
[46] Chu S, Bjorkholm J E, Ashkin A and Cable A 1986 Experimental-observation of optically trapped atoms *Phys. Rev. Lett.* **57** 314
[47] Miller J D, Cline R and Heinzen D 1993 Far-off-resonance optical trapping of atoms *Phys. Rev.* A **47** R4567
[48] Raab E L, Prentiss M, Cable A, Chu S and Pritchard D E 1987 Trapping of neutral sodium atoms with radiation pressure *Phys. Rev. Lett.* **59** 2631
[49] Migdall A L, Prodan J V, Phillips W D, Bergeman T H and Metcalf H J 1985 1st observation of magnetically trapped neutral atoms *Phys. Rev. Lett.* **54** 2596
[50] Pritchard D E 1983 Cooling neutral atoms in a magnetic trap for precision spectroscopy *Phys. Rev. Lett.* **51** 1336
[51] Kasevich M A, Riis E, Chu S and Devoe R G 1989 Rf spectroscopy in an atomic fountain *Phys. Rev. Lett.* **63** 612
[52] Aspect A, Arimondo E, Kaiser R, Vansteenkiste N and Cohen-Tannoudji C 1988 Laser cooling below the one-photon recoil energy by velocity-selective coherent population trapping *Phys. Rev. Lett.* **61** 826
[53] Neuhauser W, Hohenstatt M, Toschek P and Dehmelt H 1978 Optical-sideband cooling of visible atom cloud confined in parabolic well *Phys. Rev. Lett.* **41** 233
[54] Vuletic V, Chin C, Kerman A J and Chu S 1998 Degenerate Raman sideband cooling of trapped cesium atoms at very high atomic densities *Phys. Rev. Lett.* **81** 5768

[55] Wineland D J, Drullinger R E and Walls F L 1978 Radiation-pressure cooling of bound resonant absorbers *Phys. Rev. Lett.* **40** 1639
[56] Jessen P S, Gerz C, Lett P D, Phillips W D, Rolston S L, Spreeuw R J C and Westbrook C I 1992 Observation of quantized motion of Rb atoms in an optical-field *Phys. Rev. Lett.* **69** 49
[57] Monroe C, Meekhof D M, King B E, Jefferts S R, Itano D J, Wineland W M and Gould P 1995 Resolved-sideband Raman cooling of a bound atom to the 3D zero-point energy *Phys. Rev. Lett.* **75** 4011
[58] Anderson M H, Ensher J R, Matthews M R, Wieman C E and Cornell E A 1995 Observation of Bose-Einstein condensation in a dilute atomic vapor *Science* **269** 198
[59] Ciaramicoli G, Marzoli I and Tombesi P 2010 From a single- to a double-well Penning trap *Phys. Rev.* A **82** 044302
[60] Nielson M and Chuang I 2000 *Quantum Computation and Quantum Information* (Cambridge: Cambridge University Press)
[61] Stolze J and Suter D 2008 *Quantum Computing* (New York: Wiley-VCH)
[62] Nakahara M and Ohmi T 2008 *Quantum Computing: From Linear Algebra to Physical Realizations* (Boca Raton, FL: CRC Press)
[63] Brennen G K, Caves C M, Jessen P S and Deutsch I H 1999 Quantum logic gates in optical lattices *Phys. Rev. Lett.* **82** 1060
[64] Saif F 2005 Classical and quantum chaos in atom optics *Phys. Rep.* **419** 207
[65] Yamakoshi T, Watanabe S, Ohgoda S and Itin A P 2016 Dynamics of fermions in an amplitude-modulated lattice *Phys. Rev.* A **93** 063637 and references therein

**IOP** Publishing

Optical Forces on Atoms

**Farhan Saif and Shinichi Watanabe**

# Chapter 2

## Characterization of optical fields

*I don't know anything about photons, but I know one when I see one*
—Roy J Glauber

For more than a millennium, light has been the driving force in developing our understanding of the laws of physics. Light is a part of the electromagnetic radiation spectrum. The electromagnetic spectrum as measured in laboratory conditions covers electromagnetic waves with wavelengths from tens of meters down to a fraction of the size of an atomic nucleus, whereas visible light is usually defined as having wavelengths in the range of 400–700 nm, that is, between infrared and ultraviolet.

From the wave-particle controversy arising from the work of Thomas Young and Issac Newton to the wave-particle duality found in the work of Albert Einstein and Louis de Broglie, light has contributed hugely to shaping our understanding of the world around us. It has led to enormous development ranging from lasers to information technology. In 1920s, Max Born, Pascual Jordon, and Werner Heisenberg [1, 2] developed the quantum mechanics of light. An important contribution came from Paul Adrian Maurice Dirac [3] as he introduced the concept of non-zero vacuum energy. According to Planck, energy is absorbed or emitted by oscillators, which have energies restricted to values $n$ times the quantum of energy, $h\nu$. It was Dirac who showed that, in fact, those energies are not $nh\nu$ but $(n + 1/2)h\nu$. All of the intervals between energy levels remained unchanged, but the quantum mechanical uncertainty principle required an additional $\frac{1}{2}h\nu$ to be present.

doi:10.1088/978-0-7503-2308-6ch2  

At present, light is shown to have many non-classical states, including number states, coherent states, squeezed states, and Schrödinger cat states [4]. The quantum states of light have been demonstrated in experiments.

## 2.1 Electromagnetic field modes in vacuum

Our classical perception of electromagnetic fields, as motivated by Michael Faraday's observations, were transcribed into a set of mathematical equations developed by James Clerk Maxwell, which are as follows:

$$\nabla \cdot \mathbf{E} = 0, \tag{2.1}$$

$$\nabla \cdot \mathbf{B} = 0, \tag{2.2}$$

$$\nabla \times \mathbf{E} = \frac{\partial \mathbf{B}}{\partial t}, \tag{2.3}$$

$$\nabla \times \mathbf{B} = \mu_0 \epsilon_0 \frac{\partial \mathbf{E}}{\partial t}. \tag{2.4}$$

The set of coupled homogeneous equations describes a classical electromagnetic field in empty space, in the absence of any sources such as charges or currents. The symbols $\mathbf{E}$ and $\mathbf{B}$ express the electric and magnetic field vectors at the space-time, $(\mathbf{r}, t)$. It is often convenient to describe a free electromagnetic field in terms of vector potential $\mathbf{A}(\mathbf{r}, t)$ in the Coulomb gauge[1], which satisfies the homogeneous wave equation

$$\nabla^2 \mathbf{A} - \frac{1}{c^2} \frac{\partial^2 \mathbf{A}}{\partial^2 t} = 0, \tag{2.5}$$

and the divergence condition

$$\nabla \cdot \mathbf{A}(\mathbf{r}, t) = 0, \tag{2.6}$$

where

$$c = \frac{1}{\sqrt{\mu_0 \epsilon_0}}.$$

In terms of vector potential the electric field and magnetic field are, respectively, written as

$$\mathbf{E}(\mathbf{r}, t) = -\frac{\partial \mathbf{A}(\mathbf{r}, t)}{\partial t}, \tag{2.7}$$

[1] Introducing the vector potential $\mathbf{A}(\mathbf{r}, t)$ as more fundamental than $\mathbf{E}(\mathbf{r}, t)$ and $\mathbf{B}(\mathbf{r}, t)$ is adequate. In particular, in the context of quantum mechanics, the vector potential is conducive to the phase of the wave function of a charged particle, as shown in the Bohm–Aharanov effect. This effect is global so that the particle feels it even if the fields have non-vanishing amplitude at a large distance away. The idea is not limited to the electric and magnetic fields, but it serves as the founding idea for the gauge field theory.

$$\mathbf{B}(\mathbf{r}, t) = \nabla \times \mathbf{A}(\mathbf{r}, t). \tag{2.8}$$

Taking the curl of equation (2.3) and substituting equation (2.4), we obtain the wave equation for the electric field **E**, similar to equation (2.5). This is the wave equation for the electric field representing wave-like propagation in vacuum. Likewise, taking the curl of equation (2.4) and substituting equation (2.3) leads to a wave equation for the magnetic field **B**.

In order to develop a mathematical description of the energy associated with the electromagnetic waves, we calculate the energy density, $\rho$, as follows:

$$\rho = \frac{1}{2}\left(\epsilon_0 \mathbf{E}^*(\mathbf{r}, t) \cdot \mathbf{E}(\mathbf{r}, t) + \frac{1}{\mu_0}\mathbf{B}^*(\mathbf{r}, t) \cdot \mathbf{B}(\mathbf{r}, t)\right).$$

It is related to the energy flux

$$\mathbf{J} = \mathbf{E} \times \mathbf{B},$$

also known as the Poynting vector, through the continuity equation

$$\frac{\partial \rho}{\partial t} + \nabla \cdot \mathbf{J} = 0.$$

Incidentally, the energy density and Poynting vector define the total energy (Hamiltonian) and the momentum as

$$H = \int_V \rho \, dV, \tag{2.9}$$

$$\mathbf{P} = \int_V \mathbf{J} \, dV. \tag{2.10}$$

### 2.1.1 Solution of the wave equation

We consider the electromagnetic field present in a cubic cavity of side length $L$, as shown in figure 2.1. The planes of the cube are made of perfectly conducting material. The solution of the wave equation (2.5) provides the electric field $\mathbf{E}(\mathbf{r}, t)$ and the magnetic field $\mathbf{B}(\mathbf{r}, t)$.

*Separation of variables.* Since equation (2.5) is composed of spatial and temporal operators, following the separation-of-variable technique we write $\mathbf{A}(\mathbf{r}, t)$ as a product of the time-dependent function $q(t)$ and space-dependent function $u(\mathbf{r})$, such that

$$\mathbf{A}(\mathbf{r}, t) = A_0 \, q(t) \, u(\mathbf{r}) \, \mathbf{e}_r, \tag{2.11}$$

where $A_0$ may be regarded as the amplitude of vector potential $\mathbf{A}(\mathbf{r}, t)$, and $\mathbf{e}_r$ is the unit vector. Thus, we obtain

$$q(t) \, \nabla^2 u(\mathbf{r}) - \frac{u(\mathbf{r})}{c^2}\frac{\partial^2 q(t)}{\partial t^2} = 0. \tag{2.12}$$

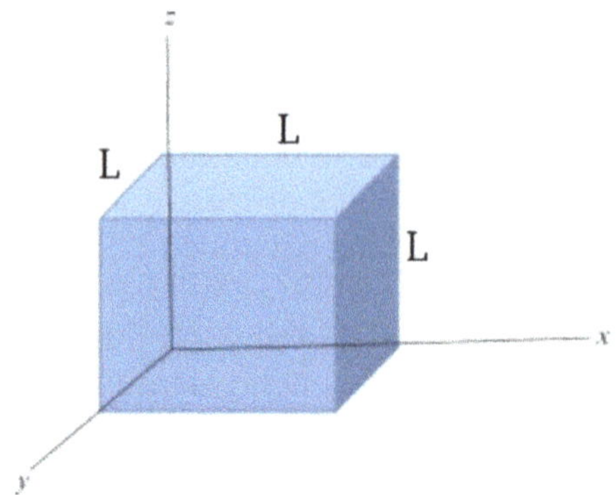


**Figure 2.1.** Electromagnetic field of frequency $\omega$ in a cubic cavity of length $L$. Each plane of the cube is made of perfectly conducting walls.

We divide by $q(t)u(\mathbf{r})$ to obtain

$$\frac{1}{u(\mathbf{r})}\nabla^2 u(\mathbf{r}) = \frac{1}{c^2}\frac{1}{q(t)}\frac{\partial^2 q(t)}{\partial t^2}. \tag{2.13}$$

The left-hand side contains a position-dependent function and operator, whereas the right-hand side is composed of a time-dependent function and operator acting in the time domain; equality is possible if both sides of equation (2.13) are equal to some constant that is not a function of position or time. We symbolically express the unknown constant as $k^2$. Therefore, we get two different equations as follows:

$$\frac{d^2 q(t)}{dt^2} + \omega^2 q(t) = 0, \tag{2.14}$$

and

$$\nabla^2 u(\mathbf{r}) + k^2 u(\mathbf{r}) = 0. \tag{2.15}$$

Here, the field frequency is written as $\omega = ck$.

The three-dimensional solution in each of the coordinates $x$, $y$, and $z$ leads to $k^2 = k_x^2 + k_y^2 + k_z^2$. Equation (2.15) is the famous Helmholtz equation, which has periodic solutions in the $x$-, $y$-, and $z$-directions. We apply the boundary conditions to solve the equation. The Coulomb gauge equation becomes

$$\nabla \cdot \mathbf{A}(\mathbf{r}, t) = q(t)\,\nabla u(\mathbf{r}) \cdot \mathbf{e}_r = 0. \tag{2.16}$$

We obtain

$$\mathbf{k} \cdot \mathbf{e}_r = k_x e_x + k_y e_y + k_z e_z = 0, \tag{2.17}$$

which implies that the propagation vector $\mathbf{k}$ is perpendicular to the unit vector $\mathbf{e}_r = (e_x, e_y, e_z)$. Since there are two linearly independent orthogonal propagation directions, there are two linearly independent polarization vectors, $\mathbf{e}_r^{(1)}$ and $\mathbf{e}_r^{(2)}$, that satisfy equation (2.17).

The vanishing of the longitudinal component at the wall of the cavity determines the vector **k** such that

$$k_x = \frac{l\pi}{L}, \tag{2.18}$$

$$k_y = \frac{m\pi}{L}, \tag{2.19}$$

$$k_z = \frac{n\pi}{L}. \tag{2.20}$$

Here, $l$, $m$, $n$ are integer numbers. Hence, we express the electric field and magnetic field components as

$$\mathbf{E}(\mathbf{r}, t) = -\sum_{l,m,n}\sum_{j=1,2} A_0^{(l,m,n)}\, \dot{q}_{l,m,n}(t)\, u_{l,m,n}(r)\, \mathbf{e}_r^{(j)}, \tag{2.21}$$

$$\mathbf{B}(\mathbf{r}, t) = \sum_{l,m,n}\sum_{j=1,2} A_0^{(l,m,n)}\, q_{l,m,n}(t)\, \nabla u_{l,m,n}(r) \times \mathbf{e}_r^{(j)}. \tag{2.22}$$

Here, the frequency of the field mode is expressed as $\omega_{l,m,n} = c\, k_{l,m,n}$.

*Electromagnetic field as harmonic oscillator.* The energy of the electromagnetic field mode is given as

$$\text{Energy} = \frac{1}{2}\int_V d^3r\left(\epsilon_0 \mathbf{E}^*(\mathbf{r}, t)\cdot \mathbf{E}(\mathbf{r}, t) + \frac{1}{\mu_0}\mathbf{B}^*(\mathbf{r}, t)\cdot \mathbf{B}(\mathbf{r}, t)\right), \tag{2.23}$$

$$= \frac{1}{2}\int_V d^3r\left(\epsilon_0 E^2(r, t) + \frac{1}{\mu_0}B^2(r, t)\right). \tag{2.24}$$

The second-order Helmholtz equation (2.15) has solutions $u_{l,m,n}(r)$, where $l$, $m$, $n$ define the modes. These are expressed in terms of trigonometric functions in general, and satisfy the property of ortho-normality, that is,

$$\frac{1}{\sqrt{V_{l,m,n}}}\frac{1}{\sqrt{V_{l',m',n'}}}\int_V d^3r\, u^*_{l,m,n}(r)\, u_{l',m',n'}(r) = \delta_{l,l'}\delta_{m,m'}\delta_{n,n'}. \tag{2.25}$$

Here, $V_{l,m,n}$ describes the effective mode volume of the mode defined by integers $l$, $m$, $n$. The definite integral over the spatial coordinate makes the Hamiltonian expression independent of $r$. Hence, we obtain the Hamiltonian description of the energy in the time domain, and simplify the above expression to

$$\begin{aligned} H &= \frac{1}{2}\sum_{l,m,n}\left(\omega^2_{l,m,n} q^2_{l,m,n}(t) + \dot{q}^2_{l,m,n}(t)\right), \\ &= \frac{1}{2}\sum_{l,m,n}\left(\omega^2_{l,m,n} q^2_{l,m,n}(t) + p^2_{l,m,n}(t)\right). \end{aligned} \tag{2.26}$$

In its present form, the Hamiltonian of the mode defined by $l$, $m$, $n$ resembles the expression for a harmonic oscillator (of unit mass) with frequency $\omega_{l,m,n}$, with $q_{l,m,n}$ and $\dot{q}_{l,m,n} = p_{l,m,n}$ acting as amplitude and momentum, respectively[2]. It is important to note that the function $q_{l,m,n}$ and its derivative $\dot{q}_{l,m,n}$ are explicitly time-dependent functions. In this respect an electromagnetic field can be compared with a harmonic oscillator; however, its nature is philosophically different as it exists in the time domain alone. Interestingly, it expresses a photon field in a single mode of frequency $\omega$ that follows harmonic oscillation in the time domain.

### 2.1.2 Modes in a one-dimensional cavity field

We solve equation (2.5) considering the electromagnetic field present in a one-dimensional cavity along the $x$-axis, with perfect mirrors of surface area $S$, at $x = 0$ and $x = L$, as shown in figure 2.2. Moreover, we apply the Coulomb gauge, that is, $\nabla \cdot u(\mathbf{r}) = 0$. The vector potential has its polarization vector in the plane perpendicular to the $x$-axis. We may express the polarization vector, $\mathbf{e}_\mu$, as $\mathbf{e}_\mu = c_y\mathbf{e}_y + c_z\mathbf{e}_z$. Here, $c_y$ and $c_z$ are constant numbers, such that $|c_y|^2 + |c_z|^2 = 1$, whereas $\mathbf{e}_y$ and $\mathbf{e}_z$ are the unit vectors along the $y$- and $z$-axes, respectively. Hence, we obtain the vector potential, leading to the electric field and magnetic field of *a single mode*, as

$$\mathbf{A}(\mathbf{x}, t) = \left(\frac{1}{LS\epsilon_0}\right)^{1/2} q(t)\sin(kx)\, \mathbf{e}_\mu, \tag{2.27}$$

$$\mathbf{E}(\mathbf{x}, t) = -\left(\frac{1}{LS\epsilon_0}\right)^{1/2} \dot{q}(t)\sin(kx)\, \mathbf{e}_\mu, \tag{2.28}$$

$$\mathbf{B}(\mathbf{x}, t) = \left(\frac{\mu_0}{LS}\right)^{1/2} \omega\, q(t)\cos(kx)\, \mathbf{e}_\nu. \tag{2.29}$$

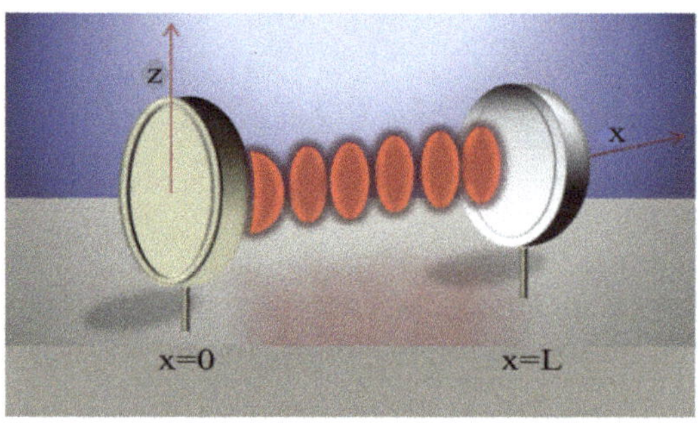


**Figure 2.2.** Electromagnetic field of frequency $\omega$ in a cavity of length $L$ with perfectly reflecting mirrors at both ends.

[2] This association of $p$ to $\dot{q}$ holds in the absence of the velocity-dependent force.

Here, $\mathbf{e}_\nu$ is the polarization vector in the $(x, y)$ plane and is perpendicular to $\mathbf{e}_\mu$, such that $\mathbf{e}_\nu \cdot \mathbf{e}_\mu = 0$. Moreover, $V = LS$ is the mode volume of the cavity, with cross-sectional area $S$.

The above discussion can be generalized to a *multi-mode* electric field that is linearly polarized along the $y$-axis. Hence, we write the corresponding expression as

$$\mathbf{E}(\mathbf{x}, t) = -\sum_j \left(\frac{1}{V_j \epsilon_0}\right)^{1/2} \dot{q}_j(t) \sin(k_j x)\, \mathbf{e}_y. \tag{2.30}$$

In addition, the *multi-mode* magnetic field in the cavity becomes

$$\mathbf{B}(\mathbf{x}, t) = \sum_j \left(\frac{\mu_0}{V_j}\right)^{1/2} \omega_j\, q_j \cos(k_j x)\mathbf{e}_z. \tag{2.31}$$

Here, $\omega_j = j\pi c/L$ denotes the frequency of the $j$th mode, and $V_j$ is the corresponding mode volume. Being the solutions of Maxwell's equation, the discrete modes, expressed by the index $j$, are classical. The classical Hamiltonian for the multi-mode electromagnetic field is given by equation (2.24). Hence, by following the same steps as for the single-mode case, and substituting the expressions for a multi-mode electric and magnetic field, we get

$$H = \frac{1}{2}\sum_j \left(\omega_j^2 q_j^2 + \dot{q}_j^2\right), \tag{2.32}$$

$$= \frac{1}{2}\sum_j \left(\omega_j^2 q_j^2 + p_j^2\right). \tag{2.33}$$

Here, $p_j = \dot{q}_j = \frac{dq_j(t)}{dt}$. Moreover, $q_j$ is merely an analog of the spatial displacement that we find in the potential energy term of a conventional harmonic oscillator Hamiltonian—at present it is a function of *time*. Hence, equation (2.33) gives the Hamiltonian of a multi-mode electromagnetic field as a function of $q_j$ and its associated conjugate momentum, $p_j$. This leads us very naturally to the quantization of the electromagnetic fields by making an analogy with the harmonic oscillator, but the comparison of the derived relationships in the two cases is non-trivial.

## 2.2 Quantization of electromagnetic field modes

The understanding that the electromagnetic fields are formally equivalent to a set of harmonic oscillators suggests an obvious procedure of quantization. We simply need a creation operator $\hat{a}_j^\dagger$ and an annihilation operator $\hat{a}_j$, and the associated number operator $\hat{N}_j = \hat{a}_j^\dagger \hat{a}_j$, to 'count photons'. We do this step-by-step and review this idea.

We simplify the Hamiltonian of a multi-mode field given in equation (2.33) by using the identity

$$r^2 + s^2 = (r + is)(r - is),$$

which helps us to factorize the equation as

$$H = \frac{1}{2}\sum_j \left(\omega_j q_j + ip_j\right)\left(\omega_j q_j - ip_j\right). \tag{2.34}$$

We further simplify the equation by labeling the factors

$$\alpha_j = \frac{1}{\sqrt{2}}\left(\omega_j q_j + ip_j\right), \tag{2.35}$$

$$\alpha_j^* = \frac{1}{\sqrt{2}}\left(\omega_j q_j - ip_j\right), \tag{2.36}$$

where $\alpha_j$ is a complex number and $\alpha_j^*$ is the conjugate of the complex number. In order to write the Hamiltonian expression for the cavity field modes, we invert equations (2.35) and (2.36) and rewrite them for $q_j$ and $p_j$, such that

$$q_j = \frac{1}{\sqrt{2}\,\omega_j}(\alpha_j^* + \alpha_j), \tag{2.37}$$

$$p_j = \frac{i}{\sqrt{2}}(\alpha_j^* - \alpha_j). \tag{2.38}$$

Hence, in terms of the complex numbers, $\alpha$ and $\alpha^*$, equation (2.33) reduces to

$$H = \frac{1}{2}\sum_j \left(\alpha_j \alpha_j^* + \alpha_j^* \alpha_j\right). \tag{2.39}$$

Since complex number $\alpha_j$ and its conjugate $\alpha_j^*$ commute, equation (2.39) can be simplified to $\sum|\alpha_j|^2$.

In a quantum mechanical description of the electromagnetic field we write $q_j$ and $p_j = \dot{q}_j$ as *operators*. Hence, the Hamiltonian of the multi-mode field reads

$$\hat{H} = \frac{1}{2}\sum_j \left(\omega_j^2 \hat{q}_j^2 + \hat{p}_j^2\right),$$

where $\hat{q}_j$ and $\hat{p}_j$ are the position and momentum operators. The corresponding commutation relations are $\left[\hat{q}_j, \hat{p}_{j'}\right] = i\hbar\hat{I}\delta_{jj'}$, $\left[\hat{q}_j, \hat{q}_{j'}\right] = \left[\hat{p}_j, \hat{p}_{j'}\right] = 0$, where $\hat{I}$ expresses the unit operator. Notice the quantization hereby depends explicitly on the constant $\hbar$ (the so-called Dirac $h$ = Planck constant divided by $2\pi = h/2\pi$). In comparison with $\alpha_j$ and $\alpha_j^*$, the corresponding factors are operators, expressed as

$$\hat{a}_j = \frac{1}{\sqrt{2\hbar\omega_j}}\left(\omega_j \hat{q}_j + i\hat{p}_j\right), \tag{2.40}$$

and

$$\hat{a}_j^\dagger = \frac{1}{\sqrt{2\hbar\omega_j}}\left(\omega_j\hat{q}_j - i\hat{p}_j\right). \tag{2.41}$$

As a result, the quantized multi-mode electromagnetic field inside the one-dimensional cavity is expressed by means of field operators $\hat{a}_j$ and $\hat{a}_j^\dagger$, called the annihilation operator and creation operator, respectively, or in short-ladder operators for the $j$th cavity mode. We define

$$\hat{q}_j = \sqrt{\frac{\hbar}{2\omega_j}}(\hat{a}_j + \hat{a}_j^\dagger), \tag{2.42}$$

$$\hat{p}_j = \frac{1}{i}\sqrt{\frac{\hbar\omega_j}{2}}(\hat{a}_j - \hat{a}_j^\dagger). \tag{2.43}$$

Hence, the quantum mechanical Hamiltonian becomes

$$\hat{H} = \frac{1}{2}\sum_j \hbar\omega_j\left(\hat{a}_j\hat{a}_j^\dagger + \hat{a}_j^\dagger\hat{a}_j\right). \tag{2.44}$$

This expression is different from equation (2.39) as the operators $\hat{a}_j$ and $\hat{a}_j^\dagger$ are related to each other through the commutation relation, $[\hat{a}_j, \hat{a}_{j'}^\dagger] = \delta_{jj'}\hat{I}$, which simplifies the above equation as

$$\hat{H} = \sum_j \hbar\omega_j\left(\hat{a}_j^\dagger\hat{a}_j + \frac{1}{2}\hat{I}\right). \tag{2.45}$$

This representation achieves one of our goals by writing the Hamiltonian expression in terms of the basic unit of energy, $\hbar\omega_j$, and counting the 'number of photons' using the number operator $\hat{N}_j = \hat{a}_j^\dagger\hat{a}_j$ for the $j$th mode.

*Electric and magnetic fields.* In terms of the ladder operators, the general expressions for the multi-mode electric field and magnetic field, as they appear in equations (2.30) and (2.31), become

$$\hat{\mathbf{E}}(\mathbf{x}, t) = i\sum_j\left(\frac{1}{V_j\epsilon_0}\right)^{1/2}\sqrt{\frac{\hbar\omega_j}{2}}\left(\hat{a}_j(t) - \hat{a}_j^\dagger(t)\right)\sin(k_jx)\,\mathbf{e}_y, \tag{2.46}$$

and

$$\hat{\mathbf{B}}(\mathbf{x}, t) = \sum_j\left(\frac{\mu_0}{V_j}\right)^{1/2}\sqrt{\frac{\hbar\omega_j}{2}}\left(\hat{a}_j(t) + \hat{a}_j^\dagger(t)\right)\cos(k_jx)\,\mathbf{e}_z. \tag{2.47}$$

*Ladder operators.* The Hamiltonian given by equation (2.45) is a product of the harmonic oscillator operators $\hat{a}_j$ and $\hat{a}_j^\dagger$. The product operator $\hat{a}_j^\dagger\hat{a}_j$ acts as a number operator since it is dimensionless and, therefore, contributes a number as it operates.

Clearly, $\hat{a}_j$ and $\hat{a}_j^\dagger$ are not Hermitian operators, that is, $\hat{a}_j^\dagger \neq \hat{a}_j$. They follow the commutation relation

$$\left[\hat{a}_j, \hat{a}_{j'}^\dagger\right] = \frac{1}{i\hbar}[\hat{q}_j, \hat{p}_{j'}] = \hat{I}\delta_{jj'}. \tag{2.48}$$

The product operator or number operator $\hat{N}_j = \hat{a}_j^\dagger \hat{a}_j$ is Hermitian, since

$$\hat{N}_j^\dagger = (\hat{a}_j^\dagger\, \hat{a}_j)^\dagger = \hat{a}_j^\dagger(\hat{a}_j^\dagger)^\dagger = \hat{a}_j^\dagger \hat{a}_j = \hat{N}_j, \tag{2.49}$$

Therefore, $\hat{N}_j$ being a Hermitian operator has real eigenvalues, $n_j$, and its eigenstates, $|n_j\rangle$, are orthogonal, that is,

$$\hat{N}_j|n_j\rangle = \hat{a}_j^\dagger\, \hat{a}_j|n_j\rangle = n_j|n_j\rangle. \tag{2.50}$$

(i) The set of operators $\left\{\hat{a}_j^\dagger, \hat{a}_j, \hat{N}_j, \hat{I}\right\}$ span a Lie algebra, which obey the following commutation relations:

$$[\hat{a}_j, \hat{I}] = [\hat{a}_j^\dagger, \hat{I}] = 0, \tag{2.51}$$

$$\left[\hat{a}_j, \hat{N}_j\right] = \hat{a}_j, [\hat{a}_j^\dagger, \hat{N}_j] = -\hat{a}_j^\dagger. \tag{2.52}$$

(ii) Following Heisenberg operator algebra, the time evolution of operator $\hat{a}_j$ reads

$$\dot{\hat{a}}_j = \frac{i}{\hbar}[\hat{H}, \hat{a}_j] = -i\omega_j\hat{a}_j, \tag{2.53}$$

$$\dot{\hat{a}}_j^\dagger = \frac{i}{\hbar}[\hat{H}, \hat{a}_j^\dagger] = i\omega_j\hat{a}_j^\dagger, \tag{2.54}$$

which is a linear system. We obtain the time evolution of operators $\hat{a}_j(t)$ and $\hat{a}_j^\dagger(t)$ as

$$\hat{a}_j(t) = \hat{a}_j(0)\mathrm{e}^{-i\omega_j t}, \tag{2.55}$$

$$\hat{a}_j^\dagger(t) = \hat{a}_j^\dagger(0)\mathrm{e}^{i\omega_j t}. \tag{2.56}$$

(iii) In order to understand the role of the harmonic oscillator operators $\hat{a}_j$ and $\hat{a}_j^\dagger$, we consider the eigenvalue equation (2.50) and operate the operator $\hat{a}_j$ such that

$$\hat{a}_j\, \hat{N}_j|n_j\rangle = \hat{a}_j\, \hat{a}_j^\dagger\hat{a}_j|n_j\rangle, \tag{2.57}$$

$$\hat{a}_j\, n_j|n_j\rangle = (1 + \hat{a}_j^\dagger\, \hat{a}_j)\hat{a}_j|n_j\rangle. \tag{2.58}$$

$$n_j\, \hat{a}_j\, |n\rangle = (1 + \hat{N}_j)\hat{a}_j|n_j\rangle. \tag{2.59}$$

We rearrange the terms to get

$$\hat{N}_j(\hat{a}_j|n_j\rangle) = (n_j - 1)(\hat{a}_j|n_j\rangle). \tag{2.60}$$

This implies $\hat{a}_j|n_j\rangle$ appears as

$$\hat{a}_j|n_j\rangle = \gamma|n_j - 1\rangle, \tag{2.61}$$

where $\gamma$ is an unknown constant. We calculate the value of $\gamma$ by projecting the dual vector $\langle n_j|$ onto the eigenvalue equation (2.50), such that

$$\langle n_j|\hat{N}_j|n_j\rangle = \langle n_j|a_j^\dagger a_j|n_j\rangle, \tag{2.62}$$

$$\langle n_j|n_j|n_j\rangle = \langle n_j - 1|\gamma^*\gamma|n_j - 1\rangle, \tag{2.63}$$

$$n_j = |\gamma|^2. \tag{2.64}$$

Therefore, we write

$$\hat{a}_j|n_j\rangle = \sqrt{n_j}|n_j - 1\rangle. \tag{2.65}$$

Hence, the operator $\hat{a}_j$ reduces the state number by 1, thus its name of the lowering operator or annihilation operator.

(iv) The smallest value of the number $n_j$ comes from the operation of $\hat{a}_j$ on $|n_j\rangle = |0\rangle$ such that

$$\hat{a}_j|0\rangle = 0, \tag{2.66}$$

which otherwise corresponds to negative energy.

(v) Similarly, we can show that

$$\hat{a}_j^\dagger|n_j\rangle = \sqrt{n_j + 1}\,|n_j + 1\rangle. \tag{2.67}$$

Hence, operators $\hat{a}_j^\dagger$ and $\hat{a}_j$, respectively, act as the raising operator and lowering operator, or alternatively the creation operator and annihilation operator. In addition, $n_j$ has positive integer values, that is, $n_j = 0, 1, 2, 3\ldots$

(vi) This reveals that in a particular $j$th mode the electromagnetic field has discrete portions of energy, $E_n^{(j)}$, that is,

$$E_n^{(j)} = \hbar\omega_j\left(n_j + \frac{1}{2}\right), \tag{2.68}$$

where

$$E_0^{(j)} = \hbar\omega_j\left(0 + \frac{1}{2}\right) \tag{2.69}$$

corresponds to vacuum energy of the $j$th mode, and

$$E_1^{(j)} = \hbar\omega_j\left(1 + \frac{1}{2}\right) \tag{2.70}$$

corresponds to one quantum of energy, which describes the particle nature of light.

*Poynting vector and momentum transferred.* The quantization of the electromagnetic fields hinges fundamentally on the dual connection between particle and wave pictures of light. This is compactly summarized by the de Broglie–Einstein formula, namely

$$p = \hbar k = \frac{2\pi\hbar}{\lambda}, \tag{2.71}$$

which relates the photon's momentum to the electromagnetic wave's wave-vector or equivalently to its wavelength, and

$$E = \hbar\omega, \tag{2.72}$$

which relates the photon's kinetic energy to the frequency of the electromagnetic wave. For electromagnetic waves in vacuum, the dispersion relation

$$\omega = ck \tag{2.73}$$

holds. If multiplied by $\hbar$ on both sides, this equation results in

$$E = cp \tag{2.74}$$

with the aid of equations (2.71) and (2.72). Of course, the relationship between field energy and field momentum had long been known as a statement of the energy conservation in vacuum, that is, the Maxwell equations lead to the continuity equation

$$\frac{\partial\rho}{\partial t} + \nabla \cdot \mathbf{J} = 0,$$

where $\rho$ defines the energy density,

$$\rho = \frac{1}{2}\left(\epsilon_0 \mathbf{E}^*(\mathbf{r},\, t)\cdot \mathbf{E}(\mathbf{r},\, t) + \frac{1}{\mu_0}\mathbf{B}^*(\mathbf{r},\, t)\cdot \mathbf{B}(\mathbf{r},\, t)\right),$$

and

$$\mathbf{J} = \mathbf{E} \times \mathbf{H}$$

is the Poynting vector. Integrated over a closed region of finite volume $V$, we get $\Delta E = (\Delta t/\Delta L)J_n$, where

$$\Delta E = \rho\Delta V,$$

and

$$J_n = \mathbf{J}\cdot\hat{n}$$

is the projection of the Poynting vector in the direction of the normal vector at the surface, and $S\Delta L = \Delta V$ relates the volume traversed by the electromagnetic wave in time, $\Delta t$, so that $E = pc$ if we simply identify $J_n = c^2 p$ as the momentum since $\Delta L/\Delta t = c$. A consequence of the quantization of the electromagnetic field, as we shall see shortly in equation (2.78), verifies $J_n = c^2 p$.

An important implication of $E = pc$ for the quantized electromagnetic field, namely photons, is the famous relativistic relationship $E = mc^2$, though it lies somewhat outside the interest of the present book. Nevertheless, let us digress from the main issue a while, considering a cart of mass $m$ capable of moving freely on a smooth surface, carrying a photon emitter and a photon receiver, both fixed firmly on the top of the cart, as shown in figure 2.3. Now suppose a photon of energy $E$ is emitted toward the receiver and is absorbed by the receiver. Throughout the process, the center-of-mass of the whole system remains unmoved, but the cart is recoiled till the photon is caught by the receiver. This means the photon has carried some residual mass, $\delta m$, from the cart and deposited it at a distance a little short of the separation between the emitter and receiver. A simple calculation leads to $E = \delta m\, c^2$ (to the lowest order in $v/c$). In short, the photon picture of the electromagnetic fields encompasses a wide range of physical phenomena ranging from special relativity to quantum mechanics.

Field quantization is best effected using the vector potential, for its simplicity, rather than **E** and **B** fields, both of which are components of a second-rank tensor constructed from the vector field. Nevertheless, we implement quantization using **E** and **B** in this book for the sake of later semi-classical treatments of the interaction of light and atoms. The quantized electromagnetic fields of a monochromatic light beam propagating freely in the $x$-direction with uniform amplitude of cross-section $S$ over a finite range of length $L$ read, in the complex representation,

$$\hat{E}_y(x,\, t) = \left(\frac{1}{LS\epsilon_0}\right)^{1/2} \sqrt{\hbar\omega}\ \hat{a}_k \mathrm{e}^{i(kx-\omega t)}, \tag{2.75}$$

and

$$\hat{B}_z(x,\, t) = -\left(\frac{\mu_0}{LS}\right)^{1/2} \sqrt{\hbar\omega}\ \hat{a}_k \mathrm{e}^{i(kx-\omega t)}. \tag{2.76}$$

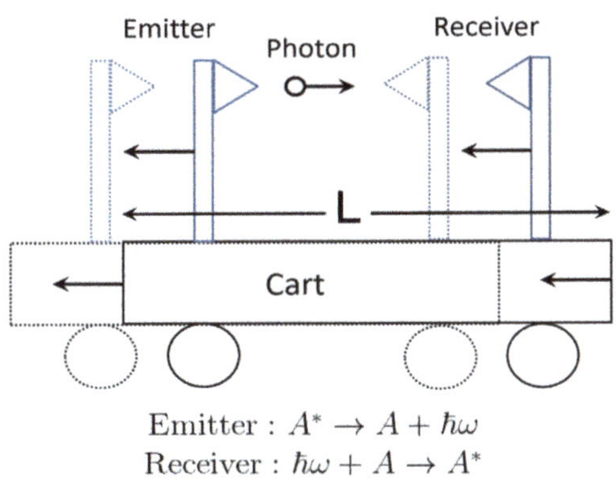


**Figure 2.3.** A gedanken for the equivalence of mass and energy, $E = mc^2$, using a moving cart equipped with a photon emitter and photon absorber. The recoil of the cart after emission of a photon suggests the loss of residual mass carried away by the photon. The formula $p = E/c$ for classical electromagnetic waves is applied.

It then holds that the energy contained in the volume $V$ is

$$\int_V \hat{\rho}\, dV = \left(\hat{N}_k + \frac{1}{2}\right)\hbar\omega, \tag{2.77}$$

where the energy-density operator is

$$\hat{\rho} = \frac{1}{4}\left(\epsilon_0\left(\hat{E}_y^\dagger \cdot \hat{E}_y + \hat{E}_y \cdot \hat{E}_y^\dagger\right) + \frac{1}{\mu_0}(\hat{B}_z^\dagger \cdot \hat{B}_z + \hat{B}_z \cdot \hat{B}_z^\dagger)\right).$$

Likewise, the Poynting vector integrated over volume $V$ yields

$$\int_V \hat{J}_z\, dV = c^2 \hat{N}_k \hbar k. \tag{2.78}$$

Thus, the above equation is **the total momentum** multiplied by $c^2$. Hence, we write the momentum operator $\hat{P}$ in number operator form, that is,

$$\hat{P} = \frac{1}{c^2}\int_V \hat{J}_z\, dV = \hat{N}_k \hbar k. \tag{2.79}$$

The momentum operator commutes with the Hamiltonian and thus acts as a constant of motion.

**Example:** In this book, we deal with optical cavities and lattices for which it is necessary to construct standing waves. Customarily, this is done by superposing the $k$ and $(-k)$ components such as

$$\hat{E}_y(x, t) \propto \left(\frac{1}{LS\epsilon_0}\right)^{1/2} \sqrt{\hbar\omega}\ (\hat{a}_k \mathrm{e}^{i(kx-\omega t)} - \hat{a}_{-k}\mathrm{e}^{i(-kx-\omega t)}), \tag{2.80}$$

which satisfies the vanishing boundary condition at $x = 0$ and $x = L$ for a selective set of values of $k$. We have $k = 2n\pi/L$, where $n = 1, 2, 3, \ldots$, provided $\hat{a}_{-k} = \hat{a}_k$. The latter operatorial equality requires an appropriate interpretation, but at least it is conducive to $\hat{N}_{-k} = \hat{N}_k$, that is, there needs to be as many photons propagating in the direction of $\mathbf{e}_x$ as in the opposite $-\mathbf{e}_x$ direction for the formation of a standing wave. Let us note to this end that equation (2.3) yields

$$B_z(x, t) \propto \frac{1}{c}\left(\frac{1}{LS\epsilon_0}\right)^{1/2} \sqrt{\hbar\omega}\ (\hat{a}_k \mathrm{e}^{i(kx-\omega t)} + \hat{a}_{-k}\mathrm{e}^{i(-kx-\omega t)})$$

so that the Poyinting vector becomes

$$\mathbf{J} \propto c^2(\hbar|k|\hat{N}_k + (-\hbar|k|)\hat{N}_{-k}) = 0,$$

with the terms involving $\mathrm{e}^{\pm 2ikx}$ integrated out to zero. Thus, there is no net field momentum inside a cavity. In what follows, we construct standing waves following an alternative procedure that proves consistent results and is useful in handling optical cavities and lattices.

## 2.3 Quantized electromagnetic fields

The quantization of the electromagnetic field as explained above is useful to understand quantum mechanical electromagnetic fields in nature. The process hinges on the non-commutativity of the **E** and **B** fields. Quantized fields can be produced, subject to experimental conditions. If the intensity (as an operator) is quantized, we observe aspects of the Einstein–de Broglie postulates. By contrast, if we quantize the phase and amplitude of the electric field and the magnetic field, then we observe their new states. Such quantized fields are utilized with remarkable control, such as in the categories of number state, coherent state, Schrödinger cat state, and squeezed state, etc discussed in the following subsections.

### 2.3.1 Fock state or number state

We express a weak electromagnetic field by a number state $|n\rangle$, also named a Fock state. Following Planck's hypothesis, an electromagnetic field is understood to be composed of discrete energy packets, photons, where in addition the energy of a packet is proportional to its frequency, $\omega$. The Hamiltonian of a single-mode field yields the energy, $E_n$, of a weak or low-intensity electromagnetic field in the photon state, $|n\rangle$, that is,

$$\hat{H}|n\rangle = \hbar\omega\left(n + \frac{1}{2}\right)|n\rangle = E_n|n\rangle, \tag{2.81}$$

where

$$\hat{H} = \hbar\omega\left(\hat{a}^\dagger\hat{a} + \frac{1}{2}\hat{I}\right). \tag{2.82}$$

The eigenstate $|n\rangle$ defines the Fock state of the field, corresponding to the energy eigenvalues, which is

$$E_n = \hbar\omega\left(n + \frac{1}{2}\right).$$

We may express the operators $\hat{a}^\dagger$, $\hat{a}$, $\hat{N}$ as the outer product of field state $|n\rangle$, that is,

$$\hat{a} = \sqrt{n}\,|n - 1\rangle\langle n|, \tag{2.83}$$

$$\hat{a}^\dagger = \sqrt{n}\,|n\rangle\langle n - 1|, \tag{2.84}$$

$$\hat{N} = n|n\rangle\langle n|. \tag{2.85}$$

Moreover, the eigenvalue of the number operator, $n$, corresponds to the number of photons in the cavity. The lowest value of $n$ being zero corresponds to the vacuum state of the electromagnetic field with the corresponding energy, $E_0 = \hbar\omega/2$, namely the zero-point energy or vacuum energy. The presence of zero-point energy in the quantum domain makes the quantum mechanical electromagnetic field different from the corresponding classical field, which gives zero energy value to the vacuum field. At the same time, zero-point energy, being the energy related to nothingness, also ignites an interesting philosophical discussion.

With the help of equation (2.46), when written for a single mode, we find the first and second moments of an electric field in the Fock state as

$$\langle \hat{E} \rangle = 0, \tag{2.86}$$

$$\langle \hat{E}^2 \rangle = c_j^2 \frac{\hbar}{\omega_j}\left(n + \frac{1}{2}\right)\sin^2(kx). \tag{2.87}$$

Equation (2.87) describes the intensity of a single-mode field. It is important to note that the intensity of an electromagnetic field is non-zero for vacuum, which is remarkably different from the zero value set for the classical counterpart. Furthermore, the intensity is linearly proportional to the photon number, $n$. For this reason, the value of $n$ also defines the intensity of a single-mode field in a Fock state.

The mathematical description of a vacuum state comes from equations (2.83) and (2.84), which provide recursion relations for $|n\rangle$. Since the recursion relation for the annihilation operator, $\hat{a}$, always changes the state in descending order, we get a cutoff condition defining the vacuum state, that is,

$$\hat{a}|0\rangle = 0. \tag{2.88}$$

As an interesting application of the recursion relation for the creation operator, $\hat{a}_j^\dagger$, we mention that the Fock state $\{|n\rangle\}$ is obtained by repeated application of the creation operator $\hat{a}^\dagger$ on the vacuum state $|0\rangle$, that is,

$$|n\rangle = \frac{(\hat{a}^\dagger)^n}{\sqrt{n!}}|0\rangle. \tag{2.89}$$

The number states satisfy the condition of ortho-normality, that is, $\langle n|n'\rangle = \delta_{n,n'}$. Therefore, these states span the Hilbert space, known as Fock space. The states obey the completeness relation, that is,

$$\sum_n |\, n\rangle\langle n \,| = \hat{I}. \tag{2.90}$$

In order to calculate the distribution function corresponding to the eigenstates in a single mode, we use equation (2.88). On substituting the definition of annihilation operator $\hat{a}$ from equation (2.40), we get a first order differential equation for the vacuum state, such that

$$\left(\omega q + \hbar\frac{\partial}{\partial q}\right)\phi_0(q) = 0. \tag{2.91}$$

Here, $\phi_0(q) = \langle q|0\rangle$. Hence, as a solution of the differential equation, we find

$$\phi_0(q) = N\exp(-q^2\omega/2\hbar), \tag{2.92}$$

where $N = \frac{1}{\sqrt[4]{\pi\hbar / \omega}}$ comes from the normalization condition.

Hence, following equation (2.89) and applying the creation operator $\hat{a}^\dagger$ successively, we obtain the higher number states $|n\rangle$, such that

$$\phi_n(q) = \frac{1}{\sqrt{n!}}\left(\omega q - \hbar\frac{\partial}{\partial q}\right)^n \phi_0(q). \tag{2.93}$$

In figure 2.4, we show the eigenfunctions $\phi_n(q)$ generated using equation (2.93) and the corresponding probability distributions $|\phi_n(q)|^2$ for the energy eigenstates. Further, in figure 2.5 the phase-space description of the vacuum state and the single-photon Fock state is shown. Engineering number states in an electromagnetic field [5, 6] and other quantum systems [7] has been performed[3].

### 2.3.2 Coherent states

We may identify a high-intensity laser source as being in a coherent state, $|\alpha\rangle$. In the following, we give three alternative definitions of the coherent state.

**Definition 1:** A coherent state is mathematically defined as the eigenstate of the annihilation operator, $\hat{a}$, that is,

$$\hat{a} \mid \alpha\rangle = \alpha \mid \alpha\rangle, \tag{2.94}$$

where $\alpha$ is a complex number. This amounts to saying that, owing to the high intensity, the omission of a photon does not change the state of the electromagnetic field. It also suggests that, when acting on $|\alpha\rangle$, we treat the operator $\hat{a}$ as a complex number $\alpha$ as if we were dealing with a classical field.

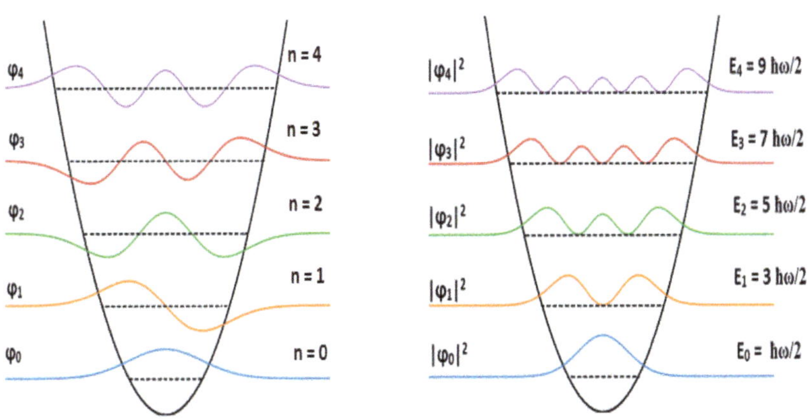


**Figure 2.4.** Harmonic oscillator in $q$-space with five eigenstates (left), and the corresponding probability distributions and energies (right).

[3] How to measure a Fock state is an interesting question ...

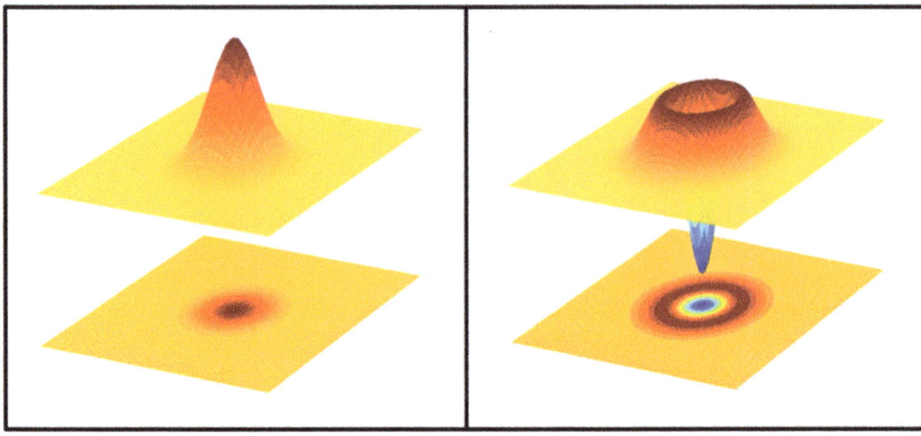

**Figure 2.5.** Wigner probability distribution functions in phase space for a vacuum state, $n = 0$ (left), and a single Fock state, that is, $n = 1$ (right).

Since any arbitrary state can be expressed in a Hilbert space as a superposition of eigenvectors, we express the coherent state in the Hilbert space spanned by the Fock states, that is,

$$|\alpha\rangle = \hat{I}|\alpha\rangle = \sum_n |n\rangle\langle n|\alpha\rangle = \sum_n c_n|n\rangle. \tag{2.95}$$

Here, $c_n = \langle n|\alpha\rangle$. We use equations (2.89) and (2.94) to obtain

$$|\alpha\rangle = \langle 0|\alpha\rangle \sum_n \frac{\alpha^n}{\sqrt{n!}}|n\rangle. \tag{2.96}$$

The quantity $\langle 0|\alpha\rangle$ is found from the normalization condition, that is,

$$\langle\alpha|\alpha\rangle = 1.$$

Therefore, we find

$$\langle 0|\alpha\rangle = \mathrm{e}^{-|\alpha|^2/2}.$$

Hence, we write the complete expression for the coherent state as

$$|\alpha\rangle = \mathrm{e}^{-|\alpha|^2/2} \sum_n \frac{\alpha^n}{\sqrt{n!}}|n\rangle. \tag{2.97}$$

We can obtain the same expression in another way, that is, by taking the Hermitian adjoint of the recursion relation for the creation operator, $\hat{a}_j^\dagger$ given in equation (2.67). We obtain

$$\langle n|\hat{a} = \sqrt{n+1}\,\langle n+1|, \tag{2.98}$$

which leads to another recursion relation,

$$\sqrt{n+1}\langle n+1|\alpha\rangle = \alpha\langle n|\alpha\rangle, \tag{2.99}$$

for the scalar products $\langle n|\alpha\rangle$, which provides

$$\langle n|\alpha\rangle = \frac{\alpha^n}{\sqrt{n!}}\langle 0|\alpha\rangle. \tag{2.100}$$

These scalar products appear as the expansion coefficients of the state $|\alpha\rangle$ in terms of Fock states, that is,

$$|\alpha\rangle = \sum_n |n\rangle\langle n|\alpha\rangle = \langle 0|\alpha\rangle \sum_n \frac{\alpha^n}{\sqrt{n!}}|n\rangle = e^{-|\alpha|^2/2} \sum_n \frac{\alpha^n}{\sqrt{n!}}|n\rangle. \tag{2.101}$$

Comparing equation (2.101) with equation (2.95), we note that the probability distribution $|c_n|^2$ is a Poisson distribution:

$$|c_n|^2 = e^{-|\alpha|^2} \frac{|\alpha|^{2n}}{n!},$$

with an average value

$$\bar{n} = \langle \hat{n} \rangle = \langle \alpha|\hat{a}^\dagger \hat{a}|\alpha\rangle = |\alpha|^2.$$

Therefore, $|\alpha|$ is related to the average value or the peak value of the distribution, such that

$$|\alpha| = \sqrt{\bar{n}}.$$

As a comment on the evolution of coherent states, it is interesting to note that the coherent state describes a state of motion close to the corresponding classical trajectory in a harmonic oscillator. It is straightforward to show that the expectation value of the position operator $\hat{q}(t) = \sqrt{\frac{\hbar}{2\omega}}(\hat{a}e^{-i\omega t} + \hat{a}^\dagger e^{i\omega t})$ executing a harmonic oscillation evolves in a coherent state as

$$\langle \hat{q}(t) \rangle = \langle \alpha \mid \hat{q}(t) \mid \alpha \rangle = \mathcal{A}\cos(\omega t - \phi), \tag{2.102}$$

where $\alpha = |\alpha| e^{i\phi}$ and $\mathcal{A}^2 = |\alpha|^2 \hbar/\omega$. Identifying $\mathcal{A}$ with the amplitude, the expectation value of the position operator for the coherent state behaves like the position of a classical particle in a corresponding oscillator. In this sense, the coherent state is considered closest to the state of the classical counterpart. We leave the proof of equation (2.102) as an exercise for the reader.

**Definition 2:** A coherent state is also generated by applying a displacement operator, $\hat{D}(\alpha)$, on the vacuum state, $|0\rangle$, of the harmonic oscillator:

$$|\alpha\rangle = \hat{D}(\alpha) \mid 0\rangle, \tag{2.103}$$

where the displacement operator $\hat{D}(\alpha)$ is defined as

$$\hat{D}(\alpha) = e^{\alpha\hat{a}^\dagger - \alpha^*\hat{a}}. \tag{2.104}$$

This implies that the coherent state is a displaced vacuum state, with the same distribution properties, such as variance.

We apply the Baker–Campbell–Hausdorff (BCH) identity

$$e^{\hat{A}+\hat{B}} = e^{-[\hat{A},\,\hat{B}]/2}\, e^{\hat{A}}\, e^{\hat{B}}, \tag{2.105}$$

on the displacement operator, which leads to

$$\hat{D}(\alpha) = e^{\alpha\hat{a}^{\dagger}-\alpha^{*}\hat{a}} = e^{|\alpha|^{2}[\hat{a}^{\dagger},\,\hat{a}]/2}\, e^{\alpha\hat{a}^{\dagger}}\, e^{-\alpha^{*}\hat{a}}. \tag{2.106}$$

Therefore, we rewrite equation (2.103) by introducing the completeness relation $(\sum_n |n\rangle\langle n| = \hat{1})$ as

$$|\alpha\rangle = \hat{1}\hat{D}(\alpha)|0\rangle, \tag{2.107}$$

$$=\sum_n |n\rangle\langle n|\hat{D}(\alpha)|0\rangle, \tag{2.108}$$

$$=e^{-|\alpha|^{2}/2}\sum_n |n\rangle\langle n|e^{\alpha\hat{a}^{\dagger}}|0\rangle, \tag{2.109}$$

$$=e^{-|\alpha|^{2}/2}\sum_n \frac{\alpha^{n}}{\sqrt{n!}}|n\rangle. \tag{2.110}$$

Here, $\hat{1}$ is the unit operator. With the help of equation (2.110) it is easy to prove that, unlike the Fock states, the coherent states are non-orthogonal, that is,

$$\langle\alpha|\beta\rangle \neq 0. \tag{2.111}$$

Here, $|\alpha\rangle$ and $|\beta\rangle$ are two different coherent states.

**Definition 3:** In an alternative way, a coherent state, $|\alpha\rangle$, is defined as a quantum state with a minimum uncertainty product, that is,

$$\Delta q \Delta p = \frac{\hbar}{2}. \tag{2.112}$$

The relation follows as we calculate the variance of position and momentum operators, expressed in terms of $\hat{a}$ and $\hat{a}^{\dagger}$, for the coherent states as

$$(\Delta q)^{2} = \frac{\hbar}{2\omega}[\langle\alpha \mid (\hat{a}^{\dagger} + \hat{a})^{2} \mid \alpha\rangle - (\langle\alpha \mid \hat{a}^{\dagger} + \hat{a} \mid \alpha\rangle)^{2}] = \frac{\hbar}{2\omega}, \tag{2.113}$$

and

$$(\Delta p)^{2} = \frac{\hbar\omega}{2}[-\langle\alpha \mid (\hat{a}^{\dagger} - \hat{a})^{2} \mid \alpha\rangle + (\langle\alpha \mid \hat{a}^{\dagger} - \hat{a} \mid \alpha\rangle)^{2}] = \frac{\hbar\omega}{2}. \tag{2.114}$$

In 1926, Erwin Schrödinger attempted to build quantum mechanical states which follow classical trajectories and obey a minimum uncertainty relationship, and eventually act as non-dispersive wave packets [8]. He succeeded for the harmonic oscillator, obtaining for the first time a one-to-one correspondence between wave packet and classical particle. In 1963, Roy J Glauber [9], and independently E C G Sudarshan [10] in the context of the quantum theory of light, introduced electromagnetic field states which resemble the classical field, and named them coherent states. As presented above, there are three definitions of these coherent states: (i) coherent states $|\alpha\rangle$ which are the eigenstates of a harmonic oscillator annihilation operator; (ii) coherent states generated by the action of a displacement operator on the vacuum state of the harmonic oscillator; and (iii) coherent states which are quantum states that minimize the uncertainty relation. These states are associated with Heisenberg–Weyl algebra in terms of the harmonic oscillator creation and annihilation operators. These three definitions are equivalent, and a coherent state obtained in this way possesses a special set of properties: (i) continuity in parameters, (ii) resolution of unity, and (iii) minimization of the uncertainty relationship, which remains at a minimum under time evolution. Non-dispersive quantum states for more general physical systems, however, remained a dream of Erwin Shrödinger.

Continuous efforts have been made to extend the notion of coherent states to other general physical systems. These generalizations are based on group theoretic approaches, thus coherent states are constructed for a variety of algebraic groups, such as Lie groups SU(1,1), and non-compact groups SO(2,1), following definitions (i) and (ii), respectively. The coherent states based on these generalizations are continuously parameterized and admit the resolution of unity, which are two of their main characteristics.

In 1996, John R Klauder suggested a new kind of coherent state suitable for systems with degenerate discrete spectra (i.e. the hydrogen atom) without an explicit dependence on any group structure [11]. In Klauder's opinion, these states possess the properties of a coherent state, as expressed above, hence the long-standing problem of finding non-dispersive wave packets for the hydrogen atom was considered solved. Paolo Bellomo and C R Stroud showed that the Klauder states are dispersive in nature for the hydrogen atom [12].

Jean Pierre Gazeau and Klauder extended the idea of temporally stable coherent states for non-degenerate discrete and continuous spectra [13]. This formalism attracted much more attention due to its independence from the groups, and these coherent states have been built for a variety of Hamiltonian systems, such as the pseudo-harmonic oscillator, the Morse potential, the Pöschl–Teller potential, the infinite square well and one-mode systems with periodic potentials [14].

The Gazeau–Klauder (GK) coherent states are defined for a general class of one-dimensional potentials defined by a power law [15]. The power-law potentials include loosely binding and tightly binding potentials, whereas the harmonic potential serves as a boundary between them. It is observed that the special construction of quantum states following the GK coherent states formalism does not overcome the inherent dispersion of a dynamical system with a non-linear energy spectrum [16, 17].

Kevin Eric Cahill and Glauber have developed coherent states for fermions [18]. Their work has been extended to develop coherent states of attractive superfluid Fermi gas to explain Bogoliubov oscillations [19].

### 2.3.3 Schrödinger cat state

The Schrödinger cat state has always been food for thought for scientists. Schrödinger coined the idea of 'his' cat state to explain the manifestation of quantum mechanics on a macroscopic scale. We may consider a Schrödinger cat state to be the superposition of two coherent states, that is,

$$|\psi\rangle = \frac{\mathcal{A}}{\sqrt{2}}(|\alpha\rangle + |-\alpha^*\rangle). \tag{2.115}$$

The constant $\mathcal{A}$ is obtained by using the normalization condition:

$$\mathcal{A}^2 = \frac{1}{1 + e^{-2|\alpha|^2}},$$

due to the fact that the two different coherent states are non-orthogonal. With $\alpha = e^{-i\varphi}$, we get a special case of

$$|\psi\rangle = \frac{\mathcal{A}}{\sqrt{2}}(|e^{-i\varphi}\rangle + |-e^{i\varphi}\rangle). \tag{2.116}$$

Here, $\varphi$ is an arbitrary phase. Total interference occurs when $\varphi = 0$ modulo $\pi$, as the values of $e^{i\varphi}$ and $e^{-i\varphi}$ coincide.

Schrödinger concocted a gedanken experiment based on a quantum mechanical phenomenon. Specifically, he proposed a cat in a box with radioactive material connected to another system that triggers poisonous gas if a radioactive particle is detected. The radioactivity, being a quantum phenomenon, creates a quantum state for the cat in either a live state, $|\text{live}\rangle$, or a dead state, $|\text{dead}\rangle$, namely $\frac{1}{\sqrt{2}}(|\text{live}\rangle + |\text{dead}\rangle)$. A macroscopic observer is capable of identifying either of the states at the same time. In other words, the cat remains in the quantum state until the observation is made. Thus, the outcome is probabilistic even if the observation is macroscopic, as long as the state of the cat is a quantum mechanical state. The coinage 'Schrödinger cat state' encapsulates the puzzling nature of quantum theory.

### 2.3.4 Squeezed state

From a simple mathematical point of view, while attempting to understand squeezed states [20, 21] it is worth noting that the uncertainty relation given in equation (2.112) does not require a unique solution for $\Delta q$ or $\Delta p$. This non-uniqueness can be expressed graphically in phase space. We make the graphical illustration by considering the scaled quantities

$$\Delta a = \sqrt{\frac{\omega}{2\hbar}}\Delta q \text{ and } \Delta b = \sqrt{\frac{1}{2\hbar\omega}}\Delta p.$$

Hence, the uncertainty relation given in equation (2.112) becomes

$$\Delta a \Delta b = \frac{1}{4}.$$

Next, we use the scaled $\Delta a$ and $\Delta b$ in defining the phase-space variables

$$Q = \Delta a \cos \theta \text{and} P = \Delta b \sin \theta. \tag{2.117}$$

A little algebra helps us to obtain

$$\frac{Q^2}{(\Delta a)^2} + \frac{P^2}{(\Delta b)^2} = 1, \tag{2.118}$$

which is *the uncertainty ellipse*, with $\Delta a$ and $\Delta b$ as the semi-major and semi-minor axes. In the case when $\Delta a \neq \Delta b$, the resulting states are called *squeezed states*. When $\Delta a = \Delta b = 1/2$, the trajectory becomes a circle, and hence a coherent state.

We show phase-space descriptions of a coherent state, Schrödinger cat state, and squeezed state in figure 2.6.

After first being observed in 1985 [22], squeezed states of light became vital in enhancing the sensitivity of interferometers. For example, they have been successfully applied in improving the sensitivity of a gravitational-wave detector [23]. In this, the squeezed state made the measurements possible below the shot-noise limit [24] and, further, it was cost effective when the other possibility of obtaining the same sensitivity was to increase the arm's length of the gravitational-wave detector so that it was much more expansive. In general, squeezed states are the optimum

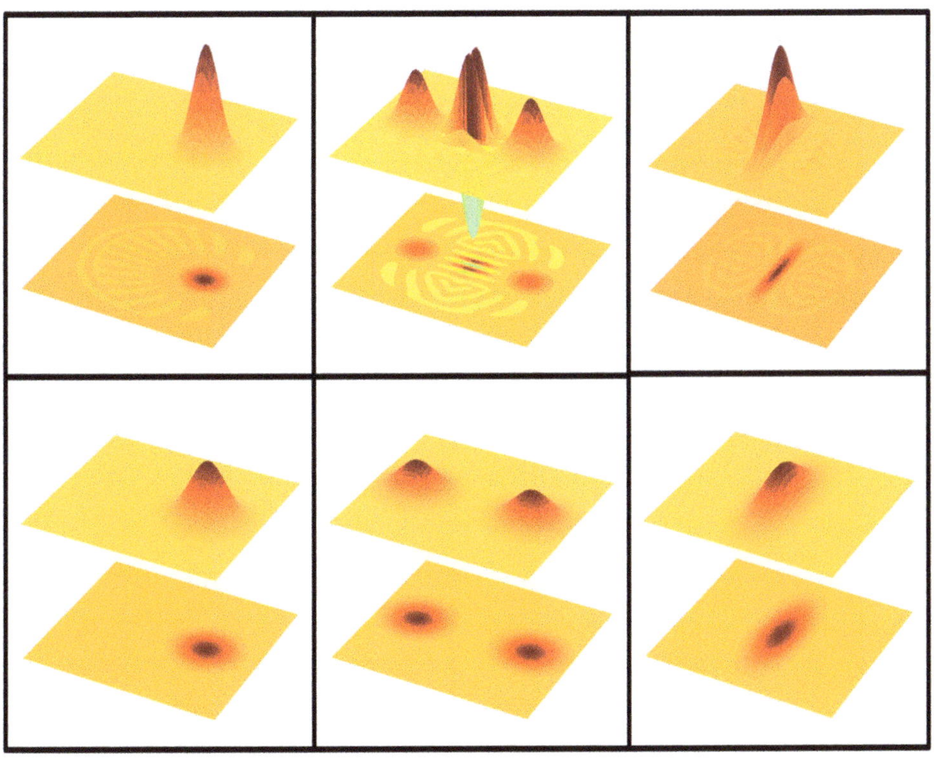

**Figure 2.6.** Wigner functions (upper column) and Husimi functions (lower column) of a coherent state (left), Schrödinger cat state (middle), and squeezed state (right).

non-classical states for laser interferometers operating with large average photon numbers.

## 2.4 Quantum distribution functions

In classical dynamics, the phase-space representation provides a comparative global picture of the dynamical system. Representation of a quantum particle in phase space is a non-trivial question. The uncertainty principle restricts the classical point-like picture of a particle and replaces it with a probabilistic cloud. Historically, various efforts have been made to make a phase-space representation possible. Below, we learn the essence of the Wigner function and the Husimi, or Q-distribution, function. These representations serve as indispensable tools for analyzing the interaction of atoms with the electromagnetic field.

### 2.4.1 Wigner function

The Wigner distribution function was introduced by Eugene Wigner in 1932 as a way to express quantum corrections to classical statistical mechanics [25, 26]. In terms of wave function, $|\Psi\rangle$, the Wigner function can be defined as

$$W(x, p) = \frac{1}{\pi\hbar}\int_{-\infty}^{\infty} \Psi^*(x - y/2)\Psi(x + y/2)e^{-ipy/\hbar}dy, \tag{2.119}$$

which expresses the quantum particle in phase space defined by the position variable $x$ and conjugate momentum variable $p$. The Wigner distribution is regarded as a quasi-probability distribution function as it may take negative values in phase space.

In figure 2.5, we show Wigner functions of a vacuum state, $n = 0$, and a single Fock state, $n = 1$. In figure 2.6, upper row, we show Wigner functions of a coherent state (left), Schrödinger cat state (middle), and squeezed state (right).

**Example:** Express a vacuum state in phase space using a Wigner distribution function.

We write the ground state of the harmonic oscillator or vacuum state as

$$\phi_0(q) = N \exp(-q^2/2b^2),$$

where $b^2 = \hbar/\omega$ and $N = \frac{1}{\sqrt[4]{\pi}b}$. The corresponding Wigner function is obtained by equation (2.119) as

$$W_0(q, p) = \frac{1}{\pi\hbar}e^{-b^2p^2/\hbar^2 - q^2/b^2}.$$

It is symmetric around the origin and displays a Gaussian distribution both in position space and in momentum space.

### 2.4.2 Husimi or Q-distribution function

The Husimi distribution function, or Q-distribution function [27, 28], expresses a quantum particle by projecting its wave function, $|\Psi\rangle$, over the coherent state, $|\alpha\rangle$, so that

$$Q(\alpha_r, \alpha_i) = \frac{1}{\pi}|\langle\alpha|\psi\rangle|^2, \tag{2.120}$$

where $\alpha_r = \text{Re}[\alpha]$ and $\alpha_i = \text{Im}[\alpha]$ are the real and imaginary parts of $\alpha$. The positive-valued probability distribution function expresses the quantum particle in the phase space made up by the real and imaginary parts of $\alpha$.

In figure 2.6, lower row, we show the Husimi function of a coherent state (left), a Schrödinger cat state (middle), and a squeezed state (right). The implication of the Schrödinger cat state is presented as two humps in the probability distribution, which implies that the state is in either of the two alternative states. In addition, the interference effect is seen in between the two humps in the corresponding Wigner function.

In addition, there are other representations such as a Sudarshan–Glauber P-distribution, Kirkwood–Rihaczek distribution, Mehta, Rivier, and Born–Jordan representations, each suitable for characterization of dynamical systems for a given problem [29–31].

**Example:** Express a vacuum state in phase space using the Husimi distribution function.

Following equation (2.120), we write the Husimi distribution function of the vacuum state as

$$Q_0(\alpha_r, \alpha_i) = \frac{1}{\pi}|\langle\alpha|0\rangle|^2, \tag{2.121}$$

where we apply

$$\langle\alpha|0\rangle = e^{-|\alpha|^2/2}$$

to obtain

$$Q_0(\alpha_r, \alpha_i) = \frac{1}{\pi}e^{-|\alpha|^2/2} = \frac{1}{\pi}e^{-(\alpha_r^2+\alpha_i^2)/2}. \tag{2.122}$$

It is symmetric around the origin and has a Gaussian distribution both in real and imaginary coordinates. We leave the phase-space illustration of the Husimi function for the vacuum state to the reader.

## Exercises

1. Show that, in a one-dimensional cavity of length $L$ and with perfectly reflecting mirrors at both ends, that the electromagnetic field $\mathbf{E}(\mathbf{x}, t)$ expressed by equation (2.28) has discrete modes with angular frequency $\omega_j = \frac{j\pi c}{L}$, where $j = 1, 2, 3, \ldots$.
2. Derive equation (2.26) from equation (2.24).
3. Show that the number operator $\hat{N}_j$ has its eigenstates as $|n_j\rangle$, whereas its eigenvalues are $n_j$.
4. Show that the expectation value of the position operator in a coherent state is

$$\langle\hat{q}(t)\rangle = \langle\alpha \mid \hat{q}(t) \mid \alpha\rangle = \mathcal{A}\cos(\omega t - \phi),$$

where $\alpha = | \alpha | e^{i\phi}$ and $\mathcal{A}^2 = | \alpha |^2 \hbar/\omega$.

5. Construct a squeezed state by considering a particular choice of $\Delta a$ and $\Delta b$. Plot it using an appropriate graphical package.
6. Calculate the Husimi distribution function for
   i. a Fock state;,
   ii. a coherent state;
   iii. a squeezed state; and
   iv. a Schrödinger cat state.
7. Show that equation (2.97) satisfies equation (2.94).
8. Show that

$$\langle 0|\alpha\rangle = e^{-|\alpha|^2/2}.$$

9. Using the BCH formula given in subsection 2.3.2, show that

$$\langle 0|\hat{D}|0\rangle = e^{-|\alpha|^2/2}.$$

10. Show that

$$(\Delta x)^2 = \frac{1}{2\omega},$$
$$(\Delta p)^2 = \frac{\omega}{2}.$$

11. Show that, in equation (2.115), we obtain

$$\mathcal{A}^2 = \frac{1}{1 + e^{-2|\alpha|^2}}.$$

## References


[1] Born M and Jordan P 1925 Zur quantenmechanik *Z. Phys.* **34** 858–88

[2] Born M, Heisenberg W and Jordan P 1925 Zur Quantenmechanik II *Z. Phys.* **35** 557–615
Born M, Heisenberg W and Jordan P 1926 Zur On quantum mechanics II *J. Phys.* **35** 557–615

[3] Dirac P A M 1927 The quantum theory of the emission and absorption of radiation *Proc. R. Soc.* A **114** 243–65

[4] Auletta G, Fortunato M and Parisi G 2009 *Quantum Mechanics* (Cambridge: Cambridge Univ. Press)

[5] Varcoe B T H, Brattke S, Weidinger M and Walther H 2000 Preparing pure photon state number state of the radiation field *Nature* **403** 743–6

[6] Bertet P, Auffeves A, Maioli P, Osnaghi S, Meunier T, Brune M, Raimond J M and Haroche S 2002 Direct measurement of the Wigner function of a one photon Fock state in a cavity *Phys. Rev. Lett.* **89** 200402

[7] Khan M M, Akram M J, Paternostro M and Saif F 2016 Engineering single-phonon number states of a mechanical oscillator via photon subtraction *Phys. Rev.* A **94** 063830

[8] Schrdöinger E 2916 Der stetige Übergang von der Mikro-zur Makromechanik *Naturwissenschaften* **14** 664
[9] Glauber R J 1963 Coherent and incoherent states of the radiation field *Phys. Rev.* **131** 2766–88
[10] Sudarshan E C G 1963 Equivalence of semiclassical and quantum mechanical descriptions of statistical light beams *Phys. Rev. Lett.* **10** 277
[11] Klauder J R 1996 Coherent state for the hydrogen atom *J. Phys. A: Math. Gen.* **29** L293
[12] Bellomo P and Stroud C R 1998 Dispersion of Klauder's temporally stable coherent states for the hydrogen atom *J. Phys. A: Math. Gen.* **31** L445
[13] Gazeau J P G and Klauder J R 1999 Coherent states for systems with discrete and continuous spectrum *J. Phys. A: Math. Gen.* **32** 123
[14] Gazeau J-P 2009 *Coherent States in Quantum Physics* (New York: Wiley-VCH)
[15] Iqbal S, Riviere P and Saif F 2010 Space-time dynamics of Gazeau–Klauder coherent states in power-law potential *Int. J. Theor. Phys.* **49** 2540–57
[16] Iqbal S and Saif F 2011 Generalized coherent states and their statistical characteristics in power-law potentials *J. Math. Phys.* **52** 082105–10
[17] Iqbal S and Saif F 2013 Gazeau-Klauder coherent states of the triangular-well potential *J. Russ. Laser Res.* **34** 77–86
[18] Cahill K E and Glauber R J 1999 Density operators for fermions *Phys. Rev.* A **59** 1538
[19] Saif F 2013 *J. Russ. Laser Res.* **34** 496
[20] Walls D F 1983 Squeezed states of light *Nature* **306** 141
[21] Loudon R and Knight P L 1987 Squeezed light *J. Mod. Opt.* **34** 709–59
[22] Slusher R, Hollberg L, Yurke B, Mertz J and Valley J 1985 *Phys. Rev. Lett.* **55** 2409
[23] Grote H, Danzmann K, Dooley K L, Schnabel R, Slutsky J and Vahlbruch H 2013 First long-term application of squeezed states of light in a gravitational-wave observatory *Phys. Rev. Lett.* **110** 181101
[24] Wheeler J A and Zurek W H 2014 *Quantum Theory and Measurement* (Princeton, NJ: Princeton University Press)
[25] Wigner E P 1932 On the quantum correction for thermodynamic equilibrium *Phys. Rev.* **40** 749–59
[26] Lougovski P, Solano E, Zhang Z M, Walther H, Mack H and Schleich W P 2003 Fresnel representation of the Wigner function: an operational approach *Phys. Rev. Lett.* **91** 010401 and references therein
[27] Husimi K 1940 Some formal properties of the density matrix *Proc. Phys. Math. Soc. Japan* **22** 264–314
[28] Leonhardt U 1997 *Measuring the Quantum State of Light Cambridge Studies in Modern Optics* (Cambridge: Cambridge University Press)
[29] Agarwal G S and Wolf E 1970 Calculus for functions of noncommuting operators and general phase-space methods in quantum mechanics. I. Mapping theorems and ordering of functions of noncommuting operators *Phys. Rev.* D **2** 2161–86
[30] Cahill K E and Glauber R J 1969 Ordered expansions in boson amplitude operators *Phys. Rev.* **177** 1857–81
[31] Cahill K E and Glauber R J 1969 Density operators and quasiprobability distributions *Phys. Rev.* **177** 1882–902

# Chapter 3

## Atom–field interaction

*'Do not Bodies and Light act mutually upon one another; that is to say, Bodies upon Light in emitting, reflecting, refracting, and inflicting it, and Light upon Bodies for heating them, and putting their parts into a vibrating motion wherein heat consists?'*

—Sir Issac Newton

Einstein's famous paper [1] reads, 'If a ray of light causes molecules hit by it to absorb or emit through an elementary process an amount of energy $h\nu$ in the form of radiation (induced radiation process), the momentum $h\nu/c$ is always transferred to the molecule, and in such a way that the momentum is directed along the direction of propagation of the ray if the energy is absorbed, and directed in the opposite direction, if the energy is emitted.' There, he presented his famous $A$ and $B$ coefficients, related to the emission and absorption of radiation, and introduced the phenomenon of spontaneous emission. Ten years later, Paul Dirac published his quantum theory of emission and absorption of radiation [2]. He also solved the mystery of the spontaneous emission of radiation, which was extended by Victor Weisskopf and Eugene Wigner [3]. The work initiated by Dirac was further explained by Werner Heisenberg and Wolfgang Pauli [4] and independently by Enrico Fermi [5] to develop the quantum theory of the absorption and emission of radiation by matter. Their work became the cornerstone of quantum electrodynamics—a full relativistic theory of charged particles interacting with the electromagnetic field.

Engineering an electromagnetic field in a finite region and its interaction with atoms led to the subject of cavity quantum electrodynamics (cavity-QED) [6]. The mode structure of a confined electromagnetic field is drastically modified by the boundary conditions. The radiative properties of an atom in confined space can thus be very different from those in the free space. For instance, spontaneous emission

doi:10.1088/978-0-7503-2308-6ch3

rates can be altered, and the energy levels may be shifted [7]. From its early achievements, cavity-QED became the workhorse of quantum optics and quantum electronics.

In this chapter, we explain the fundamental principles that are necessary for understanding the interaction of atoms with the optical field. The quantum mechanical approach to the problem is based on the quantum treatment of both the atom and the electromagnetic field. The atom here is modeled in a simplistic way, namely, it is regarded merely as a set of energy eigenstates with well-defined dipole matrix elements. No further multi-pole moments, electric or magnetic, will be considered in this book unless stated otherwise. We regard the atom as spherically symmetric so that the parity is a good quantum number. This fact leads to the selection rule that even-to-even or odd-to-odd dipole transitions are prohibited [8]. We will treat the center-of-mass motion of the atom in the optical field quantum mechanically but in the context of the effective electromagnetic field. In the semi-classical approach, however, we treat the electromagnetic field classically.

We discuss in parallel the near-resonant regime where real transitions contribute and the far-off resonant regime where virtual transitions contribute to atomic interaction with the cavity field mode. The interaction drives the temporal and spatial evolution of matter waves in the cavity field. In addition, we learn about the scattering regimes, such as Bragg scattering and Raman–Nath scattering, of the matter waves. For a general discussion of the atom–field interaction, we refer the reader to [9–15].

## 3.1 Atom–field interaction

*The two-level atom.* Let us consider an atom with a single active outer electron. Alkali atoms are a good example of such hydrogen-like atoms. In the presence of an electromagnetic field the electron may jump from a lower level to a higher level, thus developing a dipole. The two-level atom has served as an excellent pedagogical example for describing the interaction between the atom and the optical field to bring out the essence of the physics.

We label the lower level, that is, the ground state, by $|g\rangle$ and the higher level, or excited state, by $|e\rangle$, as shown in figure 3.1. These states are eigenstates of the atomic Hamiltonian operator, $\hat{H}_A$, and the corresponding eigenenergies are, respectively, $E_g$ and $E_e$, that is,

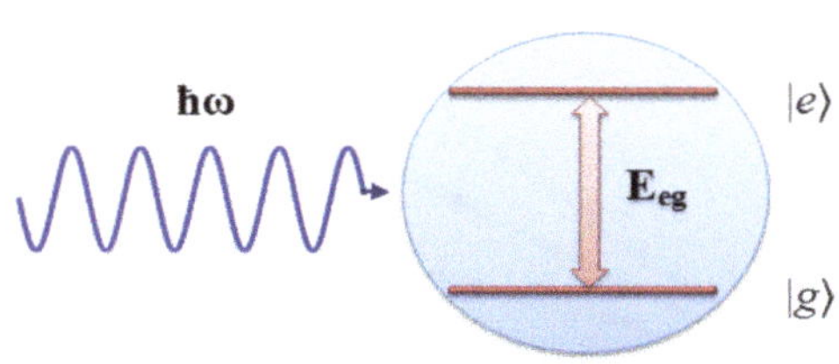


**Figure 3.1.** A two-level atom expressed by a ground state $|g\rangle$ and excited state $|e\rangle$ interacts with an electromagnetic field of frequency $\omega$.

$$\hat{H}_A|g\rangle = E_g|g\rangle,$$

$$\hat{H}_A|e\rangle = E_e|e\rangle.$$

The atomic Hamiltonian $\hat{H}_A$ can be expressed in the outer product form as we consider the completeness relation for atomic energy eigenstates, that is, $\sum_i | i\rangle\langle i | = |g\rangle\langle g| + |e\rangle\langle e| = \hat{I}$. Here, the symbol $\hat{I}$ expresses the unit operator. We write

$$\hat{H}_A = \hat{I}\hat{H}_A\hat{I} = E_e|e\rangle\langle e| + E_g|g\rangle\langle g| = E_e\hat{\sigma}_{ee} + E_g\hat{\sigma}_{gg}. \tag{3.1}$$

Here, $\hat{\sigma}_{ee} = |e\rangle\langle e|$ and $\hat{\sigma}_{gg} = |g\rangle\langle g|$.

The evolution of the atom in the single-mode electromagnetic field,

$$\mathbf{E}(\mathbf{r}, t) = E_0\, \mathbf{u}(\mathbf{r})\cos \omega t, \tag{3.2}$$

becomes time dependent due to the time dependence of the field. Here, $E_0$ and $\omega$ are the amplitude and frequency of the field, respectively. The quantity $\mathbf{u}(\mathbf{r}) = \mathbf{e}_0\, u(r)$ is the mode function, $u(r)$ describes the distribution of the optical field in space, whereas $\mathbf{e}_0$ describes the polarization vector.

*Interaction Hamiltonian.* In the presence of an electromagnetic field, a dipole is formed on an atom: between the negatively charged electron expressed by the position vector $\mathbf{r}_e$ and the positively charged nucleus of the atom expressed by the position vector $\mathbf{r}_n$ (see figure 3.2). Consequently, the interaction of the atom with the electromagnetic field is governed by the interaction Hamiltonian:

$$H_{int} = -e\mathbf{r}_e \cdot \mathbf{E}(\mathbf{r}_e, t) + e\mathbf{r}_n \cdot \mathbf{E}(\mathbf{r}_n, t). \tag{3.3}$$

The difference of sign in the two dipole moment terms is due to the negative and positive charges, respectively, associated with the displaced electron and nucleus. Here,

$$\mathbf{r}_e = \mathbf{r} + \delta\mathbf{r}_e, \qquad \mathbf{r}_n = \mathbf{r} + \delta\mathbf{r}_n.$$

The vector $\mathbf{r}$ describes the center-of-mass position vector, whereas $\delta\mathbf{r}_e$ and $\delta\mathbf{r}_n$ are given as

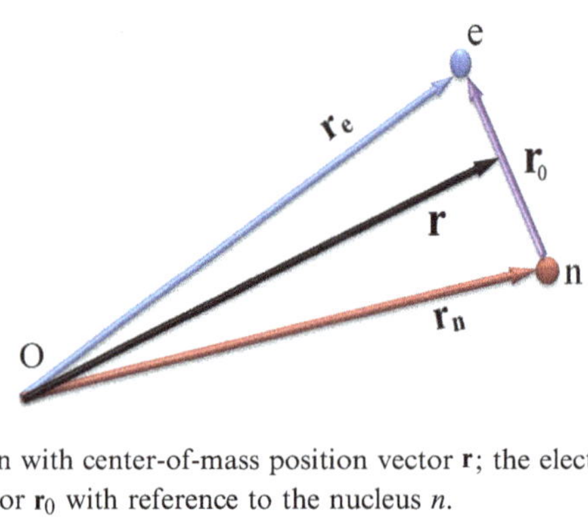


**Figure 3.2.** Atomic dipole seen with center-of-mass position vector $\mathbf{r}$; the electron $e$ in relative coordinates is expressed by the position vector $\mathbf{r}_0$ with reference to the nucleus $n$.

$$\delta \mathbf{r}_e = \frac{m_e \, \mathbf{r}_0}{m_e + m_n},$$

$$\delta \mathbf{r}_n = -\frac{m_n \, \mathbf{r}_0}{m_e + m_n}.$$

The symbols $m_e$ and $m_n$ describe the mass of the electron and the mass of the nucleus, respectively.

The wavelength of the ultraviolet, visible, or infrared radiation is measured in hundreds to thousands of nanometers. It is orders of magnitude larger than the charge distribution on an atom. Therefore, we consider that the field does not change significantly over the dimension of the atom and take $E(\mathbf{r}_e, t)$, $E(\mathbf{r}_n, t) \simeq E(\mathbf{r}, t)$, which is a *dipole approximation.* Thus, the interaction Hamiltonian becomes

$$H_{\text{int}} \simeq -e(\mathbf{r}_e - \mathbf{r}_n) \cdot \mathbf{E}(\mathbf{r}, t), \tag{3.4}$$

$$= -e\mathbf{r}_0 \cdot \mathbf{E}(\mathbf{r}, t), \tag{3.5}$$

$$= -\vec{\wp} \cdot \mathbf{E}(\mathbf{r}, t). \tag{3.6}$$

The electric field $\mathbf{E}(\mathbf{r}, t)$, however, describes the spatial profile of the field at the center-of-mass of the atom.

### 3.1.1 Semi-classical treatment

*Hamiltonian description.* Let us develop the effective Hamiltonian, which governs the atomic interaction with an optical field. We treat the atom and its center-of-mass motion quantum mechanically, whereas we consider the optical field as classical. The following Hamiltonian provides the semi-classical description of the interaction between the atom and the optical field, as

$$\hat{H} = \frac{\hat{\mathbf{p}}^2}{2m} + \hbar\omega_e|e\rangle\langle e| + \hbar\omega_g|g\rangle\langle g| - \hat{\vec{\wp}} \cdot \mathbf{E}(\mathbf{r}, t), \tag{3.7}$$

where $\hat{\mathbf{p}}$ describes the center-of-mass momentum. Here, the first term on the right-hand side corresponds to the kinetic energy of the atom. It becomes crucial when there is a significant atomic motion during the interaction of the atom with the optical field.

With the help of the completeness relation, we may express the dipole operator, $\hat{\vec{\wp}}$, as

$$\hat{\vec{\wp}} = \hat{I} \cdot \hat{\vec{\wp}} \cdot \hat{I}, \tag{3.8}$$

$$= e(|g\rangle\langle g| + |e\rangle\langle e|)\mathbf{r}_0(|g\rangle\langle g| + |e\rangle\langle e|), \tag{3.9}$$

$$= \vec{\wp}_{eg}|e\rangle\langle g| + \vec{\wp}_{ge}|g\rangle\langle e|. \tag{3.10}$$

Here, we have introduced the matrix elements of the electric dipole moment, $\vec{\wp}_{eg} \equiv e\langle e|\mathbf{r}_0|g\rangle$ and $\vec{\wp}_{ge} \equiv e\langle g|\mathbf{r}_0|e\rangle = \vec{\wp}^*_{eg}$. Moreover, we take the atom as spherically symmetric so that the diagonal elements, $\langle g|\mathbf{r}_0|g\rangle$ and $\langle e|\mathbf{r}_0|e\rangle$, vanish as the energy eigenstates have well-defined parity.

We define $\hat{\sigma}_+ \equiv |e\rangle\langle g|$ and $\hat{\sigma}_- \equiv |g\rangle\langle e|$ as the atomic raising operator and atomic lowering operator, respectively. In addition, we consider a phase, $\phi$, associated to the dipole moment, as

$$\hat{\vec{\wp}} = (\hat{\sigma}_+ e^{+i\phi} + \hat{\sigma}_- e^{-i\phi})\vec{\wp}. \tag{3.11}$$

Here, $|\vec{\wp}_{ge}| = |\vec{\wp}^*_{eg}| = \vec{\wp}$. Therefore, we obtain the interaction Hamiltonian as

$$\hat{H}_{\text{int}} = -(\hat{\sigma}_+ e^{+i\phi} + \hat{\sigma}_- e^{-i\phi})\vec{\wp} \cdot \mathbf{E}(\mathbf{r}, t), \tag{3.12}$$

which leads us to write

$$\hat{H} = \frac{\hat{\mathbf{p}}^2}{2m} + \hbar\omega_e|e\rangle\langle e| + \hbar\omega_g|g\rangle\langle g| - \vec{\wp} \cdot \mathbf{E}(\mathbf{r}, t)\,(\hat{\sigma}_+ e^{+i\phi} + \hat{\sigma}_- e^{-i\phi}). \tag{3.13}$$

*Time-dependent Schrödinger equation.* The evolution of the atom in the time-dependent electromagnetic field is expressed by the time-dependent Schrödinger wave equation, that is,

$$i\hbar\frac{\partial|\psi\rangle}{\partial t} = \hat{H}|\psi\rangle, \tag{3.14}$$

Here, $\hat{H}$ denotes the total Hamiltonian, given in equation (3.13). In order to solve the time-dependent Schrödinger equation, we express the wave function, $|\psi\rangle$, as

$$|\psi\rangle \equiv e^{-i\omega_g t}\psi_g(r, t)|g\rangle + e^{-i(\omega_g+\omega)t}\psi_e(r, t)|e\rangle. \tag{3.15}$$

Here, $\psi_g = \psi_g(r, t)$ and $\psi_e = \psi_e(r, t)$ are the probability amplitudes in the ground state and excited state, respectively.

We substitute equation (3.15) into the time-dependent Schrödinger equation, given in equation (3.14). After a little operator algebra, we finally get the coupled equations for the probability amplitudes, $\psi_g$ and $\psi_e$, expressed as

$$i\hbar\frac{\partial\psi_g}{\partial t} = \frac{\hat{p}^2}{2m}\psi_g - \frac{1}{2}\hbar\Omega_R u(r)(1 + e^{-2i\omega t})e^{+i\phi}\psi_e, \tag{3.16}$$

$$i\hbar\frac{\partial\psi_e}{\partial t} = \frac{\hat{p}^2}{2m}\psi_e - \hbar\Delta\psi_e - \frac{1}{2}\hbar\Omega_R u(r)(1 + e^{-2i\omega t})e^{-i\phi}\psi_g. \tag{3.17}$$

Here, $\Omega_R \equiv |\vec{\wp} \cdot \mathbf{e}_0 E_0/2\hbar|$, and $\Delta \equiv \omega - \omega_{eg}$ is the measure of the detuning of the external field frequency, $\omega$, away from the atomic transition frequency, expressed as $\omega_{eg} = \omega_e - \omega_g$.

In *the rotating wave approximation*, we eliminate the rapidly oscillating terms in the coupled equations (3.16) and (3.17). We assume that the probability amplitudes, $\psi_e$ and $\psi_g$, do not change appreciably over the time scale and approximate

$$1 + e^{-2i\omega t} \sim 1.$$

In a rigorous mathematical way, we perform *averaging* of equations (3.16) and (3.17) over a period, $t \equiv 2\pi/\omega$. This eliminates the explicit time dependence in the coupled equations, and we obtain

$$i\hbar\frac{\partial\psi_g}{\partial t} = \frac{\hat{p}^2}{2m}\psi_g - \hbar\Omega_R u(r)e^{+i\phi}\psi_e, \tag{3.18}$$

$$i\hbar\frac{\partial\psi_e}{\partial t} = \frac{\hat{p}^2}{2m}\psi_e - \hbar\Delta\psi_e - \hbar\Omega_R u(r)e^{-i\phi}\psi_g. \tag{3.19}$$

*3.1.1.1 Resonant and near-resonant regimes*

We consider that $\Delta = 0$ and $\Delta \approx 0$, respectively. We further simplify equations (3.18), and (3.19) by considering a very small center-of-mass momentum. Hence, we get the set of coupled differential equations as

$$i\hbar\frac{\partial\psi_g}{\partial t} = -\hbar\mu(r)e^{+i\phi}\psi_e, \tag{3.20}$$

$$i\hbar\frac{\partial\psi_e}{\partial t} = -\hbar\mu(r)e^{-i\phi}\psi_g, \tag{3.21}$$

where $\mu(r) = \Omega_R\, u(r)$. It is easy to see that the probability amplitudes $\psi_g$ and $\psi_e$ follow oscillator equations, that is,

$$\frac{\partial^2\psi_g}{\partial t^2} = -\mu^2(r)\psi_g, \tag{3.22}$$

$$\frac{\partial^2\psi_e}{\partial t^2} = -\mu^2(r)\psi_e. \tag{3.23}$$

This implies that the probabilities of finding the electron in the ground state $|g\rangle$ and in the excited state $|e\rangle$ oscillate with a position-dependent frequency, $\mu(r) = \Omega_R u(r)$. However, if we consider that over the excited state and the ground state the electric field remains constant, the oscillation frequency, $\mu$, becomes constant. For example, in the microwave regime we may take $u = 1$, therefore the oscillation frequency becomes $\mu = \Omega_R$.

*3.1.1.2 Far-off resonant regime*

We consider a large detuning between the field frequency, $\omega$, and the atomic transition frequency, $\omega_{eg}$, such that $|\Delta| \gg |\mu|$. This implies that no appreciable change occurs in the population of the excited state. It is named *secular approximation*. Hence, we take

$$\dot{\psi}_{\rm e} \approx 0, \quad \text{and} \quad \psi_{\rm e}' \approx 0.$$

Here, $\dot{\psi}_{\rm e}$ and $\psi_{\rm e}'$ are first derivatives of the probability amplitude, $\psi_{\rm e}$, with respect to time and position, respectively. We consider that the change in $\psi_{\rm e}$ with respect to time and position becomes vanishingly small. Therefore, equation (3.19) provides the probability amplitude $\psi_{\rm e}$ as

$$\psi_{\rm e}(r,\,t) \approx -\frac{\Omega_{\rm R}}{\Delta}u(r){\rm e}^{-i\phi}\psi_{\rm g}(r,\,t). \tag{3.24}$$

Substitution of the expression for $\psi_{\rm e}$ in equation (3.18) leads to

$$i\hbar\frac{\partial\psi_{\rm g}}{\partial t} \cong \frac{\hat{p}^2}{2m}\psi_{\rm g} + \frac{\hbar\Omega_{\rm R}^2}{\Delta}u^2(r)\psi_{\rm g}. \tag{3.25}$$

The time-dependent Schrödinger wave equation effectively controls the dynamics of the atom in the optical field. We note that under the condition of a large atom–field detuning, the atom stays in its electronic ground state $\psi_{\rm g}$ and *observes* the effective optical potential, $\frac{\hbar\Omega_{\rm R}^2}{\Delta}u^2(r)$.

*The effective Hamiltonian.* Equation (3.25) leads us to the effective Hamiltonian, $H_{\rm eff}$, such as

$$H_{\rm eff} \equiv \frac{\hat{p}^2}{2m} + \frac{\hbar\Omega_{\rm R}^2}{\Delta}u^2(r). \tag{3.26}$$

Here, the first term describes the center-of-mass kinetic energy, whereas the second term indicates an effective potential as seen by the atom. The spatial variation in the potential enters through the mode function of the electromagnetic field, $u$. Fascinatingly, we note that by a proper choice of the mode function almost any potential for the atom can be made.

The second term in equation (3.26) leads to an effective force $F = -2(\hbar\Omega_{\rm R}^2/\Delta)u\nabla u$, where $\nabla u$ describes gradient of $u$. Hence, the interaction of the atom with the electromagnetic field exerts a position-dependent gradient force on the atom, which is directly proportional to the *square* of the Rabi frequency, $\Omega_{\rm R}^2$, and inversely proportional to the detuning, $\Delta$. The optical force can, therefore, be made repulsive or attractive by controlling the sign of the detuning. An atom, therefore, experiences a repulsive force as it interacts with a blue detuned optical field, for which the field frequency is larger than the transition frequency, $\Delta > 0$. However, there is an attractive force on the atom if it finds a red detuned optical field, which has a frequency smaller than the atomic transition frequency, $\Delta < 0$. Thus, in principle, we can construct any atom optics component and apparatus for matter waves, analogous to classical optics.

Attractive and repulsive optical gradient force is seen at work in the development of optical tweezers at Bell Labs. This seminal work was recognized with the award of the Nobel Prize in 2018 to Arthur Ashkin [16–18]. This became a vital instrument in trapping nano-sized particles, DNA structures, and biological samples. The radiation pressure force led to the development of traps for atoms and ions. By using two evanescent wave fields with different decay constants, and with one exerting attractive force on atoms and the other repulsive behavior, the arrangement has been used to trap cold atoms near the surface of an atomic mirror and optical nanofibers. The bi-modal surface traps provide efficient trapping conditions for cold atoms.

### 3.1.2 Quantum mechanical treatment

We express the total Hamiltonian that describes the interaction of an atom and field quantum mechanically as

$$\hat{H} = \hat{H}_{\mathrm{A}} + \hat{H}_{\mathrm{F}} + \hat{H}_{\mathrm{int}}. \tag{3.27}$$

*Two-level atom.* Here, $\hat{H}_{\mathrm{A}}$ describes the Hamiltonian of the two-level atom, in the absence of the center-of-mass motion. We may express $\hat{H}_{\mathrm{A}} = \left(E_{\mathrm{e}}\hat{\sigma}_{\mathrm{ee}} + E_{\mathrm{g}}\hat{\sigma}_{\mathrm{gg}}\right)$ as

$$\hat{H}_{\mathrm{A}} = \frac{1}{2}\left(E_{\mathrm{e}} - E_{\mathrm{g}}\right)\left(\hat{\sigma}_{\mathrm{ee}} - \hat{\sigma}_{\mathrm{gg}}\right) + \frac{1}{2}\left(E_{\mathrm{e}} + E_{\mathrm{g}}\right)\left(\hat{\sigma}_{\mathrm{ee}} + \hat{\sigma}_{\mathrm{gg}}\right).$$

It is interesting to note that $\hbar\omega_{\mathrm{eg}} = \left(E_{\mathrm{e}} - E_{\mathrm{g}}\right)$ is the transition energy, whereas $\left(\hat{\sigma}_{\mathrm{ee}} + \hat{\sigma}_{\mathrm{gg}}\right) = \hat{I}$ acts as the unit operator. The constant-energy term, that is, the mean energy $\left(E_{\mathrm{e}} + E_{\mathrm{g}}\right)/2$, will be suppressed by choosing this to be the zero of the atom's energy. We adopt the following simplifying notations of Pauli's matrices:

$$\hat{\sigma}_z = \hat{\sigma}_{\mathrm{ee}} - \hat{\sigma}_{\mathrm{gg}} = | e\rangle\langle e | - | g\rangle\langle g |.$$

Hence, the above Hamiltonian becomes

$$\hat{H}_{\mathrm{A}} = \frac{1}{2}\hbar\omega_{\mathrm{eg}}\hat{\sigma}_z + \frac{1}{2}(E_{\mathrm{e}} + E_{\mathrm{g}})\hat{I}. \tag{3.28}$$

*Quantized electromagnetic field.* In section 2.2, we expressed a multi-mode electric field, $\hat{E}$, in terms of the creation and annihilation operators. Therefore, as in equation (2.46), we write the electromagnetic field:

$$\hat{\mathbf{E}}(\hat{\mathbf{r}}, t) = +i\sum_j E_j\left(\hat{a}_j - \hat{a}_j^{\dagger}\right)\mathbf{u}_j(\hat{r}). \tag{3.29}$$

Here, $E_j = c_j\left(\hbar\omega_j/2\right)^{1/2}$. Moreover, $\mathbf{u}_j(\hat{r}) = u_j(\hat{r})\mathbf{e}_j$, where $u_j$ expresses the general spatial profile of the $j$th field mode and $\mathbf{e}_j$ is the corresponding polarization vector. We have obtained the multi-mode electromagnetic field Hamiltonian in equation (2.45), that is,

$$\hat{H}_{\mathrm{F}} = \hbar\sum_j\omega_j\left(\hat{a}_j^{\dagger}\hat{a}_j + \frac{1}{2}\hat{I}\right), \tag{3.30}$$

whereas each summation term expresses a single mode of the electromagnetic field.

*Interaction Hamiltonian.* As we substitute equation (3.29) in equation (3.12), we obtain

$$\hat{H}_{\text{int}} = -(\hat{\sigma}_+ e^{+i\phi} + \hat{\sigma}_- e^{-i\phi})\vec{\wp} \cdot \hat{\mathbf{E}}(\hat{\mathbf{r}}, t), \tag{3.31}$$

$$= -i\hbar \sum_j \mu_j u_j(r)(\hat{\sigma}_+ e^{+i\phi} + \hat{\sigma}_- e^{-i\phi})\left(\hat{a}_j - \hat{a}_j^\dagger\right), \tag{3.32}$$

where $\mu_j = |\vec{\wp} \cdot \mathbf{e}_j E_j / \hbar|$. We consider the field profile $u_j(r)$ as constant in the microwave regime, and take it as unity. However, in the case where the de Broglie wavelength is of the order of the wavelength of light, the light interacts with spatially varying charge distributions, making $u(r)$ significant. On substituting $\phi = \pi/2$, we get

$$\hat{H}_{\text{int}} = \hbar \sum_j \mu_j u(r)(\hat{\sigma}_+ - \hat{\sigma}_-)\left(\hat{a}_j - \hat{a}_j^\dagger\right), \tag{3.33}$$

*Hamiltonian formulation.* Therefore, substituting the expressions for $\hat{H}_A$, $\hat{H}_F$, and $\hat{H}_{\text{int}}$ from equations (3.28), (3.30), and (3.32), respectively, in equation (3.27) we write the total Hamiltonian, which is

$$\hat{H} = \frac{1}{2}\hbar\omega_{\text{eg}}\hat{\sigma}_z + \sum_j \hbar\omega_j \hat{a}_j^\dagger \hat{a}_j + \hbar \sum_j \mu_j (\hat{\sigma}_+ - \hat{\sigma}_-)\left(\hat{a}_j - \hat{a}_j^\dagger\right), \tag{3.34}$$

$$= \frac{1}{2}\hbar\omega_{\text{eg}}\hat{\sigma}_z + \sum_j \hbar\omega_j \hat{a}_j^\dagger \hat{a}_j + \hbar \sum_j \mu_j \left(\hat{\sigma}_+ \hat{a}_j - \hat{\sigma}_+ \hat{a}_j^\dagger - \hat{\sigma}_- \hat{a}_j + \hat{\sigma}_- \hat{a}_j^\dagger\right). \tag{3.35}$$

The terms corresponding to unit operators introduce only a constant-energy shift, and for this reason we have ignored them. In equation (3.35) above, the term $\hat{\sigma}_+ \hat{a}_j^\dagger$ represents the transition of an electron from ground to excited state and the creation of a photon, whereas $\hat{\sigma}_- \hat{a}_j$ indicates de-excitation of the electron and absorption of a photon; both are unphysical phenomena. The two terms obviously violate energy conservation in this two-state framework and hence are dropped hereafter, which is rotating wave approximation.

Thus, the interaction of a two-level atom with an $n$-mode field is expressed by the Hamiltonian, that is,

$$\hat{H} = \frac{1}{2}\hbar\omega_{\text{eg}}\hat{\sigma}_z + \sum_{j=1}^{n} \hbar\omega_j \hat{a}_j^\dagger \hat{a}_j + \hbar \sum_{j=1}^{n} \mu_j \left(\hat{\sigma}_+ \hat{a}_j + \hat{\sigma}_- \hat{a}_j^\dagger\right). \tag{3.36}$$

Whereas, the interaction of a single two-level atom with a single-mode field is expressed as

$$\hat{H} = \frac{1}{2}\hbar\omega_{\text{eg}}\hat{\sigma}_z + \hbar\omega \hat{a}^\dagger \hat{a} + \hbar\mu(\hat{\sigma}_+ \hat{a} + \hat{\sigma}_- \hat{a}^\dagger). \tag{3.37}$$

*3.1.2.1 Resonant and near-resonant regimes*

Here, we consider the resonant regime, that is, $\Delta = 0$, and the near-resonant regime, where $\Delta \approx 0$. The electromagnetic field interacts with the atom and causes transition of electrons between the electronic states of the atom. The interaction regime is seen at work, for example, in developing experimentally feasible systems in quantum information and at the same time in light amplification through stimulated emission of radiation or lasers.

For the description of the atom–field interaction dynamics, we express the Hamiltonian in the *interaction picture*, given as

$$\hat{V} = e^{\frac{i\hat{H}_0 t}{\hbar}} \hat{H}_\mathrm{I} e^{\frac{-i\hat{H}_0 t}{\hbar}}. \tag{3.38}$$

Here, $\hat{H}_0 = \frac{1}{2}\hbar\omega_{eg}\hat{\sigma}_z + \hbar\omega\hat{a}^\dagger\hat{a}$, which is the sum of the atomic Hamiltonian and the electromagnetic field Hamiltonian. Moreover, $\hat{H}_\mathrm{I} = \hbar\mu(\hat{\sigma}_+\hat{a} + \hat{\sigma}_-\hat{a}^\dagger)$ expresses the interaction term.

As we substitute the explicit expressions of $\hat{H}_0$ and $\hat{H}_\mathrm{I}$, and apply the identity

$$e^{\alpha\hat{A}}\hat{B}e^{-\alpha\hat{A}} = \hat{B} + \alpha[\hat{A}, \hat{B}] + \frac{\alpha^2}{2!}[\hat{A}, [\hat{A}, \hat{B}]] + \cdots, \tag{3.39}$$

the interaction Hamiltonian becomes

$$\hat{V} = \hbar\mu(\hat{\sigma}_+\hat{a}e^{i\Delta t} + \hat{\sigma}_-\hat{a}^\dagger e^{-i\Delta t}), \tag{3.40}$$

where $\Delta = \omega - \omega_{eg}$ is the atom–field detuning. Equation (3.40) is the Jaynes–Cummings Hamiltonian, which provides the quantum mechanical description of the interaction between a two-level atom and a single-mode field.

Let us note that in the interaction picture the state vectors are transformed from state vectors $\{|S(t)\rangle\}$ of the original Schrödinger picture into state vectors $\{|I(t)\rangle\}$ of the interaction picture, represented symbolically by

$$|I(t)\rangle = e^{\frac{i\hat{H}_0 t}{\hbar}}|S(t)\rangle.$$

For the resonant case, $\Delta = 0$, the interaction Hamiltonian reduces simply to

$$\hat{V} = \hbar\mu(\hat{\sigma}_+\hat{a} + \hat{\sigma}_-\hat{a}^\dagger). \tag{3.41}$$

*3.1.2.2 Far-off resonant regime*

In this regime, we consider large detuning such that $|\Delta| \gg |\mu\sqrt{n+1}\,|$. The coupling between the field and the atom $\mu$ cannot induce a transition, but renormalizes the energies of the system. The following unitary transformation reduces the Hamiltonian to a more manageable form to the lowest order in the parameter $\frac{\mu}{\Delta}$, namely

$$\hat{U} = e^{i\frac{\mu}{\Delta}(\hat{a}\hat{\sigma}_+ - \hat{a}^\dagger\hat{\sigma}_-)}. \tag{3.42}$$

Applying perturbation theory with respect to $\mu/\Delta$, we transform the Hamiltonian $\hat{H}$, equation (3.37), into the effective Hamiltonian, that is,

$$\hat{H}_{\text{eff}} = \hat{U}\hat{H}\hat{U}^\dagger = -\frac{\hbar\mu^2}{2\Delta} - \frac{\hbar}{2}\left(\omega_{\text{eg}} + \frac{\mu^2}{\Delta}\right)\hat{\sigma}_z + \hbar\left(\omega - \frac{\mu^2}{\Delta}\hat{\sigma}_z\right)\hat{a}^\dagger\hat{a}. \tag{3.43}$$

This implies that the effective cavity frequency now depends on the atomic state, and equivalently the atomic energy splitting depends on the number of photons in the cavity. The term proportional to $\hat{\sigma}_z\hat{a}^\dagger\hat{a}$ is called the AC-Stark shift, while the 'vacuum' shift of the atomic energy splitting, $\hat{\sigma}_z\mu^2/\hbar\Delta$, is called the Lamb shift. This implies that atomic transition is shifted by $\frac{\mu^2}{\Delta}(n + 1/2)$. Incidentally, the constant-energy term in equation (3.43) may simply be ignored.

### 3.1.3 Time-dependent wave function

In general, we express the initial state of an atom and field as a tensor product, that is,

$$|\psi(t = 0)\rangle = |\psi_{\text{A}}\rangle \otimes |\psi_{\text{F}}\rangle, \tag{3.44}$$

where we may consider a two-level atom in the state

$$|\psi_{\text{A}}\rangle = c_{\text{e}}|e\rangle + c_{\text{g}}|g\rangle,$$

and the single-mode electromagnetic field as a superposition of number states, that is,

$$|\psi_{\text{F}}\rangle = \sum_n w_n|n\rangle.$$

As a result, equation (3.44) becomes

$$|\psi(t = 0)\rangle = \sum_{n=0}^{\infty} c_{\text{e},n}(t = 0)|e, n\rangle + c_{\text{g},n}(t = 0)|g, n\rangle, \tag{3.45}$$

where

$$c_{\text{e},n}(t = 0) = c_{\text{e}}w_n,$$

and

$$c_{\text{g},n}(t = 0) = c_{\text{g}}w_n.$$

The interaction of the atom with the electromagnetic field modifies the probability amplitudes, $c_{\text{e},n}$ and $c_{\text{g},n}$, as a function of time. This leads us to develop an understanding that the atom–field combined state after the interaction time, $t$, may be written as

$$|\psi(t)\rangle = \sum_{n=0}^{\infty} c_{e,n}(t)|e, n\rangle + c_{g,n}(t)|g, n\rangle, \tag{3.46}$$

where $c_{e,n}(t)$ and $c_{g,n}(t)$ are to be calculated.

The time-evolved wave function $|\psi(t)\rangle$ has important information about the interacting system, which can be obtained by applying the time transformation operator $\hat{U}$, that is,

$$\hat{U} = \exp\left(\frac{-i\hat{V}t}{\hbar}\right).$$

The Hamiltonian operator $\hat{V}$, given in equation (3.41), expresses the physical system of the atom and the field, whereas the operator $\hat{U}$ describes the transformation of the physical system beyond the initial state $|\psi(t = 0)\rangle$. Hence, we write

$$|\psi(t)\rangle = \hat{U}|\psi(0)\rangle. \tag{3.47}$$

$$= \sum_{n=0}^{\infty} c_{e,n}(t = 0)\hat{U}|e, n\rangle + c_{g,n}(t = 0)\hat{U}|g, n\rangle. \tag{3.48}$$

The time transformation operator $\hat{U}$ obtained in appendix A acts as a time machine.

*In the resonant case.* We express the atom–field interaction Hamiltonian by equation (3.41) for $\Delta = 0$. Hence, equation (3.48) leads us to write $|\psi(t)\rangle$ as

$$|\psi(t)\rangle = \sum_{n=0}^{\infty} c_{e,n}(t)|e, n\rangle + c_{g,n}(t)|g, n\rangle, \tag{3.49}$$

where

$$c_{e,n}(t) = \cos(\Omega_n t)c_{e,n}(0) - i\sin(\Omega_n t)c_{g,n+1}(0), \tag{3.50}$$

$$c_{g,n+1}(t) = \cos(\Omega_n t)c_{g,n+1}(0) - i\sin(\Omega_n t)c_{e,n}(0). \tag{3.51}$$

Here, the Rabi frequency is expressed as

$$\Omega_n = \mu\sqrt{n + 1}.$$

Equations (3.50) and (3.51) provide the probability amplitudes as functions of interaction time, $t$.

**Example:** For an atom initially in the excited state and field in the vacuum state, we write

$$|\psi(t)\rangle = \cos(\Omega_0 t)|e, 0\rangle - i\sin(\Omega_0 t)|g, 1\rangle. \tag{3.52}$$

We have considered $c_e = 1$ and $c_g = 0$, and $w_{n=0} = 1$. This implies that the probability of finding the electron in the excited state, $|\cos(\Omega_0 t)|^2$, and the probability of finding the electron in the ground state, $|\sin(\Omega_0 t)|^2$, vary periodically, such that their sum at a time of interaction $t$ is equal to one.

*In the general case.* When $\Delta \neq 0$, in general, the atom–field system is described by the interaction Hamiltonian given in equation (3.40), which reveals

$$c_{e,n}(t) = e^{-i\Delta t/2}\left\{\left(\cos(\Omega_n t) + i\frac{\Delta}{2\Omega_n}\sin(\Omega_n t)\right) c_{e,n}(0) - i\frac{\mu\sqrt{n+1}}{\Omega_n}\sin(\Omega_n t)c_{g,n+1}(0)\right\}, \tag{3.53}$$

$$c_{g,n+1}(t) = e^{i\Delta t/2}\left\{\left(\cos(\Omega_n t) - i\frac{\Delta}{2\Omega_n}\sin(\Omega_n t)\right) c_{g,n+1}(0) - i\frac{\mu\sqrt{n+1}}{\Omega_n}\sin(\Omega_n t)c_{e,n}(0)\right\}. \tag{3.54}$$

Here, the general Rabi frequency is expressed as

$$\Omega_n = \sqrt{\left(\frac{\Delta}{2}\right)^2 + \mu^2(n+1)}\,. \tag{3.55}$$

*In the far-off resonant regime.* We consider the presence of large detuning as compared with atom–field coupling, such that $\Delta \gg \mu\sqrt{n+1}$, or we may write $\frac{\mu\sqrt{n+1}}{\Delta} \ll 1$. Therefore, equation (3.55) becomes

$$\Omega_n = \frac{\Delta}{2}\sqrt{1 + \frac{4\mu^2(n+1)}{\Delta^2}} \approx \frac{\Delta}{2}\left(1 + \frac{2\mu^2(n+1)}{\Delta^2}\right). \tag{3.56}$$

In the presence of the binomial expansion of $\Omega_n$, we obtain

$$\frac{\mu\sqrt{n+1}}{\Omega_n} = \frac{\mu\sqrt{n+1}}{\frac{\Delta}{2}\left(1 + \frac{2\mu^2(n+1)}{\Delta^2}\right)} < \frac{2\mu\sqrt{n+1}}{\Delta} \ll 1. \tag{3.57}$$

In the far-off resonant regime, therefore, the atom–field wave function after an interaction time $t$ becomes

$$|\psi(t)\rangle = \sum_{n=0}^{\infty}\left\{e^{i\varphi_{e,n}(t)}\, c_{e,n}(0)|e, n\rangle + e^{-i\varphi_{g,n}(t)}\, c_{g,n}(0)|g, n\rangle\right\}. \tag{3.58}$$

Here,

$$\varphi_{e,n}(t) = \frac{\mu^2(n+1)}{\Delta}t, \tag{3.59}$$

$$\varphi_{g,n}(t) = \frac{\mu^2 n}{\Delta}t. \tag{3.60}$$

Equation (3.58) explains the parametric dependence of the system and time evolution of the field and atom due to their interaction. The atom–field combined wave function, given in equation (3.58), displays non-separability, which is a distinct characteristic of the two partite system. It emerges as a consequence of interaction, whereas before the interaction and at time $t = 0$, the atom–field combined wave function was separable. The non-separability is a fascinating feature of time evolution that introduces a correlation between the atom and field, that is, the entanglement of the two quantum systems.

In 1935, Einstein, Podolsky, and Rosen (EPR) questioned the legitimacy of quantum mechanics by using entangled states [19]. Their contention was that quantum mechanics is incomplete because the common causes are not included in the quantum state description. In 1935 and 1936, Schrödinger published a two-part article in the *Proceedings of the Cambridge Philosophical Society*, in which he discussed and extended the EPR argument. Schrödinger coined the term 'entanglement' to describe correlated and interwined quantum states [20].

Today, free-space entanglement of photons has been confirmed in experiments between the Canary Islands of La Palma and Tenerife, over a distance of 143 km [21]. Entanglement provides a successful means by which to develop quantum channels and is, therefore, of interest in quantum communication [22–24], quantum cryptography [25–27], and quantum computation [28, 29]. Bipartite entanglement has been successfully generated between two electromagnetic cavities [30, 31], multi-modes of a single cavity [32, 33], and in internal [34–38] as well as in external degrees of freedom of atoms [39–41] using optical lattices. The entanglement in multi-partite states, such as Bell states, Noon states, W states, cluster states, and graph states, have also been reported [42–44]. The experimental advances in generating quantum correlations in multi-partite systems have enabled quantum networks based on cavity-QED techniques [45] and colored lasers [46].

## 3.2 Optical lattices

Starting with a single laser beam reflected by a mirror, we can make counter-propagating plane waves that interfere with each other. It is thus possible to make periodic optical potentials along the axis of their propagation in one dimension. These have regions of strong and weak electromagnetic fields with a well-defined periodicity. With some ingenuity, such as in figure 3.3, we can construct two-dimensional and three-dimensional configurations of periodic optical potentials constituting *optical lattices*. The cold atoms are embedded in these optical lattices with the dipole-induced forces. Interestingly, it is possible to construct moving optical lattices and exotic lattices, such as the Kagome lattice [47], shown in figure 3.4. The dynamics of ultracold atoms in this lattice exhibits so-called frustration, leading to degenerate ground states. These optical lattices are thus analogous to solid-state crystals with the added advantage of purity and the absence of defects. Engineering based on these optical techniques has opened up new domains of research. The controlled interaction of matter waves with optical lattices is even regarded as a prototype of electronic dynamics in solid-state materials. For

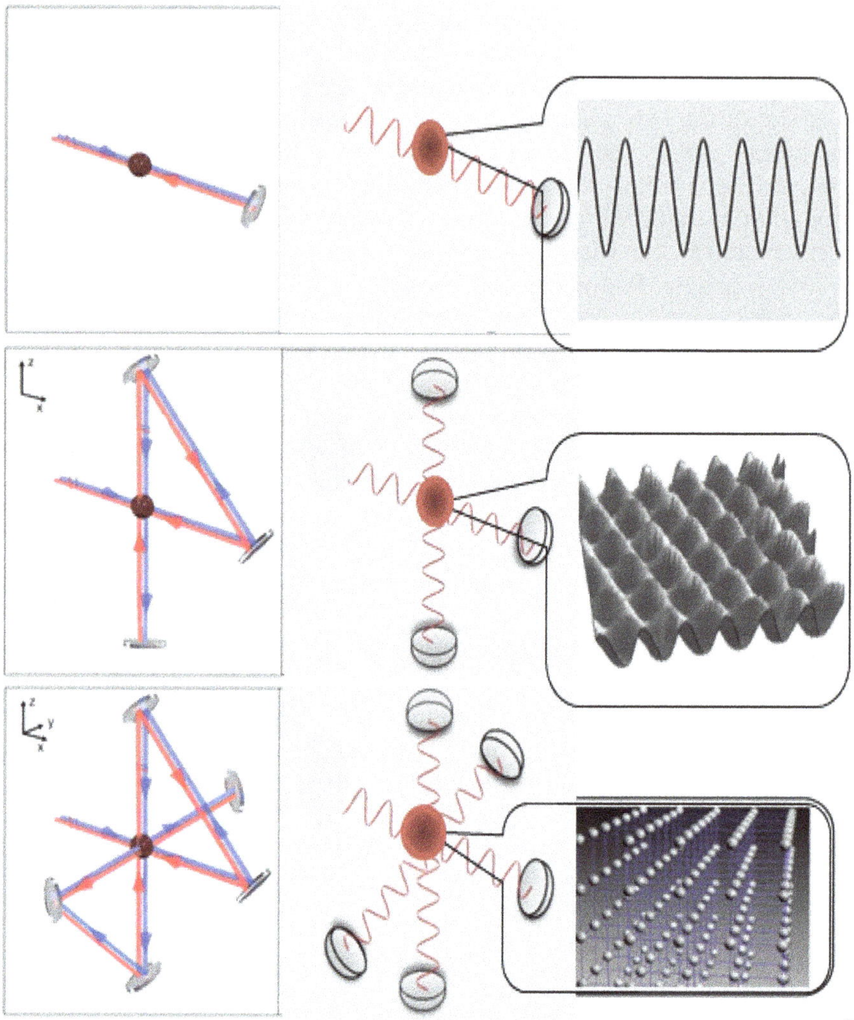

**Figure 3.3.** Schematic diagrams showing one- (top), two- (middle), and three-dimensional laser configurations (bottom) [48]. Copyright © 1969, Nature Publishing Group. With permission of Springer.

instance, cold atoms interacting in a far-off resonant regime display numerous phenomena seen in semi-conductor electronics, such as Bragg deflection and Landau–Zener transitions.

The field thus has emerging applications in which we can utilize quantum coherence to engineer quantum devices [50]. Hence, it is playing a fundamental role in new research areas such as atomtronics as well as developing sensors, interferometers, transducers, and others.

An optical lattice is obtained by counterpropagating travelling waves that are interfering with each other. The periodic potential thus generated, say along the $x$-axis, is expressed as

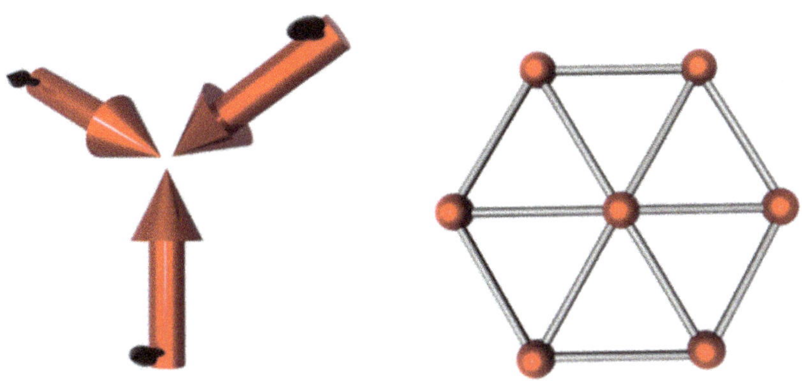

**Figure 3.4.** Three beams intersecting in a plane and enclosing angles of 120° pairwise create a two-dimensional triangular optical lattice if their polarizations is perpendicular to the plane. Adapted from [49]. © IOP Publishing Ltd. All rights reserved.

$$V(x) = \frac{V_0}{2}\cos(2k_{\mathrm{L}}x). \tag{3.61}$$

It has a periodicity of $d = \lambda/2$, which is the spacing between the two potential minima, such that $V(x + d) = V(x)$ and $k_{\mathrm{L}} = \pi/d$. Here, $V_0$ defines the maximum potential height. The periodicity of the optical lattice can be increased by tilting the counterpropagating running waves relative to each other [51, 52].

Optical lattices in real experimental conditions, however, have a finite width of size $r_0$. Hence, we write the optical force acting on an atom as a gradient of a radial potential $V(r)$ defined by a field of amplitude $V_0$ along the $x$-axis, such that

$$V(r) = \frac{V_0}{2}\exp\left[-\left(\frac{r}{r_0}\right)^2\right]\cos(2k_{\mathrm{L}}x), \tag{3.62}$$

where $r^2 = y^2 + z^2$. As discussed in subsection 3.1.1, a red detuned optical field leads to a transverse confinement for the atom.

Two-dimensional and three-dimensional optical lattices are a generalization of a one-dimensional optical lattice. For example, a two-dimensional optical lattice is created by two optical laser fields of the same intensity crossing each other in the same plane. We express the lattice potential, therefore, as

$$V(x, y) = V_0(\cos^2 kx + \cos^2 ky + 2\cos\varphi \cos kx \cos ky)$$

where $k = 2\pi/\lambda$, and $\varphi$ is the angle between the polarization vectors $e_1$ and $e_2$. A two-dimensional *square* optical lattice can be created by considering the $\varphi = \pi/2$ between the two waves. Hence, the two-dimensional optical lattice provides a two-dimensional 'egg-crate' for the atoms, whereas a change in angle between the two mutually crossing optical beams leads to higher symmetric orders. Considering two optical lattices with different periodicities, an array of double-well potential may be generated [53, 54].

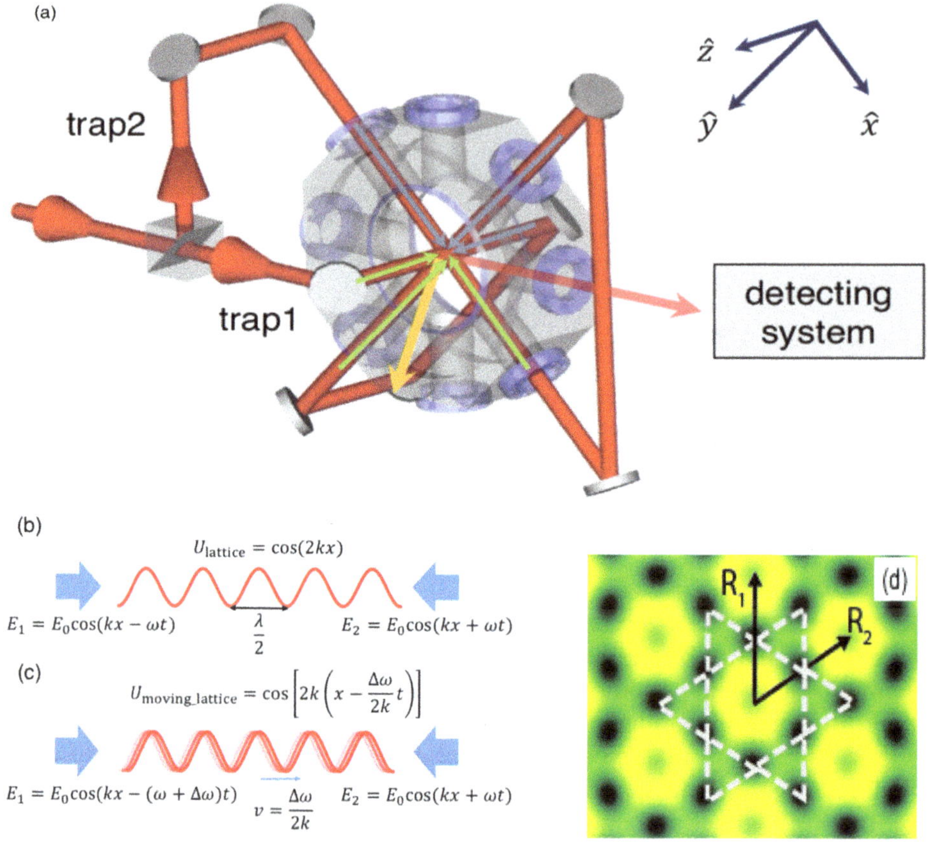


**Figure 3.5.** Three-dimensional laser configuration in a phase-stabilized magneto-optic trap. The resultant optical lattice can be made *to move* by a frequency shift $\Delta\omega$ in one arm due to acousto-optic modulator. The blue arrows represent the trap laser with frequency $\omega + \Delta\omega$, while the green arrows indicate the trap lasers with frequency $\omega$. The optical lattice can be made *static* by replacing the beam splitter and reflecting the reflected light by a mirror, therefore putting $\Delta\omega = 0$. (b) Configuration of laser beams for one-dimensional static optical lattice. We can extend this to two dimensions and three dimensions by using counterpropagating identical laser light fields. (c) The same applies for a one-dimensional moving optical lattice. Copyright © 2018, The Author(s). With permission of Springer [57]. (d) The Kagome lattice, developed by controlling the angle $\varphi$ in a two-dimensional static optical lattice. Reprinted figure with permission from [58]. Copyright © 2012 by the American Physical Society.

In a similar way, a three-dimensional optical lattice is obtained by three optical standing wave fields made up of three pairs of identical running waves with orthogonal polarization and intersecting at one point in space, as shown in figure 3.5. We may write the corresponding potential as

$$V(r) = \frac{V_0}{2}\left\{\cos(2k_x x) + \cos(2k_y y) + \cos(2k_z z)\right\}. \tag{3.63}$$

Setting the frequency of any of the constituting running waves in a pair slightly different from the other leads to the resultant optical lattice being in motion [55, 56].

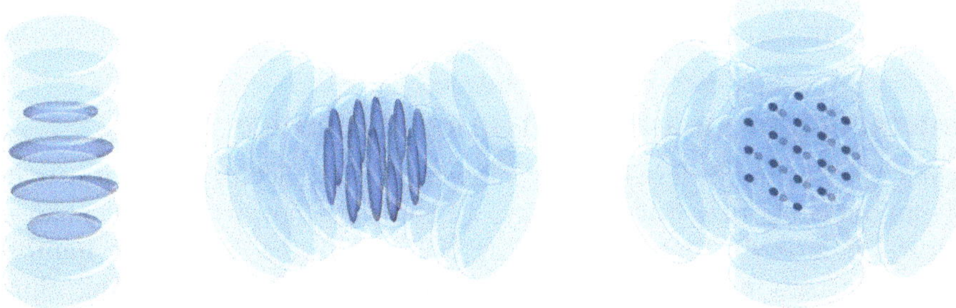

**Figure 3.6.** Cold atoms are shown in one-dimensional, two-dimensional, and three-dimensional optical lattices which appear, respectively, as disk-like structures in a one-dimensional optical lattice, a quantum wire in a two-dimensional optical lattice, and quantum dots in three-dimensional optical lattices.

**Example:** A triangular two-dimensional lattice can be generated by utilizing three travelling waves with a relative angle of $2\pi/3$, as given in reference [47] and shown in figure 3.4.

*Atom in an optical lattice.* Ultra-cold atoms placed in an optical lattice are robust against decoherence effects. At the same time, their interaction with an optical field is controllable by means of dipole force, as we discuss in the next section. Therefore, we can load a dilute Bose–Einstein condensate with a large number of atoms to manipulate the coherence effects in a highly controllable fashion. This feature makes the atom-optical system a test bed with which to realize new phenomena in numerous subjects ranging from quantum informatics to semi-conductor electronics.

An atom interacting with a one-dimensional optical lattice observes a periodic potential, following equation (3.61), where $V_0$ becomes a dipole-dependent potential well depth. Hence, there occurs a periodic array of potential wells along the one-dimensional optical lattice that are at a distance $d$ apart. The atoms are periodically confined around potential minima. However, they are free to move along the other two axes, and thus form a disc. Whereas, in a two-dimensional square optical lattice, the atoms are in an egg-tray, following a confinement in two degrees of freedom, and they are free to move along a third axis, thus forming a bundle of one-dimensional quantum wires. Such one-dimensional wires, shown in figure 3.6, have been realized experimentally. Interestingly a three-dimensional semi-conductor analog presents them as quantum dots, as being confined in three degrees of freedom. Thus, atoms in a three-dimensional optical lattice appear as an array of quantum dots separated by a distance corresponding to lattice spacing along the axes.

Atoms in an optical lattice may observe deep or shallow potential depths. The atom observes a deep optical lattice potential if the effective lattice depth, $V_0$, is of the order of a few hundred single photon recoil energies and the temperature of the atom is around recoil temperature. In this case, the dynamics of the quantum particle in the individual well is independent and one obtains multiple realizations of anharmonic oscillators. On the other hand, when the depth of the potential is just few recoil energies and the atom is at about recoil temperature, it sees a shallow potential. In this case, quantum mechanical effects caused by the spatial periodicity of the optical lattices, such as the formation of Bloch waves, become important.

## 3.3 Interaction of an atom with an optical lattice

Following the treatment developed in subsection 3.1.1, let us consider a two-level atom in an optical lattice. For simplicity, we consider an optical lattice made up by two linearly polarized counterpropagating beams with the same wave number, $k_L$, and their polarization vectors parallel to each other. The optical field, along the $x$-axis and linearly polarized along the $y$-axis, is mathematically expressed as

$$\hat{\mathbf{E}}(\hat{x}, t) = \mathbf{e}_y[\varepsilon_o \cos(k_L\hat{x})\, e^{-i\omega t} + \text{c. c. }], \tag{3.64}$$

where $\omega$ and $\varepsilon_o$ are the frequency and amplitude of the field, respectively. Note that this particular representation suggests that the mirror is placed half-a-wavelength away or an integer multiple of half-a-wavelength away from the origin so that $\mathbf{E}(x = j\lambda/2, t) = 0$, where $j$ is a positive integer.

The interacting two-level atom is defined by its ground state, $|g\rangle$, and an excited state, $|e\rangle$, separated by a transition energy, $\hbar\omega_{eg}$. Hence, the interacting laser light field, $\hat{\mathbf{E}}(\hat{x}, t)$, induces a dipole moment on the otherwise neutral atom. The evolution of the coupled atom–field system is governed by the Hamiltonian, comprising three parts, that is,

$$\hat{H} = \hat{H}_{cm} + \hat{H}_i + \hat{H}_{int}. \tag{3.65}$$

Here, $\hat{H}_{cm}$ and $\hat{H}_i$, respectively, describe the energy associated with the center-of-mass motion of the atom and its internal degrees of freedoms. The last term, $\hat{H}_{int}$, however, presents dipole interaction between the atom and the field. We explicitly write the corresponding Hamiltonian as

$$\hat{H}_{cm} = \frac{\hat{p}_x^2}{2m},$$

$$\hat{H}_i = \frac{1}{2}\hbar\omega_{eg}\hat{\sigma}_z,$$

and

$$\hat{H}_{int} = -\hat{\wp} \cdot \hat{\mathbf{E}}(\hat{x}, t) = -(\sigma_+ + \sigma_-)\vec{\wp} \cdot \hat{\mathbf{E}}(\hat{x}, t)$$

Here, we have used equation (3.12).

In the presence of sufficiently large detuning as compared with the Rabi frequency, and by applying the rotating wave approximation, we develop the effective Hamiltonian. In this regime, the probability of finding the atom in its excited state is negligible, and the spontaneous emission can be neglected. Hence, the evolution properties of the atom in the field are completely determined by the ground-state amplitude. The dynamics of an atom in its ground state are governed by the effective Hamiltonian, that is,

$$\hat{H}_{\text{eff}} = \frac{\hat{p}_x^2}{2m} + \frac{\hbar\Omega_R^2}{\Delta}\cos^2(k_L\hat{x}), \tag{3.66}$$

$$= \frac{\hat{p}_x^2}{2m} + \frac{V_0}{2}(1 + \cos(2k_L\hat{x})). \tag{3.67}$$

where $V_0 = \frac{\hbar\Omega_R^2}{\Delta}$ is the effective potential amplitude, obtained by considering $\Omega_{eg} = \Omega_{ge} = \Omega_R$. The atom in its ground state experiences the position-dependent potential energy in the far-off resonant interaction regime; the shift in energy is the AC-Stark shift.

The force on the atom is the negative of the gradiant of $H_{\text{eff}}$, that is, $F = -\nabla H_{\text{eff}}$. Hence, it consists of the product of the inverse of atom–field detuning and the derivative of the intensity profile as a function of position $I = |E(x)|^2$, such that

$$F \propto -\frac{\nabla I}{\Delta}. \tag{3.68}$$

When the laser frequency is small compared to the transition frequency, we have $(\Delta < 0)$, that is, red detuning, and atoms experience attractive force towards the locations of maximal laser intensity, whereas in the case of blue detuning, $(\Delta > 0)$, they are repelled out of the laser intensity maxima and attracted towards minimal laser intensity.

The dipole force is based on the coherent atom–photon interaction, which induces a dipole moment on the atom. The direction of the force is opposite depending on whether the induced dipole moment is in phase or out of phase by $\pi$ with respect to the electric field.

## Exercises

1. Provide a comparison of the amplitude of the vectors $\delta\mathbf{r}_e$ and $\delta\mathbf{r}_n$ in figure 3.2 with the amplitude of the vector $\mathbf{r}_0$.
2. Prove the following angular momentum commutation relations, using the above definitions of the Pauli matrices:

$$[\sigma_+, \sigma_-] = \sigma_z,$$
$$[\sigma_z, \sigma_-] = -2\sigma_-,$$
$$[\sigma_z, \sigma_+] = 2\sigma_+.$$

3. Derive equation (3.40) using the preceding equation (3.38) and

$$[\hat{H}_0, \hat{H}_I] = \hbar\mu(\sigma_+\hat{a} - \sigma_-\hat{a}^\dagger)(-i\Delta t).$$

4. Show that $\hat{U}$, given in equation (3.42), is unitary, noting $\sigma_-^\dagger = \sigma_+$ and vice versa.
5. Derive equation (3.43) following the steps below:

$$[\hat{a}\sigma_+ - \hat{a}^\dagger\sigma_-, \hat{a}^\dagger\hat{a}] = \sigma_+\hat{a} + \sigma_-\hat{a}^\dagger,$$
$$[\hat{a}\sigma_+ - \hat{a}^\dagger\sigma_-, \sigma_z] = -2(\sigma_+\hat{a} + \sigma_-\hat{a}^\dagger),$$
$$[\hat{a}\sigma_+ - \hat{a}^\dagger\sigma_-, \sigma_+\hat{a} + \sigma_-\hat{a}^\dagger] = 2(\sigma_+\sigma_- + \hat{a}^\dagger\hat{a}\sigma_z) = \hat{I} + \sigma_z + 2\hat{a}^\dagger\hat{a}\sigma_z.$$

Note $\sigma_+\sigma_- + \sigma_-\sigma_+ = \hat{I}$.

6. What is the time of interaction for a rubidium atom with an effective Rabi frequency of 1 MHz interacting with an optical lattice made up of one photon under resonance conditions? The atom is initially in its ground state.
7. Write the solution to equations (3.22) and (3.23) in the resonant regime for constant spatial profile $u(r) = 1$. The atom is prepared in the ground state at time $t = 0$.
8. Write the solution to equations (3.22) and (3.23) in the resonant regime for spatial profile $u(r) = \cos(2kx)$, where $k = 2\pi/\lambda$. The atom is prepared in the ground state at time $t = 0$.
9. Following the discussion leading to equation (3.67), calculate the force acting on the atom and explain its parametric dependence.
10. Write the number of angles between the polarization vectors of counter-propagating waves in a three-dimensional optical lattice. Plot the three-dimensional optical lattice and show the effect of a small tilt in any counterpropagating pair of running waves.

## References


[1] Einstein A 1917 Zur Quantentheorie der Strahlung *Phys. Z.* **18** 121
English translation can be found in Ter Haar D 1967 *The Old Quantum Theory* (Oxford: Pergamon)

[2] Dirac P A M 1927 The quantum theory of the emission and absorption of radiation *Proc. R. Soc. London* A **114** 243–65

[3] Weisskopf V and Wigner E 1930 Calculation of the natural brightness of spectral lines on the basis of Dirac's theory *Z. Phys.* **63** 54–73

[4] Heisenberg W and Pauli W 1929 Zur Quantentheorie der Wellenfelder *Z. Phys.* **56** 1
Heisenberg W and Pauli W 1930 Zur Quantentheorie der Wellenfelder *Z. Phys.* **59** 139

[5] Fermi E 1932 Quantum theory of radiations *Rev. Mod. Phys.* **4** 87–132

[6] Walther H, Varcoe B T H, Englert B-G and Becker T 2006 Cavity quantum electrodynamics *Rep. Prog. Phys.* **69** 1325–82

[7] Raimond J M 1996 Basics of cavity quantum electrodynamics *Quantum Optics of Confined Systems NATO ASI Series* ed M Ducloy and D Bloch 314 (Berlin: Springer) 1–46

[8] Watanabe S 1993 Atoms in static electric and magnetic fields *Review of Fundamental Processes and Applications of Atoms and Ions* ed C D Lin (Singapore: World Scientific) p 441

[9] Meystre P 2001 *Atom Optics* (Heidelberg: Springer)

[10] Schleich W P 2001 *Quantum Optics in Phase Space* (Weinheim: VCH-Wiley)

[11] Scully M O and Zubairy M S 1997 *Quantum Optics* (Cambridge: Cambridge University Press)

[12] Kazantsev A P, Surdutovich G I and Yakovlev V P 1990 *Mechanical Action of Light on Atoms* (Singapore: World Scientific)

[13] Shore B W 1990 *The Theory of Coherent Atomic Excitation* (New York: Wiley)

[14] Cohen-Tannoudji C, Diu B and Laloe F 1977 *Quantum Mechanics* (New York: Wiley)

[15] Sargent M III, Scully M O and Lamb W E 1974 *Laser Physics* (Redwood City, CA: Addison-Wesley)

[16] Ashkin A 1970 Acceleration and trapping of particles by radiation pressure *Phys. Rev. Lett.* **24** 156–9

[17] Ashkin A, Dziedzic J M, Bjorkholm J E and Chu S 1986 Observation of a single-beam gradient force optical trap for dielectric particles *Opt. Lett.* **11** 288

[18] Matthews J N A 2009 Commercial optical traps emerge from biophysics labs *Phys. Today* **62** 26

[19] Einstein A, Podolsky B and Rosen N 1935 Can quantum-mechanical description of physical reality be considered complete? *Phys. Rev.* **47** 777–80

[20] Schrödinger E 1926 Der stetige Übergang von der Mikro- zur Makromechanik *Naturwissenschaften* **14** 664

[21] Herbst T, Scheidl T, Fink M, Handsteiner J, Wittmann B, Ursin R and Zeilinger A 2015 Teleportation of Entanglement over 143 km *Proc. Natl. Acad. Sci. USA* **112** 14202–5

[22] Bennet C H, Brassard G, Crepeau C, Jozsa R, Peres A and Wootters W K 1993 Teleporting an unknown quantum state via dual classical and Einstein-Podolsky-Rosen channels *Phys. Rev. Lett.* **70** 1985

[23] Boschi D, Branca S, De Martini F, Hardy L and Popescu S 1998 Experimental realization of teleporting an unknown pure quantum state via dual classical and Einstein-Podolsky-Rosen channels *Phys. Rev.* **80** 1121

[24] Bouwmeester D, Pan J-W, Mattle K, Eibl M, Weinfurter H and Zeilinger A 1997 Experimental quantum teleportation *Nature* **390** 575

[25] Ekert A K 1991 Quantum cryptography based on Bell's theorem *Phys. Rev. Lett.* **67** 661

[26] Saif F 2001 Quantum cryptography and entanglement *Proc. Int. Conf. on Physics in Industry* ed; Anwar-ul-Haq and M Ahmad (Karachi: PCSIR)

[27] Ekert A K, Rarity J G, Tapster P R and Massimo Palma G 1992 Practical quantum cryptography based on two-photon interferometry *Phys. Rev. Lett.* **69** 1293

[28] Grover L K 1997 Quantum computers can search arbitrarily large databases by a single query *Phys. Rev. Lett.* **29** 4709

[29] Shor P W 1999 Polynomial-time algorithms for prime factorization and discrete logarithms on a quantum computer *SIAM Rev.* **41** 303–32

[30] Davidovich L, Zagury N, Brune M, Raimomd J M and Haroche S 1994 Teleportation of an atomic state between two cavities using nonlocal microwave fields *Phys. Rev.* A **50** R895

[31] El Anouz K, El Allati A and Saif F 2022 Study different quantum teleportation amounts by solving Lindblad master equation *Phys. Scr.* **97** 035102

[32] Ikram M and Saif F 2002 Engineering entanglement between two cavity modes *Phys. Rev.* A **66** 014304

[33] Rauschenbeutel A, Bertet P, Osnaghi S, Nogues G, Brune M, Raimond J M and Haroche S 2001 Controlled entanglement of two field modes in a cavity quantum electrodynamics experiment *Phys. Rev.* A **64** R050301

[34] Bogár P and Bergou J A 1996 Entanglement of atomic beams: tests of complementarity and other applications *Phys. Rev.* A **53** 49

[35] Freyberger M 1995 Simple example of nonlocality: atoms interacting with correlated quantized fields *Phys. Rev.* A **51** 3347

[36] Jaksch D, Briegel H J, Cirac J I, Gardiner C W and Zoller P 1999 Entanglement of atoms via cold controlled collisions *Phys. Rev. Lett.* **82** 1975

[37] Gerry C C 1993 Nonlocality of a single photon in cavity QED *Phys. Rev.* A **53** 4583

[38] Kneer B and Law C K 1998 Preparation of arbitrary entangled quantum states of a trapped ion *Phys. Rev.* A **57** 2096

[39] Khalique A and Saif F 2003 Engineering entanglement between external degrees of freedom of atoms via Bragg scattering *Phys. Lett.* A **314** 37–43
[40] Islam R, Ikram M and Saif F 2007 Engineering maximally entangled N-photon NOON field states using an atom interferometer based on Bragg regime cavity QED *J. Phys. B: At. Mol. Opt. Phys.* **40** 1359
[41] Islam R, Khosa A H and Saif F 2008 Generation of Bell, NOON and W states via atom interferometry *J. Phys. B: At. Mol. Opt. Phys.* **41** 035505
[42] Abbas T, Islam R, Khosa A H and Saif F 2009 Generation of cavity field cluster and GHz states using Bragg-regime atom interferometry *J. Russ. Laser Res.* **30** 267–78
[43] Islam R, Khosa A H, Saif F and Bergou J A 2013 Generation of atomic momentum cluster and graph states via cavity QED *Quantum Inf. Process.* **12** 129–48
[44] Khosa A H and Saif F 2009 Atomic cluster and graph states: an engineering proposal *J. Phys. Soc. Jpn.* **78** 114401
[45] Ritter S, Nolleke C, Hahn C, Reiserer A, Neuzner A, Uphoff M, Mucke M, Figueroa E, Bochmann J and Rempe G 2012 An elementary quantum network of single atoms in optical cavities *Nature* **484** 195–200
[46] Coelho A S, Barbosa F A S, Cassemiro K N, Villar A S, Martinelli M and Nussenzveig P 2009 Three-color entanglement *Science* **326** 823–6
[47] Becker C, Soltan-Panahi P, Kronjager J, Dorscher S, Bongs K and Sengstock K 2010 Ultra-cold quantum gases in triangular optical lattices *New J. Phys.* **12** 065025
[48] Bloch I 2005 Ultracold quantum gases in optical lattices *Nat. Phys.* **1** 23–30
[49] Becker C, Soltan-Panahi P, Kronjäger J, Dörscher S, Bongs K and Sengstock K 2010 Ultracold quantum gases in triangular optical lattices *New Journal of Physics* **12** 065025
[50] Novikova I, Xiao Y, Phillips D F and Walsworth R L 2005 EIT and diffusion of atomic coherence *J. Mod. Opt.* **52** 2381–90
[51] Peil S, Porto J V, Laburthe Tolra B, Obrecht J M, King B E, Subbotin M, Rolston S L and Phillips W D 2003 Patterned loading of a Bose-Einstein condensate into an optical lattice *Phys. Rev.* A **67** 051603(R)
[52] Hiller M, Venzl H, Zech T, Oleś B, Mintert F and Buchleitner A 2012 Robust states of ultracold bosons in tilted optical lattices *J. Phys. B: At. Mol. Opt. Phys.* **45** 095301
[53] Sebby-Strabley J, Anderlini M, Jessen P S and Porto J V 2006 Lattice of double wells for manipulating pairs of cold atoms *Phys. Rev.* A **73** 033605
[54] Anderlini M, Sebby-Strabley J, Kruse J, Porto J V and Phillips W D 2006 Controlled atom dynamics in a double-well optical lattice *J. Phys. B: At. Mol. Opt. Phys.* **39** S199–210
[55] Browaeys A, Häffner H, McKenzie C, Rolston S L, Helmerson K and Phillips W D 2005 Transport of atoms in a quantum conveyor belt *Phys. Rev.* A **72** 053605
[56] Bruderer M, Klein A, Clark S R and Jaksch D 2008 Transport of strong-coupling polarons in optical lattices *New J. Phys.* **10** 033015
[57] Chong K O, Kim J-R, Kim J, Yoon S, Kang S and An K 2018 Observation of a non-equilibrium steady state of cold atoms in a moving optical lattice *Commun. Phys.* **1** 25
[58] Jo G-B, Guzman J, Thomas C K, Hosur P, Vishwanath A and Stamper-Kurn D M 2012 Ultracold atoms in a tunable optical kagome lattice *Phys. Rev. Lett.* **108** 045305
[59] Schrödinger E 1935 Discussion of probability relations between separated systems *Proc. Camb. Philos. Soc.* **31** 555–63
Schrödinger E 1936 Discussion of probability relations between separated systems *Proc. Camb. Philos. Soc.* **32** 446–51

**IOP** Publishing

Optical Forces on Atoms

**Farhan Saif and Shinichi Watanabe**

# Chapter 4

## Scattering of atoms by light waves

*'The atoms or elementary particles … form a world of potentialities or possibilities rather than one of things or facts.'*

—Werner Heisenberg

Let us consider the scattering of ultra-cold atoms exposed to a periodic optical potential. We explain Kapitza–Dirac scattering, Bragg diffraction, and Talbot scattering of atoms, and note that the external degrees of freedom or momentum states of the atom play a significant role in the scattering.

### 4.1 Kapitza–Dirac scattering

The interaction between ultra-cold atoms and an optical lattice for a short interaction time leads to a diffraction regime, named Kapitza–Dirac scattering. Specifically, Kapitza–Dirac scattering operates in the Raman–Nath regime. This is to say that the interaction time of the particle with the light field is sufficiently short in duration that the motion of the particles with respect to the light field can be neglected.

For simplicity, we consider that the cold atom is initially in the ground state, whereas the initial momentum along the field, that is, the $x$-axis, is $p_0 = \hbar k_0$. Therefore, the initial wave packet may be described as

$$\psi(x, 0) = \frac{e^{-ik_0x}}{\sqrt{2\pi}}. \tag{4.1}$$

We consider that the atoms experience the optical lattice only during a very short transit time, $t$, and approximate the time profile by a simple on- and off-duty square pulse[1].

[1] One may consider a more general profile so as to define $t$ in terms of the time integral of the profile.

doi:10.1088/978-0-7503-2308-6ch4  

We assume that the optical lattice imparts a potential energy which is much bigger than the kinetic energy. Therefore, in the presence of large atom–field detuning, the time propagator $\hat{U}(\hat{x}, t) = \exp(-iH_{\text{eff}}t/\hbar)$. The atomic dynamics is controlled by the potential energy term given in equation (3.67). Hence,

$$\hat{U}(\hat{x}, t) \simeq \exp\left\{-i\frac{V_0}{2\hbar}t\right\} \exp\left\{-i\frac{V_0 \cos(2k_{\text{L}}\hat{x})t}{2\hbar}\right\}, \tag{4.2}$$

$$= \exp\left\{-i\frac{V_0}{2\hbar}t\right\} \sum_{\nu=-\infty}^{\infty} (-i)^{\nu} J_{\nu}\left(\frac{V_0 t}{2\hbar}\right) \mathrm{e}^{2i\nu k_{\text{L}}\hat{x}}. \tag{4.3}$$

Here, we have introduced the Jacobi–Anger identity, that is,

$$\mathrm{e}^{-i\lambda\cos\theta} = \sum_{\nu=-\infty}^{\infty} (-i)^{\nu} J_{\nu}(\lambda) \mathrm{e}^{i\nu\theta}. \tag{4.4}$$

Hence, we obtain

$$\psi(x, t) = \hat{U}(\hat{x}, t)\psi(x, 0), \tag{4.5}$$

$$= \frac{1}{\sqrt{2\pi}}\left\{-i\frac{V_0}{2\hbar}t\right\} \sum_{\nu=-\infty}^{\infty} (-i)^{\nu} J_{\nu}\left(\frac{V_0 t}{2\hbar}\right) \mathrm{e}^{-i(k_0 - 2\nu k_{\text{L}})x}, \tag{4.6}$$

$$= \frac{1}{\sqrt{2\pi}}\{-i\Omega t\} \sum_{\nu=-\infty}^{\infty} (-i)^{\nu} J_{\nu}(\Omega t) \mathrm{e}^{-i(k_0 - 2\nu k_{\text{L}})x}. \tag{4.7}$$

Here, $\Omega = \frac{V_0}{2\hbar}$. Now, recognizing this as a Fourier series expansion in $k$-space, we observe that after the interaction over a short duration $t$, the cold atoms experience $\nu$th order scattering such that the momentum after scattering becomes $\hbar k_0 - 2\nu\hbar k_{\text{L}}$, with the corresponding scattering amplitude $[J_{\nu}(\Omega t)]^2$. The experimental observation of Kapitza–Dirac scattering is shown in figure 5.9(a). The lower panels display the experimentally observed features of this regime under three different conditions.

Historically, it was Pyotr Kapitza and Paul Dirac who in 1933 predicted the scattering of 'matter by a grating of light' rather than that of 'light by a grating of matter', interchanging their earlier roles. They considered in particular the scattering of electrons from standing light waves [17]. However, the experimental verification required an intense coherent light source, which was not possible at the time, and thus encountered difficulties. It was experimentally realized by Bucksbaum *et al*, using intense optical waves in 1988 [1]. In 1986, the Kapitza–Dirac scattering of atoms was realized experimentally using thermal atomic beams as atomic sources [2, 3]. The diffraction of a well-collimated beam was observed after passage through a standing wave, as shown in figure 4.1(a).

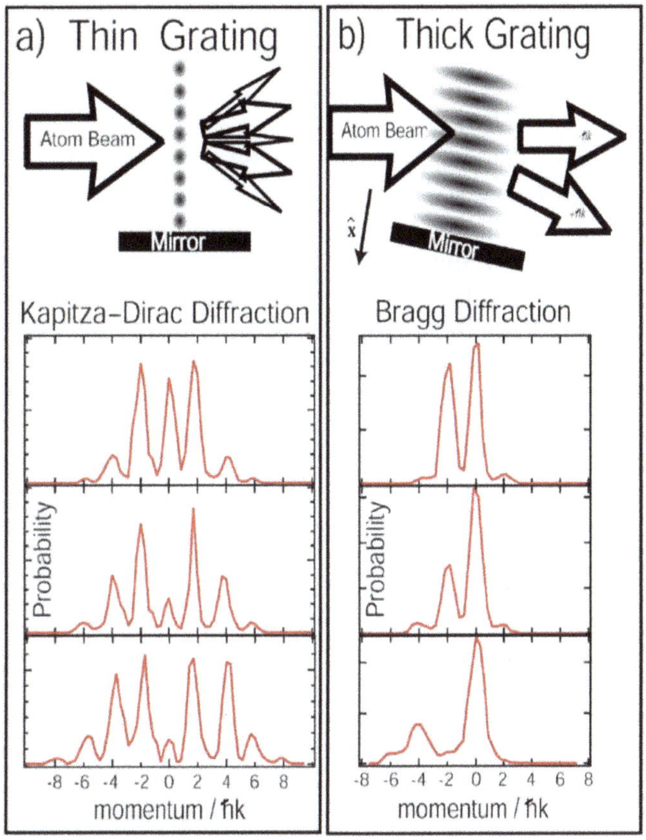


**Figure 4.1.** Comparison between diffraction from thin and thick grating: (a) Kapitza–Dirac scattering, where several diffraction peaks are seen symmetrically about $0 \times \hbar k$ under three different scenarios for $\Omega t = 1.69$ (top), 2.33 (middle), and 2.84 (lower) [2]; and (b) Bragg scattering regime. An example of a beam splitter is shown under three different conditions. Note the mirror is tilted so that the vertical momentum component has marked peaks at $0 \times \hbar k$ and $-2 \times \hbar k$ for $\Omega t = 3.8$ and 6.8, while the third frame shows second-order Bragg scattering, therefore a peak at $-4 \times \hbar k$ instead of $-2 \times \hbar k$ and at $0 \times \hbar k$. Reprinted figure with permission from [4]. Copyright (1988) by the American Physical Society.

## 4.2 Bragg scattering

Let us consider that an incident atom has a well-defined momentum $\mathbf{p}_{\text{in}} = l_0 \hbar k$. It then interacts with an optical standing wave field and comes out in the momentum state $\mathbf{p}_{\text{out}}$, which is

$$\mathbf{p}_{\text{out}} = \mathbf{p}_{\text{in}} + l\hbar k.$$

Here, $l_0$ and $l$ are integers. Energy conservation requires that

$$\frac{|\mathbf{p}_{\text{in}}|^2}{2m} = \frac{|\mathbf{p}_{\text{out}}|^2}{2m}.$$

Here, $m$ denotes the mass of the atom. Substituting the initial and final momentum into the above expression, we get the resonance condition for the Bragg regime, that is,

$$\frac{l(l + 2l_0)}{2m}\hbar^2 k^2 = 0.$$

This equation has two solutions: $l = 0$ corresponds to the undeflected atomic beam, that is, $\mathbf{p}_{out} = \mathbf{p}_{in}$, and $l = -2l_0$ corresponds to the deflected atomic beam, that is, $\mathbf{p}_{out} = -\mathbf{p}_{in}$. Therefore, we find the conservation of momentum $|\mathbf{p}_{out}| = |\mathbf{p}_{in}|$.

In summary, we can say that Bragg scattering is an energy-conserving multi-photon process through which an atomic de Broglie wave packet is coherently split into two counterpropagating momenta states due to its interaction with the oaptical lattice. Thus, an atom with a transverse initial momentum of $P_{in} = l\hbar k$, with $k$ being the wave number of the field, exits the interaction zone in superposition of the initial momentum state $P_{in} = l\hbar k$ and the state with momentum $P_{out} = -l\hbar k$. The integer $l$ is the order of the Bragg diffraction and implies an evident exchange of $2l$ photons during the atom–field interaction.

The relevant Hamiltonian for the problem has already appeared as equation (3.7), which is equivalent to the following:

$$H_I = \frac{P_x^2}{2m} + \frac{\hbar\Delta}{2}\sigma_z + \hbar\mu\cos(\mathbf{k}\cdot\mathbf{x})[\sigma_+ a + a^+\sigma_-],$$

Here, $\mu$ the atom–field coupling constant, $a(a^\dagger)$ are the annihilation (creation) operators of the single-mode electromagnetic field, and $\Delta$ is the detuning of atomic frequency with the cavity field mode. Similarly $\sigma_+$, $\sigma_-$, and $\sigma_z$ are, respectively, the atomic raising, lowering, and inversion operators. $P_x$ and $x$ are the momentum and position operators for the center-of-mass motion of the atom along the $x$-axis, respectively. Let us assume that a two-level atom, initially in ground state $|g\rangle$, interacts with a cavity field in the Fock state of some large number, $N$, of photons, $|N\rangle$. The state vector of the atom interacting with the cavity field at any arbitrary time, $t$, may be expressed as

$$|\Psi_{AF}(t)\rangle = \psi_g(x, t)|N, g\rangle + \psi_e(x, t)|(N-1), e\rangle, \tag{4.8}$$

The equations of motion are

$$i\hbar\frac{\partial\psi_g}{\partial t} = -\frac{\hbar^2}{2m}\frac{\partial^2\psi_g}{\partial x^2} + \frac{\hbar\Delta}{2}\psi_g + \hbar\mu\cos(kx)\psi_e, \tag{4.9}$$

$$i\hbar\frac{\partial\psi_e}{\partial t} = -\frac{\hbar^2}{2m}\frac{\partial^2\psi_e}{\partial x^2} - \frac{\hbar\Delta}{2}\psi_e + \hbar\mu\cos(kx)\psi_g. \tag{4.10}$$

For large detuning $\Delta \gg \mu, \frac{\hbar^2 k^2}{2m}$, and the atom initially in the ground state, we eliminate the excited state adiabatically, that is, by dropping the derivatives of $\psi_e(t)$ in equation (4.10) as small compared to the rest, so that $\psi_e = \frac{2\mu}{\Delta}\cos(kx)\psi_g$ and

$$i\hbar\frac{\partial\psi_g}{\partial t} = -\frac{\hbar^2}{2m}\frac{\partial^2\psi_g}{\partial x^2} + \frac{\hbar\Delta}{2}\psi_g + 2\hbar\frac{\mu^2}{\Delta}\cos^2(kx)\psi_g, \tag{4.11}$$

For a general quantum mechanical approach to the Bragg scattering please, see [5].

Let us focus on the case of the first-order Bragg diffraction $l = 1$, namely imparting a momentum $2\hbar k$ to the ground state, as is most commonly implemented experiment. Then, $\psi_g$ may be written as

$$\psi_g(x, t) = C^{P_0}(t)e^{ikx} + C^{P_{-2}}(t)e^{-ikx},$$

so that we have

$$i\frac{\partial C^{P_0}}{\partial t} = \left(\frac{\hbar k^2}{2m} + \frac{\Delta}{2} + \frac{\mu^2}{\Delta}\right)C^{P_0}(t) + \frac{\mu^2}{2\Delta}C^{P_{-2}}(t), \tag{4.12}$$

$$i\frac{\partial C^{P_{-2}}}{\partial t} = \left(\frac{\hbar k^2}{2m} + \frac{\Delta}{2} + \frac{\mu^2}{\Delta}\right)C^{P_{-2}}(t) + \frac{\mu^2}{2\Delta}C^{P_0}(t). \tag{4.13}$$

Imposing the initial conditions $C^{P_0}(0) = 1$ and $C^{P_{-2}}(0) = 0$ yields

$$C^{P_0}(t) = \exp\left\{-i\left(\frac{\hbar k^2}{2m} + \frac{\Delta}{2} + \frac{\mu^2}{\Delta}\right)t\right\}\cos(\beta t), \tag{4.14}$$

$$C^{P_{-2}}(t) = \exp\left\{-i\left(\frac{\hbar k^2}{2m} + \frac{\Delta}{2} + \frac{\mu^2}{\Delta}\right)t\right\}\sin(\beta t), \tag{4.15}$$

where $\beta = \frac{\mu^2}{2\Delta}$. In figure 4.2 the probability of finding the atom in momentum states $P^{(0)} = |C^{P_0}|^2$ and $P^{(-2)} = |C^{P_{-2}}|^2$ is plotted, that is,

$$P^{(0)} = |C^{P_0}(t)|^2 = \cos^2(\beta t) = \frac{1}{2}(1 + \cos(\Omega t)), \tag{4.16}$$

$$P^{(-2)} = |C^{P_{-2}}(t)|^2 = \sin^2(\beta t) = \frac{1}{2}(1 - \cos(\Omega t)), \tag{4.17}$$

that oscillate with frequency $\Omega = 2\beta$.

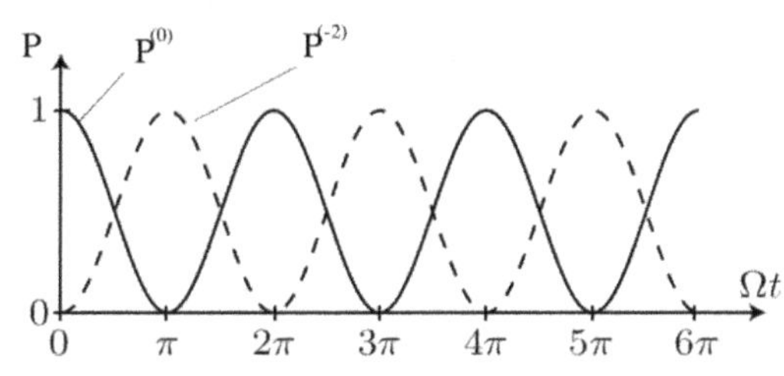


**Figure 4.2.** Bragg scattering is shown vs time. We see that the probability of finding the atom in momentum states $P^{(0)} = |C^{P_0}|^2$ and $P^{(-2)} = |C^{P_{-2}}|^2$ oscillate with frequency $\Omega = 2\beta$.

These very simple expressions explicitly show that off-resonant Bragg diffraction is capable of dealing with all aspects of atom optics. For example, interaction for $t = \frac{\pi}{2\Omega}$ yields a symmetric atomic de Broglie wave beam splitter. Similarly, interaction for $t = \frac{\pi}{\Omega}$ corresponds to an atomic mirror under first-order off-resonant Bragg diffraction.

Bragg scattering may be regarded as a multi-photon energy transfer, scattering conserving Raman process. This implies that an atom exits from the interaction region with the same initial kinetic energy as it enters. Large uncertainty in the interaction time $\Delta t$ corresponds to a minute uncertainty in energy, i.e. $\Delta E \cong \hbar/\Delta t$, such that an atom with initial momentum $l\hbar k$ with some integer $l$ gets effectively constrained into two adjacent equal-energy momentum states, $-l\hbar k$ and $l\hbar k$. This forbids a population from moving to other momentum states with a different energy. Since the energy difference with the neighbouring states is of the order of $\hbar\omega_{\text{rec}}$, so the Bragg regime is characterized by the condition $\omega_{\text{rec}} > \Omega_{\text{R}}$, where $\Omega_{\text{R}}$ is the Rabi frequency.

This is also often explained in a more geometric way. In Bragg diffraction, the condition for constructive interference of atomic de Broglie waves requires that the angle of incidence to the standing wave plane must be one of the $n$th order scattering angles $\theta_n$ that satisfies Bragg's relation, i.e. $\lambda \sin\theta_n = n\lambda_{\text{dB}}$, where $\lambda$ is the wavelength of the standing wave field and $\lambda_{\text{dB}}$ is the de Broglie wavelength of the atomic wave. The momentum transfer results only in the discrete initial values of atomic momentum along the **k** vector of the field. It is important to note that the interaction only reverses the direction (and hence changes atomic momentum from $l\hbar k$ to $-l\hbar k$) but does not alter the magnitude of the momentum. *Hence energy and momentum both are conserved in atomic Bragg diffraction.*

## 4.3 Talbot effect

The Talbot effect in classical optics, observed in 1836 by Henry Fox Talbot, is a near-field diffraction effect [6]. As a laterally periodic coherent wave is incident on a diffraction grating, the grating image repeats at regular distances away from the grating plane. This regular distance is called the Talbot length. At fractions of the Talbot length, we find fractions of the grating images. The phenomenon has applications in optical metrology, photo-lithography, optical computers, as well as in data compression, electron optics, and microscopy.

In matter-wave optics the effect is seen using atoms with non-relativistic [7] and relativistic velocities, displaying self-imaging at Talbot lengths. Here, the atoms pass a periodic optical lattice along the $x$-axis, acting as a diffraction grating. Hence, the atomic wave function appears as

$$|\psi\rangle = \sum_j f(x - x_j)\,|x\rangle. \tag{4.18}$$

Here, the function $f$ describes the distribution of atoms along the optical diffraction grating, at the $j$th diffration grating element. The propagating atoms weave Talbot carpets as a result of delicate interference between eigenmodes or wavelets.

The Schrödinger equation controls the dynamics of a non-relativistic particle, whereas the one-dimensional Dirac equation treats a relativistic particle [8].

In figure 4.3, we show the Talbot self-imaging of matter waves moving with non-relativistic and relativistic velocities. The space-time dynamics for non-relativistic matter waves is defined by the Schrödinger equation. The self-imaging takes place at a Talbot length,

$$z_{\mathrm{T}} = 2\lambda_{\mathrm{DB}}^{-1}s^2, \tag{4.19}$$

whereas fractional imaging occurs at a fraction of the distance. Here, $s$ is the spacing between grating elements, whereas $\lambda_{\mathrm{DB}}$ is the de Broglie wavelength of the matter waves. The de Broglie wavelength of the atom is $\lambda_{\mathrm{DB}} = 2\pi\hbar/mv$ as it moves with a non-relativistic velocity, $v$.

Matter waves with relativistic velocities display self-imaging in complete contrast to non-relativistic particles. In the relativistic case there are only two possible non-zero probability density lines in the space-time plane, thus exhibiting full restructuring after equal intervals of time. This leads to self-imaging of the grating at Talbot lengths (see figure 4.3, lower panel).

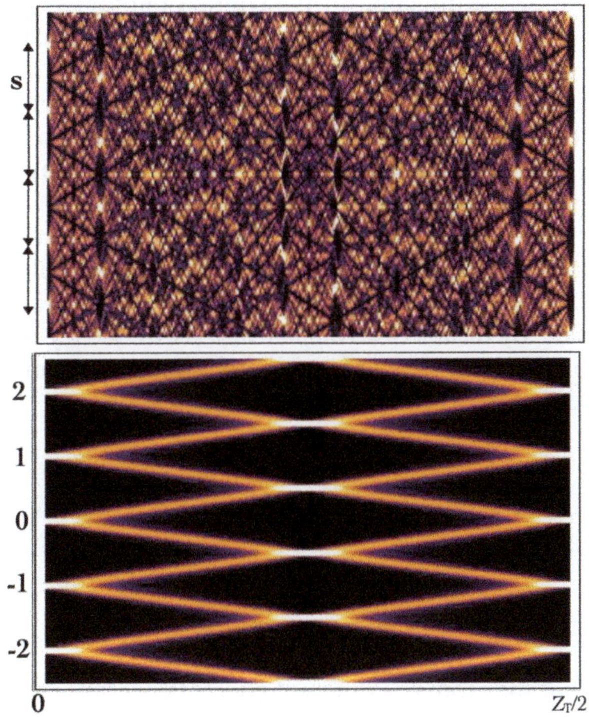


**Figure 4.3.** Talbot self-imaging for matter waves with non-relativistic velocity (upper panel). At a fraction of a Talbot length we note a fraction of the grating images. For matter waves with relativistic velocity, we find another Talbot carpet (lower panel). The Talbot length, however, is larger now as compared with the non-relativistic case due to the time-dilation effect. The figure is made for $z \in [0, z_{\mathrm{T}}/2]$. The remaining half ($z \in [z_{\mathrm{T}}/2, z_{\mathrm{T}}]$) is obtained by reflecting the first half about $z = z_{\mathrm{T}}/2$.

## 4.4 Cooling of atoms

### 4.4.1 Laser cooling

The laser cooling of atoms requires six laser beams. The six laser beams make three orthogonal pairs in which the beams in each pair point straight at each other, as shown in figure 3.5. The laser beams are made to cross at a point where an ensemble of atoms is located. The action of each incident laser beam on the atom is identical in that it leads to a cooling effect.

As the energy of the photon in the laser beam corresponds to the difference between the ground state and an excited state of the static atom, transition takes place. Hence, the atom absorbs the photon and gains energy in the direction of the absorbed photon. Once it has absorbed the photon, the atom is in an excited state. It then drops back into its ground state, emitting a photon of the same energy that it absorbed, but in a random direction.

What happens to the momentum of the atom during this process? When the atom absorbs the photon, the atom experiences a change in momentum. The atom gains the momentum in the direction of the laser beam. When the atom emits a photon, it again experiences a change in momentum. The atom achieves a recoil momentum in a direction opposite to the direction of the emitted photon. If an atom repeatedly absorbs photons from a laser beam, emitting a photon in a random direction after each absorption, the atom will experience a net change in momentum in the direction of the laser beam due to the absorption of photons. Since the emitted photons come out in all directions, the average change in momentum due to the emitted photons will be zero. Hence, the process slows down atoms that are travelling toward the laser beam.

Will this process also speed up atoms that are travelling away from the laser beam? We need a way to control whether or not an atom will absorb a photon, depending on whether the atom is moving opposite to or in the direction of the laser beam. This is possible by means of a *Doppler shift*. A frequency shift takes place due to the *movement* of the atom. The energy of the atom is the sum of its energy in its internal state $j$ and its kinetic energy, that is,

$$E = E_j + \frac{\hbar^2 k^2}{2m} = \hbar\omega_0,$$

which leads to a group velocity as

$$v_z = \frac{1}{\hbar}\frac{\partial E}{\partial k} = \frac{\hbar k}{m}.$$

The group velocity of the atom has different effects depending on whether it is in the direction of the electromagnetic field or opposite.

In the case of a static atom, it observes the frequency of an approaching photon as $\omega = ck$ or

$$\nu = c/\lambda.$$

In the case the atom moves in the direction opposite to the field, the relative velocity of the electromagnetic field rises[2], as $v_r = c - (-v_z) = c + v_z$. Hence, it observes a new shifted frequency, as

$$\nu = \frac{v_r}{\lambda} = \frac{c + v_z}{\lambda} = \left(\frac{c + v_z}{c}\right)\nu = \left(1 + \frac{v_z}{c}\right)\nu.$$

In the other case, if the atom is moving along the field, it observes a frequency shift as

$$\nu = \frac{v_r}{\lambda} = \frac{c - v_z}{\lambda} = \left(\frac{c - v_z}{c}\right)\nu = \left(1 - \frac{v_z}{c}\right)\nu.$$

We note that an atom moving in the opposite direction to the electromagnetic field observes a higher frequency, whereas an atom moving in the direction of the field experiences a smaller frequency.

We deliberately choose a frequency of the electromagnetic field, $\omega$, that is too low for a stopped atom to be absorbed. Atoms moving in a direction opposite to the photons in the laser beam will 'see' a higher frequency. If the frequency is chosen correctly, the Doppler-shifted frequency will be exactly that which can be absorbed by an atom moving toward the laser, and the atom can be slowed down. Atoms moving in the same direction as the photons in the laser beam will 'see' a lower frequency and will not absorb the photons. Therefore, we use a pair of lasers pointing in opposite directions, with one laser slowing down atoms moving in one direction while the other laser slows atoms moving in the opposite direction, as shown in figure 4.4. In this way, three pairs of orthogonal lasers can be used to slow the atoms in three dimensions. The lowest of the temperatures is achieved at the Doppler limit at which the random momentum kicks caused by photon emission limits further cooling. In early experiments on the laser cooling of atoms, a temperature of the order of the millikelvin was achieved [9]. The laser cooling technique is also known as *Doppler cooling*.

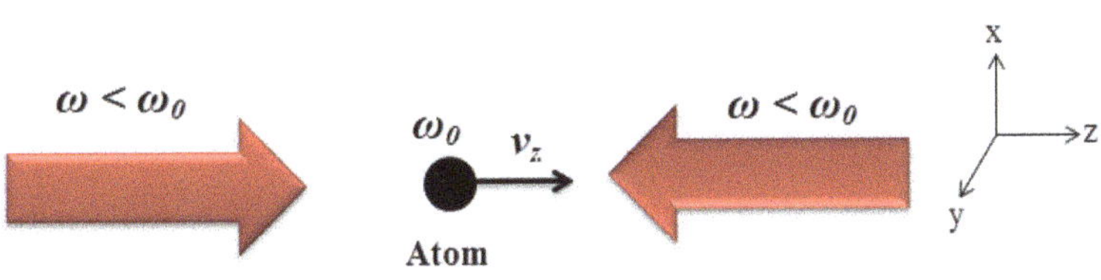


**Figure 4.4.** Doppler cooling of an atom by two laser beams propagating in opposite directions along the $z$-axis. The frequency of the lasers $\omega$ is smaller than the transition frequency $\omega_0$ of the atom moving with a velocity $v_z$. The atom moving in a direction opposite to the laser beam will 'see' a higher frequency. If the frequency is chosen correctly, the Doppler-shifted frequency will be exactly that which can be absorbed by the atom moving toward the laser and the atom can be slowed down.

[2] The relative velocity appears to be larger than $c$, however the conclusion is unaffected by this argument.

### 4.4.2 Sisyphus cooling

Sub-Doppler cooling is performed by the *Sisyphus cooling* and *evaporative cooling* techniques. In the Sisyphus cooling technique, an atom interacts with two counter-propagating laser beams with orthogonal polarization, as shown in figure 4.5. The two interfering beams create a standing wave with a polarization gradient that alternates between circular and linear polarizations. Through optical pumping, the atom can be induced to decelerating optical dipole forces more strongly than accelerating optical dipole forces. Atoms moving through the potential landscape along the direction of the standing wave lose kinetic energy as they move to a potential maximum, at which point optical pumping moves them to a lower-energy state, thus ridding them of the potential energy they carried. Hence, due to spatial modulation of the light and optical pumping rates, the atom ascends more than descends in its energy diagram [10, 11]. Following the Sisyphus cooling technique, it is possible to cool the atoms to a microkelvin scale. The lowest temperature limit is achieved at *the recoil limit*, where the momentum kick caused by photon emission is large enough relative to its velocity [12, 13].

### 4.4.3 Evaporative cooling

Evaporative cooling is based on the evaporation of higher-energy atoms from the trapping potential [14, 15], the same as the evaporation of energetic water molecules from a cup of tea, as shown in figure 4.6. The confining magnetic field makes a quadratic potential for the atoms. The atoms within the magnetic field move with a broad distribution of their velocities, and thus their kinetic energies. The evaporation of the high-velocity atoms takes place in a controllable fashion as the strength of the magnetic field is slowly decreased. In this way, the average velocity of the remaining atoms is lowered. Hence, these remaining atoms have less average energy, resulting in smaller average temperatures. Evaporative cooling can reduce the temperature to sub-microkelvin scales.

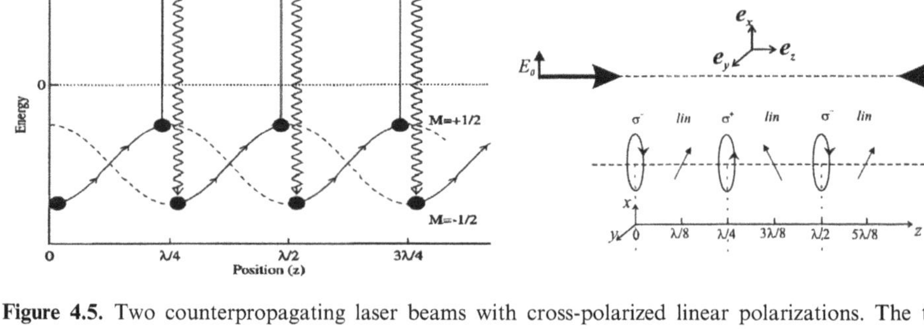


**Figure 4.5.** Two counterpropagating laser beams with cross-polarized linear polarizations. The two fields exhibit a polarization gradient with a $\lambda/2$ periodicity. Light shifts for a $|g = 1/2\rangle \longrightarrow |e = 3/2\rangle$ transition. The spatial modulation of the light shifts originates from the polarization gradient. The extrema of the light shift correspond to points where the polarization is circular. Reprinted from [12], Copyright (2001), with permission from Elsevier.

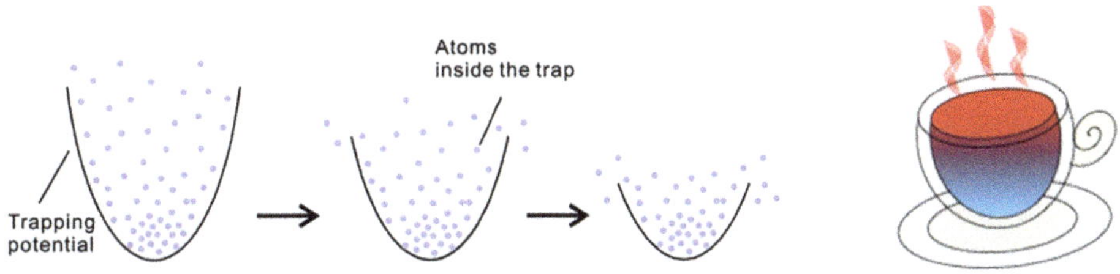


**Figure 4.6.** Evaporative cooling is based on the evaporation of higher-energy atoms from a trap, hence the remaining atoms have less energy, resulting in lower average temperatures. Credit left: [16] Copyright © 2021, The Author(s), under exclusive licence to Springer Science Business Media, LLC, part of Springer Nature. With permission of Springer.

## References

[1] Bucksbaum P H, Schumacher D W and Bashkansky M 1988 High-intensity Kapitza–Dirac effect *Phys. Rev. Lett.* **61** 1182

[2] Gould P L, Ruff G A and Pritchard D E 1986 Diffraction of atoms by light: the near-resonant Kapitza–Dirac effect *Phys. Rev. Lett.* **56** 827–30

[3] Cronin A, Schmidmayer J and Pritchard D 2009 Optics and interferometry with atoms and molecules *Rev. Mod. Phys.* **81** 1051

[4] Martin P J, Oldaker B G, Miklich A H and Pritchard D E 1988 Bragg scattering of atoms from a standing light wave *Phys. Rev. Lett.* **60** 515

[5] Marte M and Stenholm S 1992 Multiphoton resonances in atomic Bragg scattering *Appl. Phys.* B **54** 443–50

[6] Talbot H-F 1836 Facts relating to optical science. No. IV. *London, Edinburgh Dublin Phil. Mag. J. Sci.* **9** 401–7

[7] Mark M J, Haller E, Danzl J G, Lauber K, Gustavsson M and Nägerl H-C 2011 Demonstration of the temporal matter-wave Talbot effect for trapped matter waves *New J. Phys.* **13** 085008

[8] Saif F 2012 Talbot effect with matter waves *Laser Phys.* **22** 1874–8 and references therein

[9] Wineland D J and Itano W M 1979 Laser cooling of atoms *Phys. Rev.* A **20** 1521–40

[10] Dalibard J and Cohen-Tannoudji C 1985 Dressed-atom approach to atomic motion in laser light: the dipole force revisited *J. Opt. Soc. Am.* B **2** 1707–20

[11] Wineland D J, Dalibard J and Cohen-Tannoudji C 1992 Sisyphous cooling of a bound atom *J. Opt. Soc. Am.* B **9** 32–4

[12] Grynberg G and Robilliard C 2001 Cold atoms in dissipative optical lattices *Phys. Rep.* **355** 335

[13] Phillips W D 1998 Laser cooling and trapping of neutral atoms *Rev. Mod. Phys.* **70** 721

[14] Davis K B, Mewes M O and Ketterle W 1995 An analytical model for evaporative cooling of atoms *Appl. Phys.* B **60** 155–9

[15] Vogt C, Woltmann M, Herrmann S, Lämmerzahl C, Albers H, Schlippert D and Rasel (PRIMUS) E M 2020 Evaporative cooling from an optical dipole trap in microgravity *Phys. Rev.* A **101** 013634 and references therein

[16] Cao H 2021 Refrigeration Below 1 Kelvin *J. Low Temp. Phys.* **204** 175–205

[17] Kapitza P and Dirac P 1933 The reflection of electrons from standing light waves *Math. Proc. Camb. Phil. Soc.* **29** 297–300

**IOP** Publishing

Optical Forces on Atoms

**Farhan Saif and Shinichi Watanabe**

# Chapter 5

# Band formation in optical lattices

*'Because the theory of quantum mechanics could explain all of chemistry and the various properties of substances, it was a tremendous success. But still there was the problem of the interaction of light and matter.'*

—Richard P Feynman

In optical lattices, we observe spatial periodicity of the potential, which manifests itself in the formation of bands in energy space. Therefore an atom, while moving with a momentum, is allowed to occupy discrete energy values corresponding to these bands. Fascinatingly, rigorous mathematical treatment proves that the inter-band spaces increase as we increase either the intensity of the optical lattice or the coupling between the atom and the optical lattice, or both. Next, we study the characteristics of atomic dynamics in optical lattices.

## 5.1 Quantized dynamics

The interaction of a cold atom with an optical lattice is described by the three-dimensional Hamiltonian, given as

$$H = -\frac{\hbar^2}{2m}\left(\frac{\partial^2}{\partial x^2} + \frac{\partial^2}{\partial y^2} + \frac{\partial^2}{\partial z^2}\right) + V(r), \tag{5.1}$$

where we may substitute $V(r)$ from equation (3.63). Following the separation-of-variable technique, we write the general solution of the time-independent Schrödinger wave equation as

$$\psi(x, y, z) = \psi_1(x)\psi_2(y)\psi_3(z).$$

doi:10.1088/978-0-7503-2308-6ch5  

Further simplifications lead to the equation for each $\psi$, that is,

$$\sum_{s=x,y,z}\left[-\frac{\hbar^2}{2m}\frac{\partial^2}{\partial s^2}+V(s)=E\right]\psi(s). \tag{5.2}$$

Equation (5.2) implies that in each of the coordinates the wave function of the atom follows the time-independent Schrödinger wave equation. For instance, we consider the dynamics along the $x$-direction and write the eigenvalue equation as

$$\left(-\frac{\hbar^2}{2m}\frac{\partial^2}{\partial x^2}+\frac{V_0}{2}\cos(2k_{\rm L}x)\right)\psi(x)=E\psi(x). \tag{5.3}$$

Substituting $\bar{x}=k_{\rm L}x$ in the above equation, we find

$$-\frac{\partial^2\psi(\bar{x})}{\partial\bar{x}^2}+\frac{V_0}{2E_{\rm R}}\cos(2\bar{x})\psi(\bar{x})=\frac{E}{E_{\rm R}}\psi(\bar{x}),$$

where $E_{\rm R}=\frac{\hbar^2k_{\rm L}^2}{2m}$ is the recoil energy gained by the atom.

In its most general form, the above equation expresses the standard Mathieu equation, that is,

$$\frac{\partial^2\psi_\alpha(x)}{\partial x^2}+[\varepsilon_\alpha-2q\cos(2x)]\psi_\alpha(x)=0. \tag{5.4}$$

For simplicity, here and in the later discussion, we have removed the bar on variable $x$. Here, $q=\frac{V_0}{4E_{\rm R}}$ corresponds to the scaled potential height of the optical lattice, whereas the Mathieu characteristic parameter $\varepsilon_\alpha=\frac{E_\alpha}{E_{\rm R}}$ corresponds to the energy of the $\alpha$th band of the lattice. Both $q$ and $\varepsilon_\alpha$ are expressed in units of recoil energy, $E_{\rm R}$.

We may develop a pedagogic understanding of the solution, $\psi_\alpha(x)$, of the above Mathieu equation in the simplest way by considering a quantum particle of mass $m$ moving along the $x$-axis with a momentum $p$ for a negligibly small potential height, that is, $q\approx 0$. The quantum particle is described by a plane wave,

$$\exp\left\{-\frac{ipx}{\hbar}\right\}=\exp\{-ikx\},$$

which is the eigenstate of the momentum operator and energy operator, with the corresponding eigenvalue of momentum and energy, that is,

$$p=\hbar k \quad\text{and}\quad E(k)=\frac{\hbar^2k^2}{2m},$$

respectively. The particle, in its motion along the $x$-axis, observes a periodic potential of a not-so-small depth. For this reason, it gets a higher probability at the potential minima, appearing with the periodicity of the potential $d$. The larger the depth of the potential, the larger the probability of finding the particle at potential minima. Hence, we express the quantum particle mathematically by defining a function $u_k(x)$ periodic in space, such that $u_k(x) = u_k(x + d)$, and writing the plane wave together with it, that is, $\psi_{\alpha,k}(x) = u_{\alpha,k}(x)\mathrm{e}^{-ikx}$. The corresponding eigenenergy $E$ is a function of the wave vector $k$, such that $E = E_\alpha(k)$. The energy degeneracy of the atom is removed because of the tunneling effect.

In a formal way, however, we write the solution of the time-independent Schrödinger wave equation for a periodic potential following the Bloch theorem.

## 5.2 Bloch theorem

The Bloch theorem provides general solutions to a general class of spatially periodic potentials, such as in equation (5.4). We prove the Bloch theorem in three steps.

(a) First, let us consider a unitary operator, $\hat{T}$, which provides a translational displacement such that

$$\hat{T}\psi_{\alpha,k}(x) = \psi_{\alpha,k}(x + d). \tag{5.5}$$

Hence, a quantum operator, $\hat{O}$, is displaced over the space as

$$\hat{O} \longrightarrow \hat{T}\hat{O}\hat{T}^{-1},$$

for this reason, we write the Hamiltonian $\hat{H}(x)$ displaced over the lattice constant $d$ by the displacement operator $\hat{T}$ as

$$\hat{T}\hat{H}\hat{T}^{-1} = \hat{H}(x + d).$$

Since

$$\hat{H}(x + d) = \frac{\hat{p}^2}{2m} + \hat{V}(x + d) = \frac{\hat{p}^2}{2m} + \hat{V}(x) = \hat{H}(x),$$

this helps to write the above equation as

$$\hat{T}\hat{H}\hat{T}^{-1} = \hat{H}(x + d) = \hat{H}(x),$$

or

$$\hat{T}\hat{H}(x) = \hat{H}(x)\hat{T},$$

leading to a simple commutation relation, that is,

$$[\hat{T}, \hat{H}(x)] = 0. \tag{5.6}$$

(b) The general commutation relation between the displacement operator, $\hat{T}$, and the Hamiltonian operator, $\hat{H}$, indicates that they have common eigenstates. Hence, we conclude that the eigenvalue equation $\hat{H}\psi_{\alpha,k}(x) = E\psi_{\alpha,k}(x)$ implies a set of orthonormal eigenstates, $\psi_{\alpha,k}(x)$, which are spanning a Hilbert space. As a consequence, these states are also the eigenstates of the displacement operator $\hat{T}$, that is,

$$\hat{T}\psi_{\alpha,k}(x) = \psi_{\alpha,k}(x + d) = \mathrm{e}^{ikd}\psi_{\alpha,k}(x). \tag{5.7}$$

Here, we have considered that the energy operator $\hat{H}$ is Hermitian, hence, the corresponding eigenvalues are real. In addition, the operator $\hat{T}$ is unitary and has complex eigenvalues, $e^{ikd}$, with a corresponding absolute value as one.

(c) This helps to define a function, $u_{\alpha,k}(x) = e^{-ikx}\psi_{\alpha,k}(x)$, which has the property that

$$u_k(x + d) = e^{-ik(x+d)}\psi_{\alpha,k}(x + d) = e^{-ik(x+d)}\hat{T}\psi_{\alpha,k}(x) = e^{-ikx}e^{-ikd}e^{ikd}\psi_{\alpha,k}(x) = u_{\alpha,k}(x).$$

Thus, we find the eigenstate of the periodic Hamiltonian to be

$$\psi_{\alpha,k}(x) = e^{ikx}u_{\alpha,k}(x), \tag{5.8}$$

which are the Bloch states.

In his own words, Felix Bloch said of the states $\psi_{\alpha,k}(x)$, 'These are the de Broglie waves which are modulated in the rhythm of lattice structure'. For this reason, proof of the Bloch theorem helps us to understand the non-trivial nature of Bloch states and adds insight to our understanding of their physical nature. Thus, looking at the Bloch solution, we may explain it as describing the evolution of a quantum particle, or a plane wave with momentum $\hbar k$ in free space; when it evolves in a periodic potential it rides over a track provided by the periodicity of the potential, $u_{\alpha,k}(x)$, which periodically modifies the dynamics of the plane wave.

## 5.3 Energy bands

In a heuristic way, we may understand the effect of the periodicity of the potential by requiring $e^{ikd} = 1$, which is true for $k + 2jk_L$ values, where $j$ is an integer. We express the energy in $k$-space as

$$E_j(k) \approx \frac{\hbar^2(k + 2jk_L)^2}{2m}, \tag{5.9}$$

and we, therefore, find successive parabolic relationship between the energy and the wave number $k$ appearing with successive $j$ values, where $j = 0, \pm1, \pm2, \ldots$, as shown in figure 5.1.

Hence, superposition of reciprocal parabola contributes to periodicity over one unit cell $\{-k_L, +k_L\}$ that depicts a *band structure* in energy space, called a *Brillouin zone*. We plot parabolic expressions for $j = 0, \pm1, \pm2, \ldots$ and considering one Brillouin zone we find a plot, as shown in figure 5.1, made up by reciprocal parabola.

The band number $\alpha$ can be related to the parabola number $j$, such that for each $j$ value there exists two bands, that is, $\alpha = 2|j|$ and $2|j| - 1$. For example, band numbers 1 and 2 are made up by the parabola centered at $j = \pm1$, and similarly band numbers 3 and 4 originate from the parabola centered at $j = \pm2$. The band number $\alpha$ is a positive integer. We connect number $j$ with initial velocity as

$$\frac{\hbar^2(k + 2jk_L)^2}{2m} = \frac{p_0^2}{2m} = \frac{1}{2}mv_0^2. \tag{5.10}$$

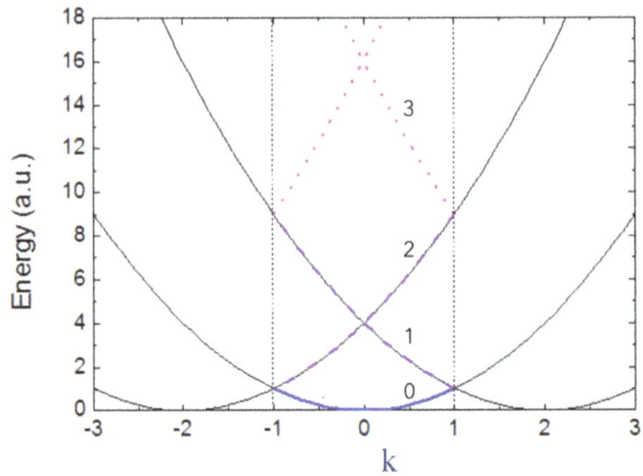


**Figure 5.1.** Energy bands vs quasi-momentum $k$ as multiples of $k_L = \pi/d$; the band spectrum is obtained by the superposition of the parabolic energy spectrum centered around reciprocal points $E_j(k) = \frac{\hbar^2(k+2jk_L)^2}{2m}$.

This implies that a particle with a certain initial velocity, and therefore a certain initial momentum, corresponds to a certain band in energy space. In addition, $k$ can take any value from $\left[-\frac{\pi}{d}, +\frac{\pi}{d}\right]$, which is the Brillouin zone for an optical lattice. In Bloch theory, $\hbar k$ behaves very similar to momentum. Taking $\hbar$ as a constant multiplier, we name $k$ as the quasi-momentum. It remains conserved over the lattice. The quasi-momentum $k$ is related to the initial momentum as $p_0 = mv_0 = \hbar k + 2j\hbar k_L$.

*Band non-parabolicity and effective mass.* In general, we may express the energy of a periodic potential $E_\alpha(k)$ by a Maclaurin series as

$$E_\alpha(k) = E_\alpha(k=0) + \frac{1}{2!}k^2\frac{\partial^2 E_\alpha(k)}{\partial^2 k}\bigg|_{k=0} + \frac{1}{4!}k^4\frac{\partial^4 E_\alpha(k)}{\partial^4 k}\bigg|_{k=0} + \cdots, \tag{5.11}$$

Here, energy is zero for $k = 0$, whereas symmetry along the $k$-axis requires the disappearance of odd-order terms in the expansion. Therefore, equation (5.11) becomes

$$E_\alpha(k) = \frac{1}{2!}k^2 E_\alpha^{''}(0) + \frac{1}{4!}k^4 E_\alpha^{''''}(0) \ldots, \tag{5.12}$$

where $E_\alpha^{''}(0)$ and $E_\alpha^{''''}(0)$ correspond to second- and forth-order derivatives of energy with respect to $k$, calculated at $k = 0$. For a periodic potential of negligible amplitude, that is, $q \approx 0$, the higher-order terms can be neglected:

$$E_\alpha(k) \approx \frac{1}{2}k^2 E_\alpha^{''}(0), \tag{5.13}$$

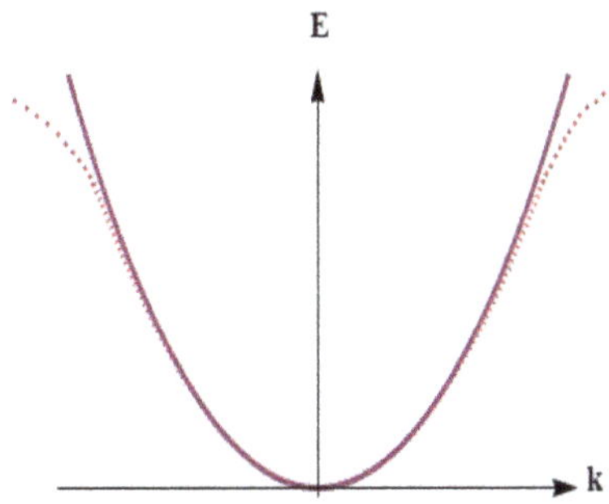


**Figure 5.2.** Non-hyperbolicity is plotted as a function of $k$ for a fixed value of $r_a$; the thick line corresponds to $r_a = 0$, while the dotted line expresses $r_a \neq 0$.

which can be equated with the energy of a free particle, that is, $\frac{\hbar^2 k^2}{2m^*}$. For this reason, we define an effective mass associated with the dynamics of an atom in an optical lattice as

$$m^* = \hbar^2 / E_\alpha^{''}(0) = \hbar^2 \left( \frac{\partial^2 E_\alpha(k)}{\partial^2 k} \bigg|_{k=0} \right)^{-1}. \tag{5.14}$$

For a free particle, that is, $q = 0$, the effective mass $m^*$ becomes the rest mass of the quantum particle, $m$.

However, in the presence of non-vanishing amplitude, $q$, the next higher-order term in equation (5.12) becomes significant. This consideration introduces a non-linearity in energy, as shown in figure 5.2, that is,

$$E_\alpha(k) = \epsilon_{\alpha 1} k^2 + \epsilon_{\alpha 2} k^4 = \frac{\hbar^2 k^2}{2m^*}(1 + r_a k^2), \tag{5.15}$$

where $r_a = \epsilon_{\alpha 2} / \epsilon_{\alpha 1}$ and

$$\epsilon_{\alpha 1} = \frac{1}{2!} E_\alpha^{''}(0) = \frac{\hbar^2}{2m^*} \text{ and } \epsilon_{\alpha 2} = \frac{1}{4!} E_\alpha^{''''}(0).$$

In equation (5.14), we express the effective mass $m^*$, which is a function of $k$ and may obtain positive and negative values. The equation (5.15) explains a symmetric deviation from the parabolic behavior.

*Energy bands of the optical lattice.* Equation (5.4) states that for a fixed value of $q$, there are a countably infinite number of solutions, $\psi_\alpha(x)$, labeled by the quantum number $\alpha$. However, only for specific values of the Mathieu characteristic parameter $\varepsilon_\alpha$ will the solutions be periodic with periods of $\pi$ or $2\pi$. These parameter values, respectively, are denoted by $a_\alpha$ and $b_\alpha$ for even and odd values of $\alpha$. Because of the intrinsic parity of the potential, the solutions can be characterized as being even, $ce_\alpha(x, q)$, or cosine-like for even values of $\alpha$ with $\alpha \geqslant 0$, whereas they are odd, $se_\alpha(x, q)$, or sine-like for odd values of $\alpha$ with $\alpha \geqslant 1$. The corresponding characteristic Mathieu parameter provides an understanding of the width of a band in energy space. The lower edge of the $\alpha$th band is

$$E_\alpha^{(\mathrm{lower})} = a_\alpha(q)E_\mathrm{R}, \tag{5.16}$$

whereas the upper edge of the $\alpha$th band is

$$E_\alpha^{(\mathrm{upper})} = b_{\alpha+1}(q)E_\mathrm{R}. \tag{5.17}$$

Here, in the deep optical lattice limit, that is, $q \gg 1$, the width of an energy band is defined as [1]

$$\delta\varepsilon_\alpha = (b_{\alpha+1} - a_\alpha)E_\mathrm{R} \simeq \frac{2^{4\alpha+5}\sqrt{\frac{2}{\pi}}\, q^{\frac{\alpha}{2}+\frac{3}{4}} \exp(-4\sqrt{q})}{\alpha!}, \tag{5.18}$$

which shows that in the deep optical lattice ($q \gg 1$ limit), energy bands are realized as degenerate energy levels as the band width is negligible. For the lowest-energy band, the width of a band in the energy domain is defined as

$$\delta E_0 = \frac{4}{\sqrt{\pi}} E_\mathrm{R}(V_o/E_\mathrm{R})^{\frac{3}{4}} \exp(-2\sqrt{V_o/E_\mathrm{R}}). \tag{5.19}$$

## 5.4 Bloch states for optical lattices

Following the Bloch theorem, the eigenfunctions of the Hamiltonian given in equation (5.4) appear as a product of a plane wave and a periodic function, that is,

$$\psi_{\alpha,k}(x) = u_{\alpha,k}(x)\, \mathrm{e}^{-ikx}, \tag{5.20}$$

where the function $u_{\alpha,k}(x)$ is periodic in $x$-space, such that

$$u_{\alpha,k}(x) = u_{\alpha,k}(x + d). \tag{5.21}$$

Here, $d$ describes the lattice spacing, and the function $u_{\alpha,k}(x)$ is the Mathieu function. Hence, the whole position space is stratified in unit cells corresponding to lattice sites. For a fixed potential height $q$ and eigenvalue $\varepsilon_\alpha$, we have a large number of solutions, $u_{\alpha,k}(x)$, for the second-order differential equation (5.4), identified by the quantum number $\alpha$.

The periodicity of the Mathieu equation in position space also places the same binding on the solution $\psi(x)$, which in turn restricts the wave number $k$ and requires that $\psi_{\alpha,k} = \psi_{\alpha,k+2j\pi/d}$ for any displacement over a unit cell of size $d$. Since $j$ takes any integer value, that is, $j = 1, 2, 3, \ldots$, moreover $\psi_{\alpha,k}$ describes an infinite set of momentum states periodically appearing at $k + jk_\mathrm{L} = k + j\pi/d$. We conclude that the Bloch states of the optical lattice, $|\alpha, k\rangle$, are defined as

$$\psi_{\alpha,k}(x) = \langle x|\alpha,\, k\rangle = \frac{1}{\sqrt{2\pi}}u_{\alpha,k}(x)\mathrm{e}^{-ikx}. \tag{5.22}$$

Moreover, $\alpha$ expresses the band number, that is, $\alpha = 1, 2, 3, \ldots$, also known as the $s, p, d, f, \ldots$ bands, respectively.

As discussed above, the function $u_{\alpha,k}$ satisfies the boundary condition, that is, $u_{\alpha,k}(x + d) = u_{\alpha,k}(x)$, where the parameter $d$ is the lattice constant. The periodicity condition implies that all the shifted solutions expressed as $\exp\{i(k + 2\pi j/d)x\}$ are solutions of the corresponding time-independent Schrödinger equation. Hence, we express $|\alpha,\, k\rangle$ as a superposition in $j$ space or momentum space as

$$|\alpha,\, k\rangle = \sum_j C_j^{(\alpha)}(k)|p_k = k + 2j\pi/d\rangle. \tag{5.23}$$

Here, $\langle x|p_x\rangle = \exp(ixp_x/\hbar)$ is the eigenstate of the momentum operator, with $C_j^{(\alpha)}((k)$ as the superposition coefficient or probability amplitude.

Hence, Bloch states have certain characteristics: (i) the Bloch states, being a product of a periodic function and a running wave, are *non-local* in position space; (ii) they are local in momentum space, with a definite $k$ value; (iii) they are *not* the eigenstates of the momentum operator, which implies that they do not have well-defined momentum (except for $u(x)$ = constant); and (iv) the plane waves are special kind of Bloch states, with $u(x)$ = constant.

## 5.5 Wannier states

Bloch orthonormal basis states in quasi-momentum space, $\psi_{\alpha,k}$, have their counterpart as Wannier states, or Wannier functions (WFs), in the position space, $w_\alpha(x - x_l)$, as shown in figure 5.3. Both are connected by a Fourier coefficient, that is,

$$w_\alpha(x - x_l) = \frac{V}{2\pi}\int_{\mathrm{BZ}} dk\ \psi_{\alpha,k}(x)\ \mathrm{e}^{-ikx_l}, \tag{5.24}$$

where $V$ is the unit cell volume, $x_l = ld$, and $l$ is an integer. The integration runs over the Brillouin zone. The inverse transformation reads

$$\psi_{\alpha,k}(x) = \sum_l w_\alpha(x - x_l)\ \mathrm{e}^{ikx_l}. \tag{5.25}$$

The Wannier states are mutually orthonormal, that is,

$$\int_V w^*_{\alpha'}(x - x_{l'})\ w_\alpha(x - x_l)\ dx = \delta_{\alpha,\alpha'}\ \delta_{l,l'}, \tag{5.26}$$

therefore, they span a Hilbert space. Both states are invariant under translation: the Bloch states have translational symmetry in position space, $x \longrightarrow x + d$, and the Wannier states in momentum space, $k \longrightarrow k + 2\pi/d$. Bloch states extend over the whole coordinate space in the same way that Wannier states extend over the

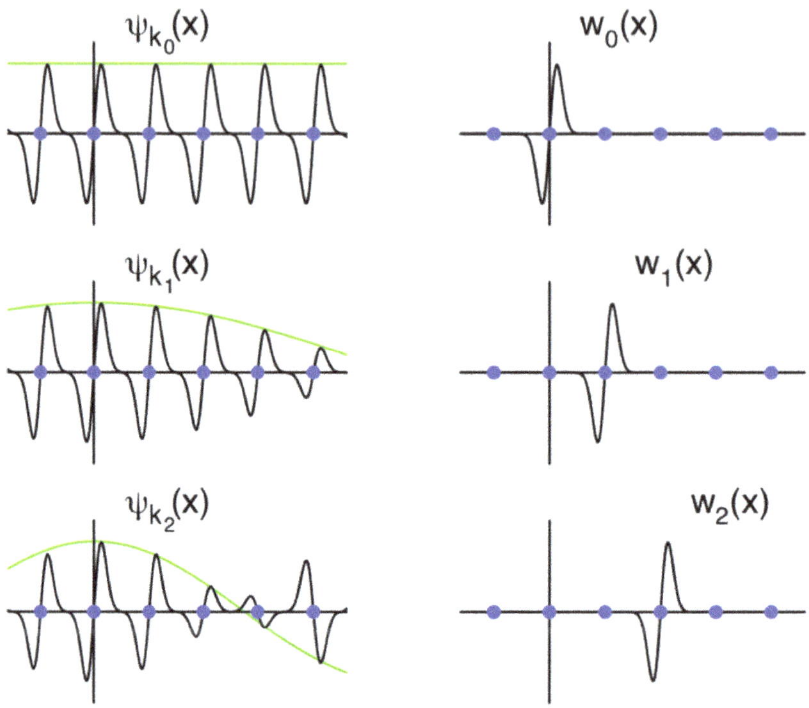


**Figure 5.3.** Transformation from Bloch functions to Wannier functions (WFs). Left: Real-space representation of three of the Bloch functions $e^{ikx}u_{\alpha,k}(x)$ associated with a single band in one dimension, for three different values of wave vector $k$. Right: WFs associated with the same band, forming periodic images of one another. Blue dots indicate atoms; green curves indicate envelopes $e^{ikx}$ of the Bloch functions. The two sets of Bloch functions at every $k$ in the Brillouin zone and WFs at every lattice vector span the same Hilbert space. Reprinted figure with permission from [4]. Copyright (2012) by the American Physical Society.

whole $k$-space. In addition, Wannier states are localized in position space as Bloch states are in $k$-space. For these reasons, the two states have similar properties but in different spaces [3].

These considerations also lead to Bloch functions as the solution made up by the linear combination of atomic de Broglie waves. The WFs, as the Fourier transform of Bloch functions, become more localized with increasing potential depth, as shown in figure 5.4. We plot two WFs for $V_0 = 3E_R$ and $V_0 = 10E_R$. For a small potential height the atom has a larger tendency to tunnel to neighboring sites. The side lobes in the Wannier function for $V_0 = 3E_r$ indicate small tunneling to the next neighbouring lattice minima due to tunneling. These are suppressed for higher values of potential $V_0 = 10E_R$, showing a diminishing value of tunneling, as shown in figure 5.4.

## 5.6 Formation of bands

Bands in the energy space are obtained by recalling the Bloch theorem, as explained in section 5.2. We note that the Schrödinger equation with a periodic potential

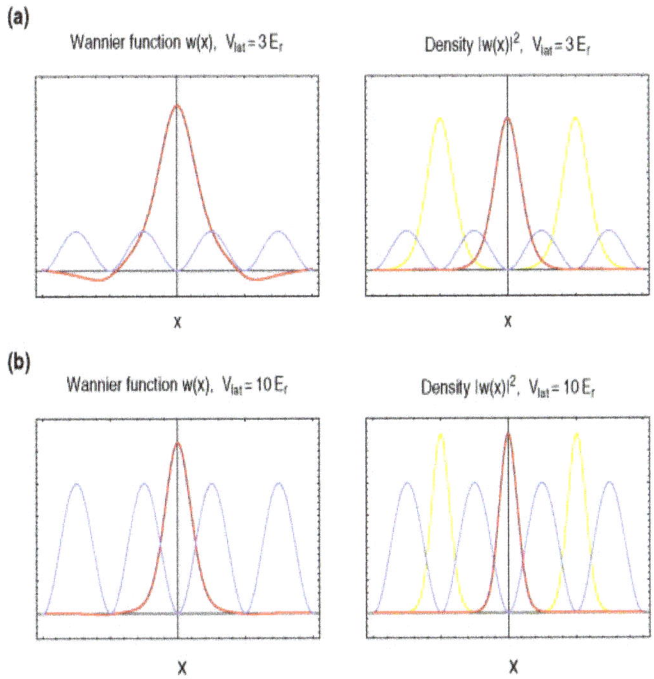


**Figure 5.4.** (a) Wannier functions for $V_0 = 3E_R$ and $V_0 = 10E_R$, and (b) their probability density $|w^2(x)|$ shown as a function of optical lattice sites. Reproduced with permission from [2].

$$V(x) = V(x + d)$$

has its solution defined in the form

$$\psi_{\alpha,k}(x) = u_{\alpha,k}(x)\, e^{ikx},$$

which is an eigenfunction of the displacement operator $T$ for any value of quasi-momentum $k$. Since $T$ commutes with $H$, the function $\psi_{\alpha,k}(x)$ is also the eigenstate of $H$ that places a restriction on attributing values to $k$. For example, eigenstates of $H$ with the periodic potential $V$ defined over a ring with $N$ number of sites with lattice spacing $d$, shown in figure 5.5, have the property

$$\psi_{\alpha,k}(x) = \psi_{\alpha,k}(x + Nd).$$

Applying the Bloch theorem, we note that a straightforward substitution of $\psi_{\alpha,k}(x) = u_{\alpha,k}(x)e^{ikx}$ yields

$$e^{iNkd} = 1,$$

that implies $Nkd = 2\alpha\pi$, where $\alpha$ is referred to as the band index and takes on integer values, namely $\alpha = 0, 1, 2, \ldots, N - 1$. This implies

$$|k|d/2\pi = \alpha/N,$$

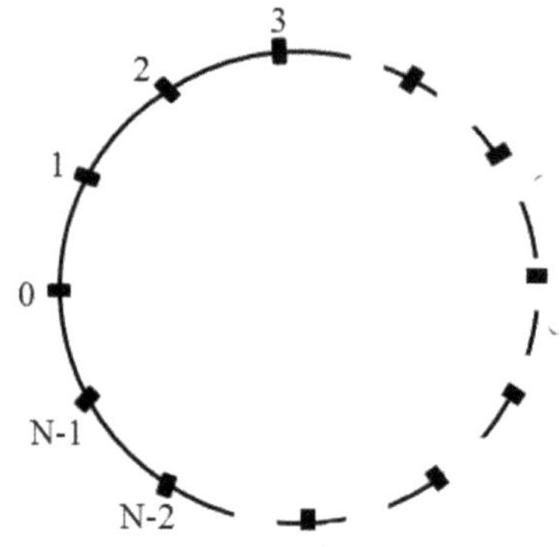


**Figure 5.5.** We show the periodic potential $V$ defined over a ring with $N$ number of sites with lattice spacing $d$, such that $\psi(x) = \psi(x + Nd)$.

that is, for $N \gg 1$, the distance between successive $k$ values diminishes and $k$ behaves like a quasi-continuous variable. The periodicity of the potential, however, responds differently in energy space as it leads to the formation of a band spectrum, with certain unacquirable values for energy that define forbidden values of energy or energy gaps between bands.

We consider an atom of mass $m$ in an optical lattice, expressed by the time-independent Schrödinger equation:

$$\left(-\frac{\hbar^2}{2m}\frac{\partial^2}{\partial x^2} + V - E\right)\psi = 0. \tag{5.27}$$

Here, $V(x) = V(x + d)$ defines the periodic lattice, $E$ is the corresponding energy, and $\psi$ expresses the corresponding Bloch functions. We understand the Bloch functions by saying that the density of the atoms propagating over the periodic lattice reflects the periodic feature of the lattice, therefore

$$|\psi_{\alpha,k}(x)|^2 = |\psi_{\alpha,k}(x + d)|^2.$$

This equation is valid up to phase, or

$$\psi(x) = u(x)e^{i\kappa(x)},$$

where $u(x)$ is periodic with periodicity $d$, whereas $\kappa(x)$ is a real function of $x$. In this case, the potential $V$ becomes constant for $d$ approaching infinity, we have $u(x)$ as a constant, and $\kappa(x) = kx$, leading to a plane-wave solution. The Bloch theorem states that this plane-wave-type feature is quite general and that the solution is locally modulated by $u(x)$.

### 5.6.1 Kronig–Penney model

Let us work out a concrete example where $V$ is constant over a finite region, with a periodicity of a finite $d$, that is,

$$V(x) \equiv \begin{cases} V_0 & 0 < x < d/2, \\ 0 & d/2 < x < d. \end{cases} \tag{5.28}$$

This is a particular case of the so-called Kronig–Penney model. Hence, within the limits of this model, we express the corresponding potential as made up of periodically appearing wells and barriers, as displayed in figure 5.6. The wave function valid in a well, $\psi_{\rm w}(x)$, thus satisfies

$$\left(-\frac{\hbar^2}{2m}\frac{\partial^2}{\partial x^2} - E\right)\psi_{\rm w}(x) = 0, \tag{5.29}$$

the wave function $\psi_{\rm w}(x)$ is in general expressed as

$$\psi_{\rm w}(x) = {\rm e}^{ikx}(A{\rm e}^{i(k_{\rm w}-k)x} + B{\rm e}^{-i(k_{\rm w}+k)x}) = {\rm e}^{ikx}u_{\rm w}(x), \tag{5.30}$$

such that $u_{\rm w}(x)$ is periodic with periodicity $d$ and

$$\frac{\hbar^2 k_{\rm w}^2}{2m} = E. \tag{5.31}$$

Moreover, the wave function for the barrier, $\psi_{\rm b}(x)$, is

$$\left(-\frac{\hbar^2}{2m}\frac{\partial^2}{\partial x^2} - (\epsilon - V_0)\right)\psi_{\rm b}(x) = 0, \tag{5.32}$$

where

$$\psi_{\rm b}(x) = {\rm e}^{ikx}(C{\rm e}^{i(k_{\rm b}-k)x} + D{\rm e}^{-i(k_{\rm b}+k)x}) = {\rm e}^{ikx}u_{\rm b}(x), \tag{5.33}$$

such that $u_{\rm b}(x)$ is periodic with periodicity $d$ and

$$\frac{\hbar^2 k_{\rm b}^2}{2m} = E - V_0. \tag{5.34}$$

*Boundary conditions.* Recall that we have a ring of $N$ lattice sites. Representing the periodicity of the solution $\psi(x) = \psi(x + Nd)$, say at a single point, $x = 0$, that it demands the continuity of the function and its derivative in the following form:

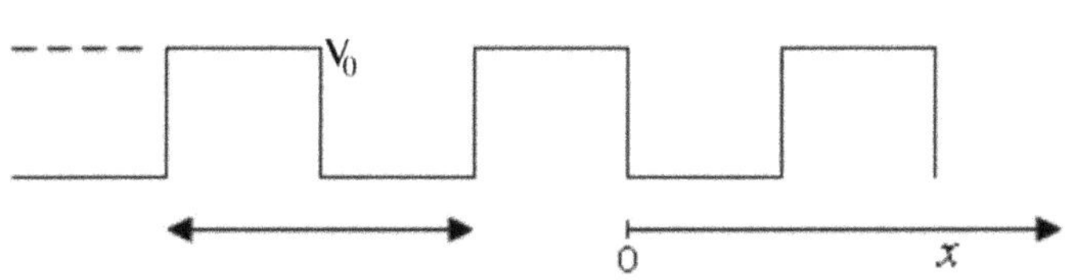


**Figure 5.6.** The periodic potential is shown along the $x$-axis with a periodicity of lattice-site spacing $d$. The potential changes from its minimum value of $V = 0$ to the maximum value $V = V_0$.

$$\psi(0^+) = \psi(Nd - 0), \qquad \left.\frac{d\psi}{dx}\right|_{x=0^+} = \left.\frac{d\psi}{dx}\right|_{x=Nd-0}.$$

Here, $0^+$ means infinitesimally to the right of $x = 0$, thus pertaining to $\psi_w$, while $x = Nd - 0$ means infinitesimally to the left of $x = 0$ but after going around the ring once, thus pertaining to $\psi_b$. From these, we obtain $Nkd = 2\alpha\pi$ with $\alpha = 0, 1, 2, \dots, N - 1$ as before, and

$$A + B = C + D, \qquad k_w(A - B) = k_b(C - D),$$

or, in matrix notation,

$$[K]\begin{pmatrix} A \\ B \end{pmatrix} = [L]\begin{pmatrix} C \\ D \end{pmatrix},$$

where

$$[K] = \begin{pmatrix} 1 & 1 \\ k_w & -k_w \end{pmatrix}, \quad [L] = \begin{pmatrix} 1 & 1 \\ k_b & -k_b \end{pmatrix}.$$

Likewise, for the periodicity pertaining to $u_w$ and $u_b$, consider the other boundary at $x = d/2$ for $u_w$ and at $x = -d/2$ for $u_b$. We have

$$u_w(d/2) = u_b(-d/2), \qquad \left.\frac{du_w}{dx}\right|_{x=\frac{d}{2}} = \left.\frac{du_b}{dx}\right|_{x=-\frac{d}{2}},$$

thus,

$$[M]\begin{pmatrix} A \\ B \end{pmatrix} = [N]\begin{pmatrix} C \\ D \end{pmatrix},$$

where

$$[M] = e^{ikd/2}\begin{pmatrix} e^{ik_w d/2} & e^{-ik_w d/2} \\ (k_w - k)e^{ik_w d/2} & -(k_w + k)e^{-ik_w d/2} \end{pmatrix},$$

$$[N] = e^{-ikd/2}\begin{pmatrix} e^{-ik_b d/2} & e^{ik_b d/2} \\ (k_b - k)e^{-ik_b d/2} & -(k_b + k)e^{ik_b d/2} \end{pmatrix}.$$

*Dispersion relation.* In order to get a non-trivial solution of the set of equations, it is clear that the matrix $[K]^{-1}[L] - [M]^{-1}[N]$ must be singular, thus

$$\det([\tilde{K}][L] - [\tilde{M}][N]) = 0,$$

where

$$[\tilde{K}] = \begin{pmatrix} -k_w & -1 \\ -k_w & 1 \end{pmatrix}$$

and

$$[\tilde{M}] = e^{-ikd/2}\begin{pmatrix} -(k_w + k)e^{-ik_wd/2} & -e^{-ik_wd/2} \\ -(k_w - k)e^{ik_wd/2} & e^{ik_wd/2} \end{pmatrix}.$$

It is left for the reader to verify the following equation as an exercise:

$$\cos k_wd/2 \cos k_bd/2 - \eta \sin k_wd/2 \sin k_bd/2 = \cos kd, \quad (5.35)$$

where

$$\eta = \frac{k_w^2 + k_b^2}{2k_wk_b}.$$

The wave numbers $k_w$ and $k_b$, being functions of energy, in equation (5.35) express a dispersion relation between energy $E$ and $k$.

*Basic characteristics.* (i) The absolute value of the right-hand side of equation (5.35) is limited by 1, while on the left-hand side the term $\eta$ may have an arbitrarily large value. The mathematical expression for $\eta$ elaborates that it is a function of $k_b$ and $k_w$, which are functions of the energy $E$. For this reason, only those values of energy are allowed which satisfy equation (5.35), whereas the values of energy which do not satisfy equation (5.35) are not allowed to be picked up by the atom; these are regarded as forbidden energy values and appear as energy gaps.

(ii) In the case of vanishing potential height $V_0$, $k_w \approx k_b$, which implies that both sides of equation (5.35) become equal, and

$$E = \frac{\hbar^2 k^2}{2m},$$

thus, we have a free particle.

(iii) For $V_0 > E$, we find that the wave vector in the barrier region becomes imaginary, that is, $k_b = i\kappa_b$, depicting tunneling behavior. We write

$$\frac{\hbar^2\kappa_b^2}{2m} = V_0 - E.$$

Hence, substituting the trigonometric identity $\cos(i\theta) = \cosh\theta$ and $\sin(i\theta) = i\sinh\theta$, the dispersion relation becomes

$$\cos k_wd/2 \cosh \kappa_bd/2 - \eta \sin k_wd/2 \sinh \kappa_bd/2 = \cos kd, \quad (5.36)$$

where

$$\eta = \frac{k_w^2 - \kappa_b^2}{2k_w\kappa_b}. \quad (5.37)$$

(iv) In the case of a very large value of $V_0$, we may, therefore, write $V_0 - E \approx V_0$. Equation (5.37) leads to $\eta \approx -\frac{1}{2}\sqrt{\frac{V_0}{E}}$. This implies that, for very large value of $V_0$, $\eta$ picks up a large value. Therefore, equation (5.36) is satisfied in two ways: for $\sinh \kappa_bd/2$ approaching zero, which implies a trivial situation in which the lattice spacing $d$ approaches zero; and for $\sin k_w d/2$ approaching zero, which implies $k_w d/2 = n\pi$. This reveals $k_w = n\pi/(d/2)$, and we obtain

$$E = \frac{\hbar^2 k_w^2}{2m} = \frac{n^2\pi^2\hbar^2}{2\,mL^2},$$

where $L = d/2$. We find that the band spectrum transforms into a discrete energy spectrum. In retrospect, we may look at the situation as follows. For a ring of $N$ sites, each eigenenergy, specified by $n$ in the case of $V_0 \gg 1$, is $N$-fold degenerate. As $V_0$ is lowered, degeneracy is lifted due to tunneling, and thus each group of degenerate energy levels splits into a band of $N$ levels. This is the basis of the tight-binding picture of energy bands, to be discussed next.

## 5.7 Band formation for optical lattices: tight-binding picture

Let us treat the case of an optical lattice represented by a periodic potential $V(x) = V_0 \sin^2(k_L x)$, where $k_L = \frac{\pi}{d}$. According to the Bloch theorem, the general expression for the atomic wave function in the $\alpha$th band of the periodic optical lattice potential is described as

$$\psi_{\alpha,k}(x) = \frac{1}{2\pi} e^{ikx} \sum_{j=-\infty}^{+\infty} C_j^{(\alpha)} \exp(2ijk_L x) = e^{ikx}\, u_{\alpha,k}(x), \tag{5.38}$$

which satisfies the periodicity condition of period $d$. We substitute equation (5.38) in the time-independent Schrödinger wave equation (5.3), and simplify the expression by applying

$$\hat{p}|k + 2jk_L\rangle = \hbar(k + 2jk_L)|k + 2jk_L\rangle,$$

$$\hat{p}^2|k + 2jk_L\rangle = \hbar^2(k + 2jk_L)^2|k + 2jk_L\rangle,$$

and

$$\sin^2(k_L x)|k + 2jk_L\rangle = \frac{1}{2}|k + 2jk_L\rangle + \frac{1}{4}(|k + 2(j+1)k_L)\rangle + |k + 2(j-1)k_L)\rangle).$$

Hence, the Schrödinger wave equation for a particular $j$th state in the $\alpha$th band with energy $E_{\alpha,j}$ can be expressed as

$$(k + 2jk_L)^2 C_j^{(\alpha)}(k) + s(C_{j-1}^{(\alpha)}(k) + C_{j+1}^{(\alpha)}(k)) = rC_j^{(\alpha)}(k). \tag{5.39}$$

The quantity $s$ depends on the potential height $V_0$ such that

$$s = \frac{2m}{\hbar^2} V_0,$$

and $E_{\alpha,j}$ is represented via its equivalent:

$$r = \frac{2m}{\hbar^2}\left(E_{\alpha,j} - \frac{V_0}{2}\right).$$

This equation is known as the central equation in the tight-binding model. In this particular case, equation (5.39) provides a three-term recurrence relation where each $j$th term is coupled with its neighbouring $(j + 1)$th and $(j - 1)$th terms

It is interesting to note that for a vanishingly small potential height, that is, for $V_0$ approaching zero, we obtain equation (5.9). As $V_0$ gains larger values so does the $s$ parameter, giving rise to non-parabolicity and leading to energy gaps between the bands.

We write the recurrence equation (5.39) in matrix form as

$$[M][X] = 0,$$

where $[X] = [\cdots, C_{j-2}^{(\alpha)}(k), C_{j-1}^{(\alpha)}(k), C_j^{(\alpha)}(k), C_{j+1}^{(\alpha)}(k), C_{j+2}^{(\alpha)}(k), \cdots]^t$; here, $t$ corresponds to a transpose of matrix $[X]$. Moreover, the matrix $[M]$ is defined as

$$[M] = \begin{bmatrix} \ddots & & & & & \\ & (k+2(j-2)k_L)^2 - r & s & 0 & 0 & 0 \\ & s & (k+2(j-1)k_L)^2 - r & s & 0 & 0 \\ & 0 & s & (k+2jk_L)^2 - r & s & 0 \\ & 0 & 0 & s & (k+2(j+1)k_L)^2 - r & s \\ & 0 & 0 & 0 & s & (k+2(j+2)k_L)^2 - r \\ & & & & & & \ddots \end{bmatrix}.$$

Eigenvalue $r$ is usually obtained by diagonalizing the tridiagonal matrix $[M]$, starting with some suitable range of $j$ such that $-J \leqslant j \leqslant J$, and ever increasing the limit $J$ till $r$ converges to a desired degree of accuracy. We plot the band structure for the first three bands in figure 5.7 for typical values of $s$.

## Exercises

1. Explain the dispersion relation given in equation (5.35) for the case when $V = E$.

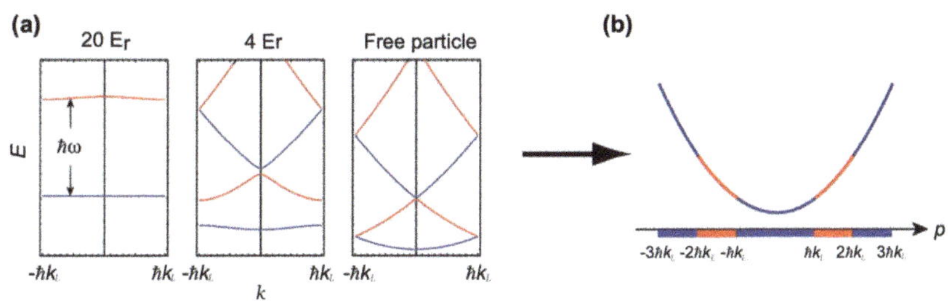


**Figure 5.7.** (a) Energy bands for different potential depths corresponding to optical lattice potential: the band gap increases as the potential depth, represented by $s$, is raised. During an adiabatic ramp down, the quasi-momentum is conserved and (b) a Bloch wave with quasi-momentum $k$ in the $n$th energy band is mapped onto a free particle with momentum $p$ in the $n$th Brillouin zone of the lattice. Reprinted figure with permission from [4]. Copyright (2008) by the American Physical Society.

2. Show that, for the periodic component $u_k(x)$ of the Bloch wave function $\varphi k = uk\ e^{ikx}$, the Schrödinger equation for a periodic potential $V(x)$ reduces to $\frac{\hbar^2}{2m}(u'' + 2iku' - k^2u) + (E - V(x)u) = 0$.
3. Using the equation obtained in exercise 2, show that in the limit $V$ approaching zero we obtain free particle eigenenergy,

$$E = \frac{\hbar^2 k^2}{2m}.$$

4. Show that a particle in free space may be expressed by a Bloch wave, following boundary conditions $u(0) = u(d)$.
5. Plot the Bloch energy bands shown in figure 5.7 for an optical lattice using Python.

## References

[1] Abramowitz M and Stegun I A 1965 *Handbook of Mathematical Functions* (New York: Dover)

[2] Rey A M 2021 Ultra-cold bosonic atoms in optical lattices: an overview *Rev. Acad. Colomb. Cienc. Exactas Fis. Nat.* **45** 666

[3] Marzari N, Mostofi A A, Yates J R, Souza I and Vanderbilt D 2012 Maximally localized Wannier functions: theory and applications *Rev. Mod. Phys.* **84** 1419

[4] Bloch I, Dalibard J and Zwerger W 2008 Many-body physics with ultracold gases *Rev. Mod. Phys.* **80** 885–964

**IOP** Publishing

Optical Forces on Atoms

**Farhan Saif and Shinichi Watanabe**

# Chapter 6

## Interacting atoms in optical lattices

*'Imagine how many aspects of nature we would miss if we lived on the surface of the Sun. Without inventing refrigerators, we would only know gaseous matter and never observe liquids or solids, and miss the beauty of snowflakes.'*

—Wolfgang Ketterle

The presence of external forces on atoms modifies the optical lattice potential. For example, the presence of an external constant force adds a linear potential, whereas a linear force on the atoms introduces a harmonic potential. The added potential limits the periodicity of the optical lattice, reshapes the atomic trajectory, and affects the overall atomic dynamics.

### 6.1 An atom in an optical lattice

In the presence of a negligibly small atom–atom interaction, we write the interaction of an ultra-cold atom with an optical lattice (OL) by means of the time-dependent Schrödinger wave equation, that is,

$$i\hbar \frac{\partial |\psi(x, t)\rangle}{\partial t} = \hat{H} |\psi(x, t)\rangle. \tag{6.1}$$

Here,

$$\hat{H} = \frac{\hat{p}^2}{2m} + V_0(\hat{x}), \tag{6.2}$$

where $V_0$ describes the one-dimensional periodic optical lattice potential seen by the ultra-cold atom and defined in equation (3.61), whereas $|\psi(x, t)\rangle$ describes the time-evolved wave function in space and time. Owing to the orthonormal property of

doi:10.1088/978-0-7503-2308-6ch6  

Wannier functions (WFs) $\langle x|\alpha, n\rangle = w_\alpha(x - x_n)$, they are orthogonal and normalizable, that is,

$$\langle \alpha, n|\alpha', n'\rangle = \int_{-\infty}^{+\infty} w_\alpha^*(x - x_n) w_{\alpha'}(x - x_{n'}) dx = \delta_{\alpha\alpha'}\delta_{nn'}. \tag{6.3}$$

The wave function, $|\psi(t)\rangle$, can be expressed as a linear combination of the WFs. Here, $\alpha$ is the band index. Hence, we write

$$|\psi(t)\rangle = \sum_{\alpha,n} |\alpha, n\rangle\langle \alpha, n|\psi(t)\rangle = \sum_{\alpha,n} c_{\alpha,n}(t)|\alpha, n\rangle,$$

where $n$ is an integer specifying the $n$th lattice site. On substituting the expansion in equation (6.1), we obtain

$$i\hbar\dot{c}_{\alpha,n} = \epsilon_{\alpha n} c_{\alpha n} - \sum_{\alpha' n' \neq n} J_{nn'}^{\alpha\alpha'} c_{\alpha',n'} \tag{6.4}$$

where

$$J_{nn'}^{\alpha\alpha'} = -\langle \alpha, n|\hat{H}|\alpha', n'\rangle.$$

Equation (6.4) expresses how the atom hops not only from one site to another but also from one band to another.

*Tight-binding approximation.* At low temperatures, atoms do not find sufficient energy for inter-band transition. This restricts the atoms to the lowest energy band. We thus drop the band index $\alpha$. Moreover, in the *tight-binding approximation*, we take into account the nearest-neighboring lattice sites. The tight-binding limit, that is, opposite to the free evolution, implies that the optical potential amplitude is large so that the cold atom is mostly bound to potential minima, and occasionally hops from one lattice site to nearest lattice site. The model is based on the consideration that the system can be expressed by the single-atom Hamiltonian near each lattice site. The corresponding eigenfunctions are the WFs that are localized in each lattice with one lattice spacing away from each other. The solution as a whole is a stationary state. For this reason, the above equation becomes

$$i\hbar\dot{c}_n = \epsilon_n c_n - J(c_{n+1} + c_{n-1}). \tag{6.5}$$

As the tunneling to neighboring sites is symmetric, we take $J_{n,n+1} \approx J_{n,n-1} = J$. The expressions for on-site energy, $\epsilon_n$, and tunneling, $J$, respectively, are

$$\epsilon_n = \langle n|\hat{H}|n\rangle = \int w^*(x - x_n)\left[\frac{\hat{p}^2}{2m} + V_0(\hat{x})\right]w(x - x_n)\, dx \tag{6.6}$$

and

$$J = -\langle n|\hat{H}|n + 1\rangle = -\int w^*(x - x_n)\left[\frac{\hat{p}^2}{2m} + V_0(\hat{x})\right]w(x - x_{n+1})\, dx. \tag{6.7}$$

*Hamiltonian in outer product.* Following the above discussion and using the completeness relation in the Wannier state basis, the Hamiltonian given in equation (6.2) can be rewritten for a single band in the tight-binding approximation as

$$\hat{H} = \epsilon \sum_n |n\rangle\langle n| - J \sum_n (|n\rangle\langle n+1| + |n+1\rangle\langle n|), \tag{6.8}$$

where we have taken the on-site energy as constant, that is, $\epsilon_n \approx \epsilon$.

*Second quantization.* We introduce the second quantization with a view to a later discussion of the Bose–Hubbard model of interacting bosonic atoms. The second quantization also provides a compact and transparent representation of the hopping of atoms from one site to another. Identical atoms are indistinguishable, and in quantum mechanics this results in symmetry under interchange of each pair of atoms. The atoms are called bosons, and those with anti-symmetry are called fermions. Symbolically, we express this as

$$\Psi_{\text{boson}}(\xi, \xi') = \Psi_{\text{boson}}(\xi', \xi),$$

$$\Psi_{\text{fermion}}(\xi, \xi') = -\Psi_{\text{fermion}}(\xi', \xi),$$

where $\xi$ and $\xi'$ represent the degrees of freedom. It follows from the above discussion that an atom consisting of an even number of fermions, such as electrons, protons, neutrons, etc, are bosons, whereas the atoms with an odd number of fermions behave as fermions. Thus, for example, potassium $^{39}$K is bosonic while $^{40}$K is fermionic. The symmetry under interchange is describable using a set of creation and annihilation operators satisfying bosonic or fermionic commutation relations.

In order to develop the second quantization, we express the Hamiltonian using the field operator for the ultra-cold atom, $\hat{\psi}_a(\mathbf{x})$ [1], such that

$$\hat{H} = \int dx\, \hat{\psi}_a^\dagger(\mathbf{x}) \left( -\frac{\hbar^2}{2m} \nabla^2 + V_0(\mathbf{x}) \right) \hat{\psi}_a(\mathbf{x}),$$

where the bosonic field operators $\hat{\psi}_a(\mathbf{x})$ and $\hat{\psi}_a^\dagger(\mathbf{x})$ are non-Hermitian operators, defined as

$$\hat{\psi}_a^\dagger(\mathbf{x}) = \sum_{\alpha,n} \hat{a}_{\alpha,\, n}^\dagger w_\alpha^*(\mathbf{x} - x_n),$$

and

$$\hat{\psi}_a(\mathbf{x}) = \sum_{\alpha,n} \hat{a}_{\alpha,n} w_\alpha(\mathbf{x} - x_n).$$

The operators $\hat{a}_{\alpha,n}$ and $\hat{a}_{\alpha,\, n}^\dagger$ are the annihilation operator and creation operator for the atom in the $\alpha$ band and at the $n$th optical lattice site [2]. They follow the conventional commutation relationship between ladder operators, namely

$$[\hat{a}_{\alpha n}, \hat{a}_{\alpha' n'}^\dagger] = \delta_{\alpha\alpha'} \delta_{nn'},$$

for bosonic atoms. Likewise,

$$\{\hat{a}_{\alpha n}, \hat{a}^{\dagger}_{\alpha' n'}\} = \hat{a}_{\alpha n}\hat{a}^{\dagger}_{\alpha' n'} + \hat{a}^{\dagger}_{\alpha' n'}\hat{a}_{\alpha n} = \delta_{\alpha\alpha'}\delta_{nn'},$$

for fermionic atoms. It then follows that

$$[\hat{\psi}^{\dagger}_{a}(\mathbf{x}), \hat{\psi}^{\dagger}_{a}(\mathbf{x}')] = \delta(\mathbf{x} - \mathbf{x}'),$$

for bosonic atoms, and

$$\{\hat{\psi}^{\dagger}_{a}(\mathbf{x}), \hat{\psi}^{\dagger}_{a}(\mathbf{x}')\} = \delta(\mathbf{x} - \mathbf{x}'),$$

for fermionic atoms.

As we consider the atoms at a very low temperature, they are in the lowest single band. For this reason, we may simplify our expression by omitting the band index $\alpha$, such that

$$\hat{H} = \sum_{n,\, n'} \hat{a}^{\dagger}_{n}\hat{a}_{n'} \int dx\, w^{*}(x - x_{n})\left(-\frac{\hbar^{2}}{2m}\nabla^{2} + V_{0}(\mathbf{x})\right)w(x - x_{n'}). \tag{6.9}$$

The expression can be rewritten in two parts: first, where $n = n'$, which corresponds to the on-site energy of the atom at the $n$th site; second, when $n \neq n'$, which corresponds to tunneling from the $n$th to $n'$ th site, respectively, mentioned in equations (6.6) and (6.7) in the tight-binding limit. Therefore,

$$\hat{H} = \epsilon\sum_{n}\hat{a}^{\dagger}_{n}\hat{a}_{n} - \sum_{n' \neq n} J_{n,n'}\hat{a}^{\dagger}_{n}\hat{a}_{n'}. \tag{6.10}$$

Considering only the next neighboring term, that is, $n \pm 1$, we get

$$\hat{H} = \epsilon\sum_{n}\hat{a}^{\dagger}_{n}\hat{a}_{n} - J\sum_{n}(\hat{a}^{\dagger}_{n}\hat{a}_{n+1} + \hat{a}^{\dagger}_{n}\hat{a}_{n-1}). \tag{6.11}$$

Here, we take $n$ as a dummy index to rewrite the above expression as

$$\hat{H} = \epsilon\sum_{n}\hat{a}^{\dagger}_{n}\hat{a}_{n} - J\sum_{n}(\hat{a}^{\dagger}_{n}\hat{a}_{n+1} + \hat{a}^{\dagger}_{n+1}\hat{a}_{n}). \tag{6.12}$$

This expression, equivalent to the outer product from equation (6.8), introduces $\hat{a}_n$ and $\hat{a}^{\dagger}_{n}$ as dynamical variables for the Bose–Hubbard model.

*Dispersion relation for arbitrary amplitude OL.* In the presence of a periodic optical potential, however, quasi-momentum or lattice momentum plays almost the same role as momentum in free space, with a constraint that it is within a Brillouin zone, that is, $[-\pi/d, +\pi/d]$. Therefore, we introduce an ansatz on the probability amplitude, $c_n$, and express it as $c_n = f^{(k)}_{n}\mathrm{e}^{-i(E+\epsilon_n)t/\hbar}$, where $f^{(k)}_{n} = \mathrm{e}^{inkd}$. The energy, $E$, in the exponent depends on $k$, that is, $E \equiv E(k)$. Thus, substituting $c_j$ into equation (6.5), we get for the lowest band

$$E(k) = -2J\cos kd, \tag{6.13}$$

which provides a dispersion relation for the non-interacting system. This further explains that

$$\Delta E = E(k = \pi/d) - E(k = 0) = 4J. \tag{6.14}$$

Hence, $k$ may take values in the range that corresponds to the energy width $\Delta E$. Moreover, $E(k)$ has translational symmetry over a period $k + 2k_L = k + 2\pi/d$. For this reason, $k$ acts like a quantum number, named *quasi-momentum*.

**Example 1: dispersion relation for small *k*:** Equation (6.13) describes the dispersion relation for a cold atom in the lowest band of the optical lattice. In the case of a very small quasi-momentum $k \approx 0$, the dispersion relation becomes

$$E(k) \approx -2J\left(1 - \frac{(kd)^2}{2!}\right) = -2J + \frac{\hbar^2}{2m^*}k^2,$$

where the second term represents the effective kinetic energy and

$$m^* = \frac{\hbar^2}{2Jd^2} = \frac{\hbar^2 k_L^2}{2\pi^2 J}.$$

Here, $k_L = \pi/d$. The dispersion relation is useful in understanding the dynamics of a quantum particle of effective mass $m^*$ moving in an optical lattice with a negligible quasi-momentum, $k$. Noting that $E = \hbar\omega$, we write the corresponding phase velocity as

$$v_{ph} = \frac{\omega}{k} = \frac{-2J}{\hbar k} + \frac{\hbar k}{2m^*},$$

and the group velocity as

$$v_g = \frac{\partial\omega}{\partial k} = \frac{\hbar k}{m^*}.$$

The effective mass $m^*$ can be obtained for the band edge $k \approx \pi/d$ by using the above dispersion relation. Note that $E(k \approx \pi/d) = -E(k \approx 0)$. This implies that the effective mass $m^*$ has the same magnitude, but is of the opposite sign at the band edge.

## 6.2 Constant external forcing

Let us consider an ultra-cold atom evolving in an optical lattice in the presence of a constant force, $F$, as shown in figure 6.1. As we discuss in this section, the quantum particle has distinct dynamical behavior in the system as compared with its counterpart evolving in the absence of constant forcing. We express the corresponding Hamiltonian as

$$\hat{H} = \frac{\hat{p}^2}{2m} + V(\hat{x}) + V_1(\hat{x}), \tag{6.15}$$

where we have included the potential energy operator as $V_1(\hat{x}) = F\hat{x}$. In the case of a neutral atom propagating in an optical lattice along the gravitational field, the

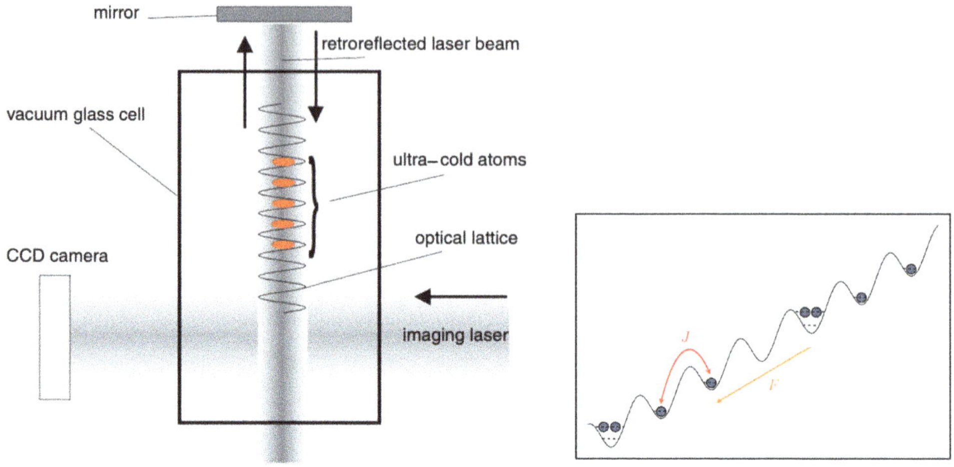


**Figure 6.1.** Left: Schematic diagram of the experiment: ultra-cold atomic dynamics is shown in a periodic optical lattice in the presence of a linear gravitational potential [3]. John Wiley & Sons. Copyright © 2004 Wiley-VCH Verlag GmbH & Co. KGaA, Weinheim. Right: Energy vs displacement plot: the atoms in the optical lattice in the presence of the linearly changing potential display a dynamics somewhat similar to that of a periodically changing ramp. Reprinted figure with permission from [4], Copyright (2016) by the American Physical Society. Image Credits: Igor Pikovski.

constant force $F$ is defined as $F = mg$, where $g$ is the gravitational acceleration. We rewrite the Hamiltonian in the Wannier states basis as

$$\hat{H} = -J \sum_{n=-\infty}^{\infty} (|n\rangle\langle n+1| + |n+1\rangle\langle n|) + Fd \sum_{n=-\infty}^{\infty} n|n\rangle\langle n|. \tag{6.16}$$

The on-site energy term provides a constant energy shift, and for this reason we exclude it. Here, the last term is obtained by assuming that the external potential remains constant over a lattice site, that is,

$$\sum_{n'}\sum_{n}|n'\rangle\langle n'|V_1(\hat{x})|n\rangle\langle n| = \sum_{n} V_1(x_n)|n\rangle\langle n| = F\ d\ \sum_{n} n|n\rangle\langle n|. \tag{6.17}$$

Following the same procedure developed in the previous section, we express the Hamiltonian given in equation (6.16) in quasi-momentum space as

$$\hat{H}(k) = -J\cos\hat{k}d + F\ \hat{x}. \tag{6.18}$$

We express the operator $\hat{x}$ in quasi-momentum space, that is,

$$\hat{x} = i\hbar\frac{\partial}{\partial p} = i\frac{\partial}{\partial k}.$$

Hence, we obtain

$$\hat{H}(k) = -J\cos\hat{k}d + iF\frac{\partial}{\partial k}. \tag{6.19}$$

### 6.2.1 Acceleration theorem

The wave-packet dynamics follows the Heisenberg equations, that is,

$$\dot{\hat{x}} = \frac{i}{\hbar}[\hat{H}, \hat{x}], \tag{6.20}$$

$$\dot{\hat{p}} = \frac{i}{\hbar}[\hat{H}, \hat{p}]. \tag{6.21}$$

Considering the correspondence principle, and owing to a close resemblance between quasi-momentum and real momentum, we write the semi-classical description of wave-packet dynamics in the lowest band of a lattice as

$$\dot{x} = \frac{1}{\hbar}\frac{\partial H}{\partial k}, \quad \hbar\dot{k} = -\frac{\partial V}{\partial x}. \tag{6.22}$$

This leads us to write the force on the quantum particle as

$$\ddot{x} = \frac{1}{\hbar}\frac{\partial^2 H}{\partial k^2}\frac{\partial k}{\partial t} = -\frac{1}{\hbar^2}\frac{\partial^2 H}{\partial k^2}\frac{\partial V}{\partial x}. \tag{6.23}$$

Equation (6.23) reveals a simple equation for the force experienced by the material wave packet, that is,

$$m^*\ddot{x} = -\frac{\partial V}{\partial x}, \tag{6.24}$$

where

$$m^* = \hbar^2\left(\frac{\partial^2 H}{\partial k^2}\right)^{-1}. \tag{6.25}$$

Hence, we can associate an effective mass with the atom, induced by the presence of the optical lattice in the presence of constant force, $F = m^*a$, where $a = -\ddot{x}$. As discussed above, the effective mass can be positive or negative; in addition, it can vary from zero to infinity. The effective mass for the system is

$$m^* = \frac{\hbar^2}{Jd^2 \cos kd} = \frac{\hbar^2 k_L^2}{\pi^2 \cos kd}. \tag{6.26}$$

Therefore, we find the effective mass changing from a positive value, $+\frac{\hbar^2 k_L^2}{J\pi^2}$, at the center of the Brillouin zone, where quasi-momentum is zero, to a negative value, $-\frac{\hbar^2 k_L^2}{J\pi^2}$, at the edge of the Brillouin zone, where $k = \pm\pi/d$.

*Time evolution.* The substitution of the Hamiltonian given in equation (6.18) in equations (6.20) and (6.21) leads to

$$\dot{\hat{x}} = \frac{i}{\hbar}[\hat{H}, \hat{x}] = Js \sin ps, \tag{6.27}$$

$$\dot{\hat{p}} = \frac{i}{\hbar}[\hat{H}, \hat{p}] = -F. \tag{6.28}$$

Here, we introduce $s = d/\hbar$. Integration of the two differential equations reveals

$$\hat{x}(t) = \hat{x}_0 + \frac{J}{F}[\cos(\omega_{\mathrm{B}}t - \hat{p}_0 s) - \cos(\hat{p}_0 s)], \tag{6.29}$$

$$\hat{p}(t) = \hat{p}_0 - Ft. \tag{6.30}$$

We simplify equation (6.29) by using the identity

$$\cos(A + B) - \cos(A - B) = -2 \sin A \sin B,$$

where $A = \omega_{\mathrm{B}}t/2$ and $B = \omega_{\mathrm{B}}t/2 - p_0 s/2$. We obtain

$$\hat{x}(t) = \hat{x}_0 - \frac{2J}{F} \sin \frac{\omega_{\mathrm{B}}t}{2} \sin\left(\frac{\omega_{\mathrm{B}}t}{2} - \hat{p}_0 s\right). \tag{6.31}$$

In terms of quasi-momentum $\hat{p} = \hbar\hat{k}$ and $\hat{x} = \hat{n}d$, we rewrite equations (6.30) and (6.31) such that

$$\hat{n}(t) = \hat{n}_0 - \frac{2\omega}{\omega_{\mathrm{B}}} \sin \frac{\omega_{\mathrm{B}}t}{2} \sin\left(\frac{\omega_{\mathrm{B}}t}{2} - \hat{k}_0 d\right), \tag{6.32}$$

$$\hat{k}(t) = \hat{k}_0 - Ft/\hbar, \tag{6.33}$$

where $\omega = J/\hbar$ and $\omega_{\mathrm{B}} = Fd/\hbar$.

## 6.3 Wannier–Stark states and the Wannier–Stark ladder

The Hamiltonian given in equation (6.19) leads to a time-independent Schrödinger wave equation in quasi-momentum space, which is

$$\hat{H}(k)\Psi = E\Psi,$$

and reveals a first-order differential equation, that is,

$$\left[-J \cos kd + iF\frac{d}{dk}\right]\Psi(k) = E\Psi(k). \tag{6.34}$$

Rearranging the terms and performing a simple integration provides

$$\log \Psi(k) = -\frac{i}{F}\left(Ek + \frac{J}{d} \sin kd\right) + \log \mathcal{N},$$

which is further simplified as

$$\Psi(k) = \mathcal{N} \exp\left\{-i\left(\frac{Ek}{F} + \frac{J}{Fd} \sin kd\right)\right\}. \tag{6.35}$$

Here, $\mathcal{N}$ is the normalization constant and is calculated from the normalization condition on $\Psi$, as $\mathcal{N} = \sqrt{\frac{d}{2\pi}}$. It is interesting to note that the $2\pi$-periodic nature of

$H(k)$ also imposes a condition of periodicity on the eigenfunctions. We, therefore, find

$$\Psi(k) = \Psi(k + 2k_L), \tag{6.36}$$

where $k_L = \pi/d$. The periodicity condition given in equation (6.36) requires that

$$e^{-2iEk_L/F} = 1.$$

Hence, a non-trivial solution is obtained as

$$2Ek_L/F = 2m\pi, \quad \text{where } m = 0, \pm1, \pm2, \pm3, \dots,$$

revealing the eigenenergies as

$$E_m = mFd, \tag{6.37}$$

and the corresponding eigenstates as

$$\Psi_m(k) = \langle k|\Psi_m\rangle = \sqrt{\frac{d}{2\pi}}\, e^{-i(mkd+\gamma \sin kd)}, \tag{6.38}$$

where $\gamma = J/Fd = \omega/\omega_B$. Thus, we get the eigenstates, $\Psi_m(k)$, and eigenenergies, $E_m$, of the system, respectively, also known as Wannier–Stark states and the Wannier–Stark ladder. Wannier–Stark states are orthonormal and thus make a complete Hilbert space. The Wannier–Stark ladder provides an energy spectrum that has an equal spacing between next-neighboring eigenenergies. These two characteristics make the system similar to a harmonic oscillator, and therefore the two systems have comparable dynamics.

## 6.4 Bloch oscillations

We consider the dynamics of a very narrowly peaked wave packet in time in an optical lattice in the presence of a linear potential. For simplicity, we consider that the initial state is $|\psi(0)\rangle = |n = 0\rangle$, and that the particle is at origin in coordinate space. The time-evolved wave packet is written as

$$|\psi(t)\rangle = \hat{U}(t)|\psi(0)\rangle.$$

On substituting equation (B.9) for the evolution operator $\hat{U}$ from appendix B, we obtain

$$\begin{aligned} |\psi(t)\rangle &= \sum_n \sum_{n'} J_{n-n'}\left(2\gamma \sin\frac{\omega_B t}{2}\right) e^{i(n-n')\left(\frac{\pi}{2}-\frac{\omega_B t}{2}\right)-in'\omega_B t}|n\rangle\langle n'|0\rangle, \\ &= \sum_n J_n\left(2\gamma \sin\frac{\omega_B t}{2}\right) e^{in\left(\frac{\pi}{2}-\frac{\omega_B t}{2}\right)}|n\rangle. \end{aligned} \tag{6.39}$$

As a confirmation of the obtained result, we note that at $t = 0$ we regain the initial state $|0\rangle$ since the zeroth-order Bessel function, $J_0$, attains a unit value when the argument becomes zero, that is, $J_0(0) = 1$, whereas all the other Bessel functions are zero, that is, $J_{n\neq 0}(0) = 0$. Therefore, equation (6.39) becomes

$$|\psi(0)\rangle = \sum_n J_n(0) e^{in\frac{\pi}{2}} |n\rangle = |0\rangle.$$

As the time evolves, the initial state is reached repeatedly whenever the argument of the Bessel function becomes zero, that is,

$$2\gamma \sin \frac{\omega_B t}{2} = 0. \tag{6.40}$$

The periodic behavior is repeatedly achieved in time for $t$ = (integer) × $T_B$, where

$$T_B = \frac{2\pi}{\omega_B}, \quad \text{and} \quad \omega_B = \frac{Fd}{\hbar}. \tag{6.41}$$

Hence, we note that as the evolution in time takes place, higher-order Bessel functions, $J_n$, grow beyond an initial zero value successively with increasing $n$. This implies that as the time increases the larger Wannier states are occupied, successively showing spreading behavior. However, at $t = T_B$, the arguments of Bessel functions become zero, thus the initial state $|0\rangle$ is recovered. The process repeats itself periodically at integer multiples of time, $T_B$.

**General solution:** As a combined effect of an optical lattice and linear potential, a quantum particle in its one-dimensional motion *oscillates* in time. We express a wave packet in general in the Wannier basis as

$$|\Psi(0)\rangle = \sum_n c_n(0)|n\rangle. \tag{6.42}$$

Here, $c_n(0) = \langle n|\Psi(0)\rangle$. The time evolution is defined by the unitary operator $\hat{U}(t) = e^{-iHt/\hbar}$ developed in appendix B, and the Hamiltonian $\hat{H}$ is defined in equation (6.19), that is,

$$\hat{U}(t) = \sum_n \sum_{n'} J_{n-n'}\left(2\gamma \sin \frac{\omega_B t}{2}\right) e^{i(n-n')\left(\frac{\pi}{2} - \frac{\omega_B t}{2}\right) - in'\omega_B t} |n\rangle\langle n'|. \tag{6.43}$$

In order to obtain the time-evolved wave packet, $|\Psi(t)\rangle$, we apply the time evolution operator. Thus, we write

$$\begin{aligned} |\Psi(t)\rangle &= \hat{U}(t)|\Psi(0)\rangle, \\ &= \sum_{n''} c_{n''}(0)\ \hat{U}(t)|n''\rangle, \\ &= \sum_{n, n'} c_{n'}(0)\ J_{n-n'}\left(2\gamma \sin \frac{\omega_B t}{2}\right) \exp\left\{ i(n-n')\left(\frac{\pi}{2} + \frac{\omega_B t}{2}\right) \right. \\ &\qquad \left. - in'\omega_B t \right\} |n\rangle, \\ &= \sum_n c_n(t)|n\rangle. \end{aligned} \tag{6.44}$$

Hence, the probability distribution function, $|c_n(t)|^2$, is obtained as

$$|c_n(t)|^2 = \left| \sum_{n'} c_{n'}(0) J_{n-n'}\left(2\gamma \sin \frac{\omega_B t}{2}\right) \exp\left\{ i(n - n')\left(\frac{\pi}{2} + \frac{\omega_B t}{2}\right) - in'\omega_B t \right\} \right|^2, \quad (6.45)$$

which displays an oscillatory dynamics as a function of time, with a frequency $\omega_B = Fd/\hbar$. The oscillatory behavior is known as Bloch oscillation, as shown in figure 6.2. It is interesting to note that the Bloch oscillation frequency, $\omega_B$, is a product of constant force $F$ and lattice spacing $d$. This indicates that the Bloch oscillation of a particle is a consequence of the combined effect of the optical lattice and the constant external force $F$. Furthermore, it is the same frequency with which a classical particle sweeps the classical phase space, as discussed previously. Moreover, $c_n(t) = \langle n|\Psi(t)\rangle$, which implies a projection of $|\Psi\rangle$ along the $n$th Wannier–Stark basis. Hence, at $T_B = 2\pi/\omega_B$, we find $|c_n(t)|^2 = |c_n(0)|^2$, that is, the original state of the wave packet. We find the same feature at integer multiples of the Bloch oscillation time period $T_B$, which is similar to the wave-packet dynamics we observe in a harmonic oscillator.

*Variance in momentum space and position space.* Let us consider a wave packet, $|\Psi\rangle$, that initially follows a Gaussian probability distribution, $W(x, p)$ [5]. It originates from a certain position, $x_0 = n_0 d$, and momentum, $p_0 = \hbar k_0$, in phase space; hence, we write

$$W(x, p, t = 0) = \frac{1}{2\pi\sigma_{x0}\sigma_{p0}} \exp\left( -\frac{(x - x_0)^2}{2\sigma_{x0}^2} - \frac{(p - p_0)^2}{2\sigma_{p0}^2} \right). \quad (6.46)$$

Here, $\sigma_{x0}$ and $\sigma_{p0}$ are initial distribution widths in position space and momentum space. The evolution of the wave packet in time is evaluated by calculating the average value of position with respect to time, that is,

$$\langle x(t)\rangle = x_0 - \frac{2J}{F} e^{-(\sigma_{p0}^2 s^2)/2} \sin \frac{\omega_B t}{2} \sin\left(\frac{\omega_B t}{2} - p_0 s\right), \quad (6.47)$$

and the average value of momentum as a function of time, i.e.

$$\langle p(t)\rangle = \langle \Psi|\hat{p}|\Psi\rangle = p_0 - Ft. \quad (6.48)$$

The time evolution of the variance in position space is obtained by calculating the first moment, $\langle x(t)\rangle$, and second moment, $\langle x^2(t)\rangle$, such that

$$\begin{aligned} \sigma_x^2(t) &= \langle x^2(t)\rangle - \langle x(t)\rangle^2 \\ &= \sigma_x^2(t = 0) + \frac{2J^2}{F^2}(1 - e^{-\sigma_{p0}^2 s^2})\sin^2 \frac{\omega_B t}{2}. \end{aligned} \quad (6.49)$$

Similarly, we obtain the variance in momentum space as a function of time, that is,

$$\begin{aligned} \sigma_p^2(t) &= \langle p^2(t)\rangle - \langle p(t)\rangle^2 \\ &= \sigma_p^2(t = 0). \end{aligned} \quad (6.50)$$

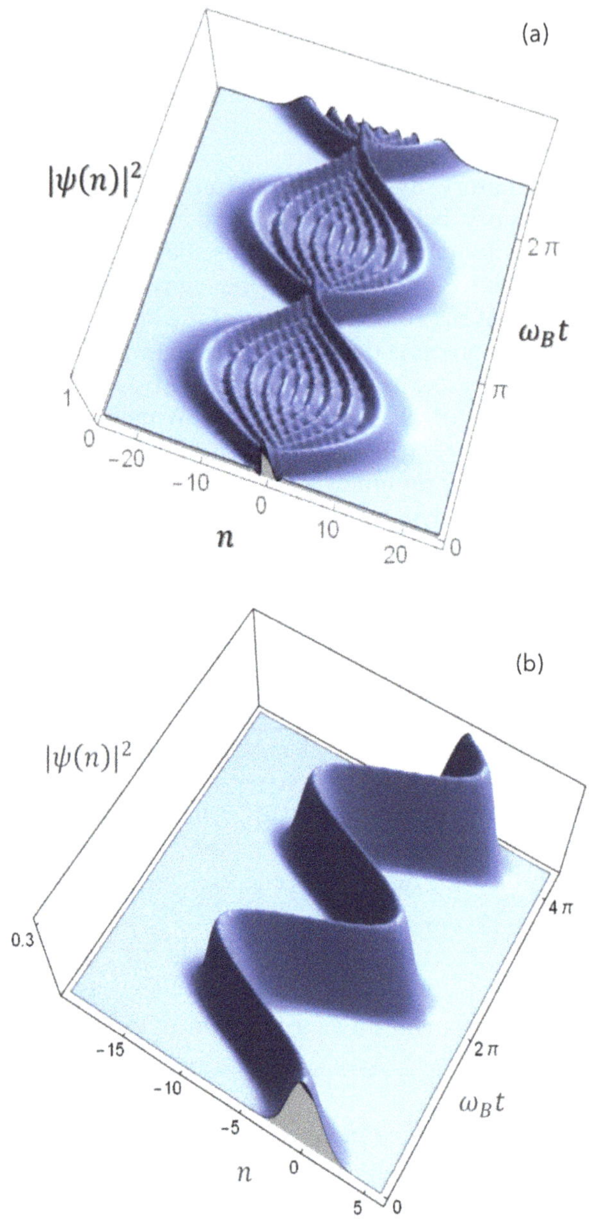


**Figure 6.2.** (a) Quantum mechanical wave-packet distribution initiating from $n = 0$ with a very narrow distribution within a lattice site showing coherent dynamics. The initial narrow wave packet spreads as a coherent superposition over many sites of the potential, labeled by $n$. (b) However, for an initial atomic wave packet that has a spread over many lattice sites, the wave packet displays minimum dispersion. In time, the oscillations are periodic with frequency $\omega_B$, and the wave packet comes back to its original state; see reference [4]. Reprinted figure with permission from [4], Copyright (2016) by the American Physical Society. Image Credits: Igor Pikovski.

Here, $\sigma_x(t=0) = \sigma_{x0} = d\Delta n$ and $\sigma_p(t=0) = \sigma_{p0} = \hbar/\sigma_{x0}$. The distribution in position space evolves in time following equation (6.49). It is interesting to note that the position distribution is a function of the initial width of the probability distribution in momentum space, $\sigma_p(t=0)$. Thus, for a broad initial distribution in momentum space, we find $e^{-\sigma_{p0}^2 s^2} = 0$. From equation (6.49), we obtain

$$\sigma_x^2(t) = \sigma_x^2(t=0) + \frac{2J^2}{F^2}\sin^2\frac{\omega_B t}{2}, \tag{6.51}$$

which shows a breathing behavior at a fixed mean position, that is, $\langle x(t)\rangle = x_0$. Whereas, for a very narrow initial momentum distribution, we note that $e^{-\sigma_p^2(0)s^2} = 1$, thus $\sigma_x^2(t) = \sigma_x^2(t=0)$. The mean position in this case becomes

$$\langle x(t)\rangle = x_0 - \frac{2J}{F}\sin\left(\frac{\omega_B t}{2}\right)\sin\left(\frac{\omega_B t}{2} - p_0 s\right), \tag{6.52}$$

thus it shows oscillatory behavior. Maintaining equation (6.49), we obtain $\sigma_x^2(t) = \sigma_{x0}^2$. Figure 6.2 shows the probability distribution function $|\Psi(t)|^2$ as a function of evolution time and the optical lattice sites, $n$.

## 6.5 Beyond the tight-binding approximation: the Landau–Zener transition

When the energy bands become quasi-degenerate, for instance $E_0(k) \approx E_1(k)$ for the lowest two bands, the tight-binding model fails. There occurs an inter-band transition which, under certain circumstances, permits an approximate description using the Landau–Zener transition model.

The atomic dynamics in the optical lattice can take place in two possible ways. First, the atom exhibits band dynamics following

$$\dot{x} = \frac{1}{\hbar}\frac{\partial \epsilon_q}{\partial k} \tag{6.53}$$

and

$$\dot{k} = -\frac{1}{\hbar}\frac{\partial \epsilon_q}{\partial x}. \tag{6.54}$$

Equation (6.53) defines the group velocity of the atomic wave packet in a band, whereas equation (6.54) expresses Newton's second law, which defines the force acting in $k$-space.

Second, the atom experiences a Landau–Zener transition when it passes an avoided crossing in the quasi-momentum space, as shown in figure 6.3. We avoid the mathematical subtleties in obtaining an explicit expression for the Landau–Zener transition in periodic lattices by introducing an innovative approach. This is of pedagogic interest as it provides a useful understanding of a different perspective on the phenomenon.

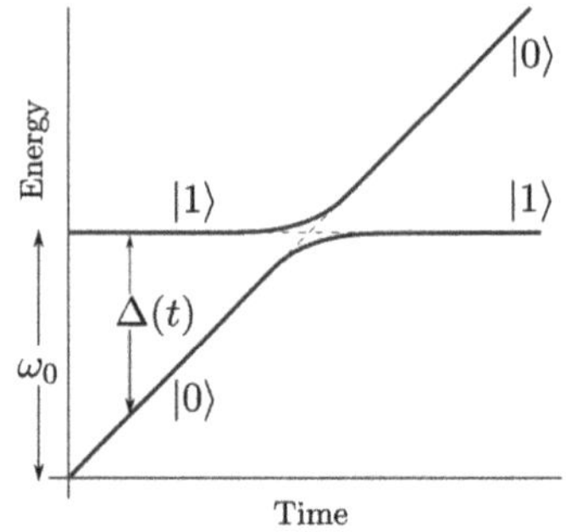


**Figure 6.3.** Energies of the dressed $|0\rangle$ and $|1\rangle$ states in the rotating-wave approximation. When the detuning is swept linearly through resonance, the probability of jumps across the avoided crossing is given by the Landau–Zener formula. Reproduced from [6]. 

Instead of quasi-momentum space, where we find the avoided crossings between two levels, we pick up the time axis and develop a two-state model for the atom with an understanding that the detuning is changing [6]. We consider that the detuning, $\Delta$, between the interaction frequency $\omega$ and the transition frequency $\omega_0$, that is, $\Delta = \omega - \omega_0$, changes linearly with time:

$$\frac{d\Delta(t)}{dt} = \dot{\Delta} = \text{constant}.$$

We consider that the atom is in a superposition of two orthogonal basis states, $|0\rangle$ and $|1\rangle$, such that we can describe the system at any time as $|\psi(t)\rangle = \psi_0|0\rangle + \psi_1|1\rangle$. We consider that at an initial time, $t_i$, the atom is at the lower-energy state, so that $\psi_0(t = t_i) = 1$ and $\psi_1(t = t_i) = 0$. With reference to the lower-state energy being zero, the higher state has energy $\hbar\omega_0$. Let the two-state system be coupled with a general time-dependent interaction; hence, the Hamiltonian of the systems, in the rotating-wave approximation, is given as

$$H = \hbar\omega_0|1\rangle\langle 1| + \hbar\Omega^* e^{i\omega t}|0\rangle\langle 1| + \hbar\Omega e^{-i\omega t}|1\rangle\langle 0|. \tag{6.55}$$

Here, $\Omega$ is the strength of the time-dependent interaction or coupling, and $\omega$ is the corresponding frequency. Our goal is to calculate the probability of finding the system in zero state ($|0\rangle$) as a function of time for linearly changing detuning $\Delta$ in time. In the case of a constant detuning, the system can be solved by the Jaynes–Cummings model, and described by the Rabi flipping formula for the two-state system. Time-varying detuning results in a constant *de-phasing* of the coherent Rabi oscillations, over a de-phasing time $\tau_D$ that corresponds with the time to accumulate a $2\pi$ phase difference. Hence, $\tau_D = \sqrt{4\pi/\dot{\Delta}}$.

We simplify the discussion by considering that the change in detuning takes place in discrete steps of small increments, $\delta t$. Over one incremental segment of time, the detuning remains constant and hence the evolution can be described by the usual Rabi flopping probability amplitudes. The time-dependent Hamiltonian, equation

(6.55), leads to the transition rate from the level $|0\rangle$ in the presence of a decay rate, $\gamma$, of the Rabi oscillations as

$$\Gamma = \Omega^2 \frac{\gamma}{\Delta^2 + \gamma^2/4}, \tag{6.56}$$

where $\gamma = 1/\tau_D$. The transition rate controls the probability of the population in state $|0\rangle$, that is, $P_0(t)$. Therefore, the probability after the segment of time $\delta t$ is

$$P_0(t + \delta t) = (1 - \Gamma\delta t)P_0(t) \approx \mathrm{e}^{-\Gamma\delta t}P_0(t).$$

The probability of being in state $|0\rangle$ from the initial time $t_i$ to a later time $t$ is given as

$$\begin{aligned} P(t) &= \exp\left(-\int_{t_i}^{t} \Gamma(t)dt\right), \\ &= \exp\left(-\Omega^2 \int_{t_i}^{t} \frac{\gamma}{\Delta(t)^2 + \gamma^2/4}dt\right), \\ &= \exp\left(-\frac{2\Omega^2}{\dot{\Delta}} \int_{\Delta(t_i)}^{\Delta t} \frac{\gamma/2}{\Delta^2 + \gamma^2/4}d\Delta\right), \\ &= \exp\left(-\frac{2\Omega^2}{\dot{\Delta}}\left[\arctan\left(\frac{\Delta(t)}{\gamma/2}\right) - \arctan\left(\frac{\Delta(t_i)}{\gamma/2}\right)\right]\right). \end{aligned} \tag{6.57}$$

From equation (6.57), it is interesting to note that for time $t = t_i$, the probability $P_0 = 1$, which implies that the atom is in its initial state. In addition, for a constant detuning, that is, $\dot{\Delta} = 0$, the transition probability $P_0$ is zero, indicating a transition to state $|1\rangle$. In the case of $\Delta(t_i) \ll \gamma/2 \ll \Delta(t)$, the entire avoided crossing region is ramped: from far below resonance, resonance, and far above resonance. In the limiting case, equation (6.57) reduces to the Landau–Zener formula, that is,

$$P_0 = \exp\left(-\frac{2\pi\Omega^2}{\dot{\Delta}}\right).$$

The parameter $\dot{\Delta}$ is related to the quasi-momentum space, such that

$$\dot{\Delta} = \frac{\partial\Delta}{\partial k}\dot{k}$$

to be evaluated at the location of the avoided crossing in quasi-momentum $k$.

*Concurrence of Landau–Zener transition and Bloch oscillation.* An ultra-cold atom in an optical lattice in the presence of a constant force is expressed by the Hamiltonian in equation (6.15). It exhibits Bloch oscillations, as shown in figure 6.2. In the presence of a unitary transformation,

$$\hat{U} = \exp\left\{-\frac{i}{\hbar}\hat{x}Ft\right\},$$

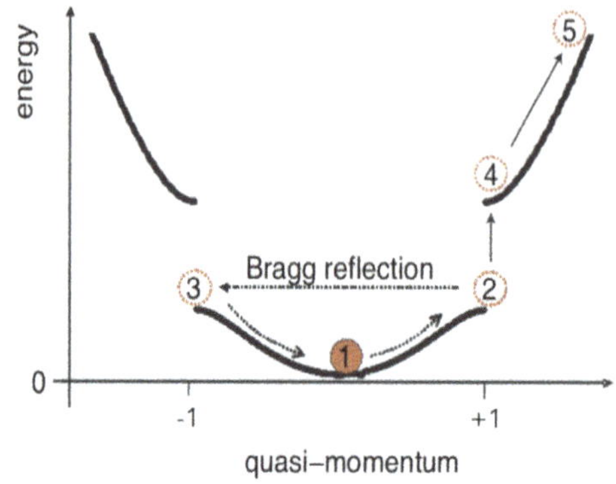


**Figure 6.4.** Band structure depiction of Bloch oscillations and Landau–Zener transition: position (1) marks the quantum particle placed in the lowest Bloch band. Following a constant force, it reaches the band edge at point (2). Following Bragg reflection, a reversal in momentum takes place and it appears at point (3) and performs Bloch oscillations. Due to the Landau–Zener transition, a fraction of the wave packet appears at point (4) and is accelerated in the higher band [3]. John Wiley & Sons. Copyright © 2004 Wiley-VCH Verlag GmbH & Co. KGaA, Weinheim.

applied to the corresponding Schrödinger equation, we obtain the transformed wave function

$$|\tilde{\psi}\rangle = \hat{U}|\psi\rangle, \tag{6.58}$$

and the Hamiltonian as

$$H = \frac{\hbar^2(k - Ft)^2}{2m} + V_0(x), \tag{6.59}$$

where $V_0(x)$ is the periodic optical potential. The first term in the Hamiltonian indicates that the quasi-momentum of the atom increases linearly in the Brillouin zone. However, on the Brillouin zone edge, it undergoes Bragg reflection and a reversal of quasi-momentum takes place. A small fraction of the wave function is transferred to the higher bands, following the Landau–Zener transition. Figure 6.4 presents the inter-band dynamics of a quantum particle following Bloch oscillations and Landau–Zener transition in the extended-zone representation. Here, point (1) marks the ultra-cold atom placed in the lowest band. Following a constant force and a linear increase in the quasi-momentum, it reaches the band edge at point (2). The Bragg reflection causes the reversal in momentum and it appears at point (3) in the Brillouin zone, and performs Bloch oscillations. Due to the Landau–Zener transition, a fraction of the wave packet appears at point (4) and is accelerated in the higher band. Hence, as a direct consequence of Landau–Zener transitions, we find the generation of entanglement between the bands [7].

## 6.6 Many-body effects of ultra-cold atoms in an optical lattice

The inter-atomic interactions between ultra-cold atoms while they are in an optical lattice give rise to a rather profound manifestation of various many-body effects.

Here, we give a sample of some notable many-body effects in these systems. As aforementioned, quantum particles with zero or integer spin (1, 2, 3, …) are bosons, whereas those with half-integer spin (1/2, 3/2, 5/2, …) are fermions. The interaction of bosons with a periodic optical lattice is described by the Bose–Hubbard model, whereas the interaction of fermions with a periodic optical lattice is expressed by the Fermi–Hubbard model. These systems introduce newer theoretical techniques and experimental advances to the study of optical forces on atoms. With the former, we consider the inter-relationship between the superfluid phase and the Mott insulator phase. With the latter, we consider Bogoliubov oscillations of fermionic coherent states.

## 6.7 Bose–Hubbard model

The presence of atom–atom interaction modifies the physical system expressed by the Hamiltonian in equation (6.2). Therefore, together with the interaction of the atom with the periodic optical lattice, we include an atom–atom interaction term, such that

$$\hat{H} = \frac{\hat{p}^2}{2m} + \hat{V}_0(x) + \frac{1}{2}\hat{V}(x - x').$$

Here, $\hat{V}_0(x)$ describes a one-dimensional optical lattice potential seen by the atom, and $\hat{V}(x - x_i)$ expresses the potential seen by the atom due to neighboring interacting atoms. The atom–atom interaction term in the second quantization becomes

$$\frac{1}{2}\hat{V}(x - x') = \frac{1}{2}\int dx\, dx' \hat{\psi}_a^{\dagger}(\mathbf{x})\hat{\psi}_a^{\dagger}(\mathbf{x}')\hat{V}(\mathbf{x} - \mathbf{x}')\hat{\psi}_a(\mathbf{x}')\hat{\psi}_a(\mathbf{x}).$$

The operators $\hat{\psi}_a(\mathbf{x})$ and $\hat{\psi}_a^{\dagger}(\mathbf{x})$ are the non-Hermitian field operators from section 6.1, defined in terms of the WFs and the creation and annihilation operators, such that

$$\hat{\psi}_a^{\dagger}(\mathbf{x}) = \sum_{\alpha,n} \hat{a}_{\alpha,\, n}^{\dagger} w_{\alpha}^{*}(\mathbf{x} - x_n),$$

and

$$\hat{\psi}_a(\mathbf{x}) = \sum_{\alpha,n} \hat{a}_{\alpha,n} w_{\alpha}(\mathbf{x} - x_n).$$

The atom–atom interaction term can be restructured by rewriting the integral expression $U_{n,n',m,m'}$, as

$$U_{n,n',m,m'} = \int dxdx' w^{*}(x - x_n)w^{*}(x' - x_n')V(\mathbf{x} - \mathbf{x}')w(x' - x_m)w(x - x_m'). \quad (6.60)$$

Thus, we obtain

$$\hat{U} = \frac{1}{2}\sum_{n,\, n',\, m,\, m'} \hat{a}_n^{\dagger}\hat{a}_{n'}^{\dagger}\hat{a}_m^{\dagger} a_{m'}^{\dagger} U_{n,n',m,m'}. \quad (6.61)$$

In the case of a non-vanishing atom–atom interaction term, we evaluate the integral expression, given in equation (6.60). A simplicity is introduced as we take the atom–atom interaction as a point interaction, that is,

$$V(\mathbf{x} - \mathbf{x}') = \frac{4\pi^2 a_s \hbar^2}{m}\delta(\mathbf{x} - \mathbf{x}'),$$

where $a_s$ is the scattering length. This approximation holds to a great extent when the scattering length is made resonantly large. This means that atoms feel the mutual force at distances of the order of $a_s$ from each other. We calculate the integral as

$$U \approx \int dx|w(x - x_n)|^4, \tag{6.62}$$

and consider only the next neighboring term, that is, $n \pm 1$; hence, we obtain

$$H = \epsilon\sum_n a_n^\dagger a_n + J\sum_n (a_n^\dagger a_{n+1} + a_n^\dagger a_{n-1}) + \frac{U}{2}\sum_n a_n^\dagger a_n^\dagger a_n a_n,. \tag{6.63}$$

Here, we take $n$ as a dummy index to rewrite the above expression as

$$H = \epsilon\sum_n a_n^\dagger a_n + J\sum_n (a_n^\dagger a_{n+1} + a_{n+1}^\dagger a_n) + \frac{U}{2}\sum_n a_n^\dagger a_n^\dagger a_n a_n. \tag{6.64}$$

In terms of the number operator $\hat{N}_n = a_n^\dagger a_n$, the last term is simplified so that

$$H = \epsilon\sum_n \hat{N}_n + J\sum_n (a_n^\dagger a_{n+1} + a_{n+1}^\dagger a_n) + \frac{U}{2}\sum_n \hat{N}_n(\hat{N}_n - 1). \tag{6.65}$$

Thus far, we have worked out the representation of the Hamiltonian, which is suitable for the number basis set $\{|\cdots, N_{-1}, N_0, N_1, \cdots\rangle\}$, where $N_j$ is the number of atoms at the $j$th site with the total number of atoms $N_{\text{total}} = \sum_n N_n$ fixed. However, the corresponding eigenvalue problem quickly becomes formidable as $N_{\text{total}}$ becomes large. Approximation schemes become desirable, and have indeed been applied under various circumstances. One example is the study of the transition between the Mott insulator state and the superfluid state. In experiments on ultra-cold gas of rubidium ($^{87}$Rb) atoms a quantum phase transition is observed. The rubidium Bose–Einstein condensate (BEC) with repulsive interactions is held in a three-dimensional optical lattice potential. As the potential depth of the lattice is increased, a transition is observed from a superfluid to a Mott insulator phase. In the superfluid phase, each atom is spread out over the entire lattice, with long-range phase coherence, whereas in the insulator phase, exact numbers of atoms are localized at individual lattice sites, with no phase coherence across the lattice [8].

### 6.7.1 Superfluid to Mott insulator

We perform the mean-field calculations using the Gutzwiller ansatz for the ground-state wave function $|\Psi_{\text{MF}}\rangle = \prod_j |\phi_j\rangle$, with $|\phi_j\rangle = \sum_{N=0}^{\infty} f_N^{(j)} |N_j\rangle$. Here, $|N_j\rangle$ denotes the

Fock state with $N$ atoms at the $j$th site, and $\hat{a}_j^\dagger \hat{a}_j = \hat{N}_j$ is the corresponding number operator [9]. We minimize the expectation value of the Hamiltonian,

$$\langle \Psi_{\text{MF}} | H | \Psi_{\text{MF}} \rangle - \mu \langle \Psi_{\text{MF}} | \sum_j \hat{N}_j | \Psi_{\text{MF}} \rangle,$$

with respect to the coefficients $f_N^{(j)}$. The Lagrange multiplier, $\mu$, enforces a given mean particle number $N = \sum_j \langle N_j \rangle$. This corresponds to a calculation in the grand canonical ensemble with chemical potential $\mu$ at temperature $T = 0$.

A Mott insulator phase is indicated by solutions in the form of single Fock states, $|\phi_j\rangle \to |N_j\rangle$. A signature of a Mott insulator phase is found as the density, $\rho_j$, attains an integer occupation number, that is,

$$\rho_j = \langle \hat{N}_j \rangle,$$

and as the fluctuations,

$$\sigma_j^2 = \frac{\langle \hat{N}_j^2 \rangle - \langle \hat{N}_j \rangle^2}{\langle \hat{N}_j \rangle},$$

approach zero (figure 6.5).

Solutions in the form of the superposition of Fock states indicate the presence of a superfluid component. In the homogeneous case ($\epsilon_j = 0$), the phase diagram in the $J - \mu$ plane consists of a series of lobes [11]. Inside the lobes (that is, for a small $J$ in comparison with the on-site repulsion energy $U$) the system is in a Mott phase; outside, it is a superfluid.

Within the mean-field decoupling approximation, the boundary between the superfluid phase and the Mott phase in the Bose–Hubbard model is obtained from the equation

$$(\mu - \epsilon)^2 - (\mu - \epsilon)[U(2N - 1) - 2DJ] + 2DJU + U^2N(N - 1) = 0. \quad (6.66)$$

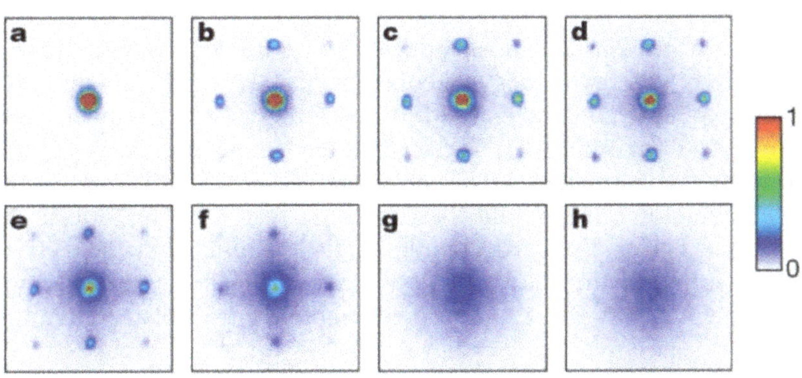


**Figure 6.5.** A transition from a superfluid state to a Mott insulator state in rubidium atoms was obtained after suddenly releasing the atoms from an optical lattice potential with different potential depths after a time of flight of 15 ms. Values of the potential depth (measured in units of recoil energy) were (a) 0, (b) 3, (c) 7, (d) 10, (e) 13, (f) 14, (g) 16, and (h) 20 [10]. Copyright © 2002, Macmillan Magazines Ltd. With permission of Springer.

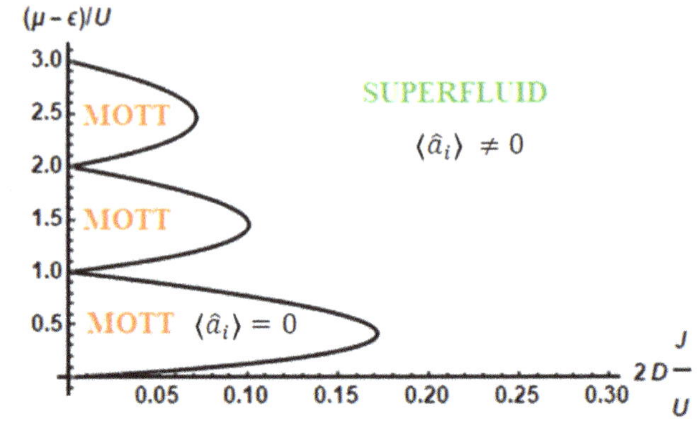


**Figure 6.6.** Mean-field phase diagram of the Bose–Hubbard model at zero temperature in $D$ spatial dimensions. Here, $J$ is the hopping energy, $\epsilon$ is the on-site energy, $U$ is the on-site interaction energy, and $\mu$ is the chemical potential. Reprinted figure with permission from [12], Copyright (2019) by the American Physical Society.

In figure 6.6, we plot the superfluid–Mott phase diagram of the Bose–Hubbard model at zero temperature in $D$ spatial dimensions, obtained from equation (6.66). Inside the lobes is the Mott phase, characterized by an integer filling number $N$ and where the expectation value of the annihilation operator for each site is equal to zero; outside the Mott lobes is the superfluid phase, characterized by a non-vanishing expectation value of the annihilation operator for each site. The tips of the lobes are critical points. The corresponding critical transitions, which occur for $N = 1$ at $[2DJ/U]_c = 0.172$ and $[(\mu - \epsilon)/U]_c = 0.414$, can be controlled by varying $U$ at fixed $J$ or varying $J$ at fixed $U$. Non-critical transitions, which are the ones not occurring at the tips, can be obtained, instead, at fixed $J$ and $U$, by changing the effective chemical potential $\mu - \epsilon$ [12–14].

## 6.8 Fermi–Hubbard model

Experimental advances in achieving optical lattices for cold atoms have made it possible to study the dynamics of coherent matter waves, quantum phases, entanglement, collapse and revival, Mott insulators, the superfluidity of BECs, and atomic Bose–Fermi mixtures. These achievements make such systems a testing ground to study the characteristics of cold Fermi gases [15, 16].

The interaction of a Fermi ultra-cold atomic gas with an optical lattice is governed by the Fermi–Hubbard Hamiltonian. Assuming that the optical lattice is sufficiently deep and the gas is dilute enough so that only nearest-neighboring terms contribute, we may express the single-band Fermi–Hubbard Hamiltonian as

$$\begin{aligned} H - \mu N = &-\mu \sum_n \left(\hat{c}^\dagger_{\uparrow n}\hat{c}_{\uparrow n} + \hat{c}^\dagger_{\downarrow n}\hat{c}_{\downarrow n}\right) + U\sum_n \hat{c}^\dagger_{\uparrow n}\hat{c}^\dagger_{\downarrow n}\hat{c}_{\uparrow n}\hat{c}_{\downarrow n} \\ &- \sum_{\langle n,m\rangle_r} J_r\left(\hat{c}^\dagger_{\uparrow n}\hat{c}_{\uparrow m} + \hat{c}^\dagger_{\downarrow n}\hat{c}_{\downarrow m}\right). \end{aligned} \tag{6.67}$$

In obtaining the Hamiltonian, we assume that excitation to the excited levels of the individual lattice well does not take place due to the very low temperature. The operator $\hat{c}_{\sigma,n}$ annihilates a fermion with spin $\sigma$, and the operator $\hat{c}^{\dagger}_{\sigma,n}$ creates it at the site $n$. Moreover, $N$ describes the number of atoms and $\langle n, m\rangle_r$ expresses all the nearest-neighbor pairs in the $r$ direction of the three-dimensional optical lattice. We have taken the number of atoms in the two species as equal so that the chemical potential $\mu$ for the two different spins is the same. The parameter $U$ describes the interaction strength, and $J_r$ controls the tunneling in the $r$ direction. Assuming the lattice well to be a quadratic potential, we have

$$U = \prod_r (8a/\lambda\sqrt{\pi})(2s_r E_R^3 \hbar^2/m\lambda^2)^{1/4}, \tag{6.68}$$

$$J_r = E_R \exp\left(-\pi^2\sqrt{\frac{s_r}{4}}\right)[(\pi^2 s_r/4 - \sqrt{s_r}/2) - (s_r/2)(1 + \exp(-\sqrt{s_r}))]. \tag{6.69}$$

Here, $\hbar$ is the Planck's constant divided by $2\pi$, $a$ is the scattering length, $\lambda$ is the wavelength of the lattice, and $E_R = 4\pi^2\hbar^2/2m\lambda^2$ is the recoil energy. The parameter $s_r$ defines the scaled lattice depth in the corresponding direction, $r$.

In momentum space, the Fermi–Hubbard Hamiltonian becomes

$$\begin{aligned} H = &\sum_\kappa (\epsilon_\kappa - \mu)(\hat{c}^{\dagger}_{\uparrow\kappa}\hat{c}_{\uparrow\kappa} + \hat{c}^{\dagger}_{\downarrow\kappa}\hat{c}_{\downarrow\kappa}) \\ &+ \frac{U}{M}\sum_{\kappa, q, \kappa'} \hat{c}^{\dagger}_{\uparrow\kappa'+\kappa+q}\hat{c}^{\dagger}_{\downarrow\kappa'-\kappa+q}\hat{c}_{\downarrow\kappa'+\kappa}\hat{c}_{\uparrow\kappa'-\kappa}, \end{aligned} \tag{6.70}$$

where $M$ indicates the number of lattice sites, and $\epsilon_\kappa = 2\sum_r J_r[1 - \cos(\kappa_r d)]$ reveals a symmetry, that is, $\epsilon_{-\kappa} = \epsilon_\kappa$.

*Heisenberg equations of motion.* The Hamiltonian leads us to study the evolution of the field operators, $\hat{c}^{\dagger}_{\uparrow-\kappa}$ and $\hat{c}^{\dagger}_{\downarrow\kappa}$, by means of the Heisenberg equations of motion which, on linearization, can be written as

$$\dot{\hat{c}}^{\dagger}_{\uparrow-\kappa} = -\frac{i}{\hbar}\left\{(\epsilon_\kappa - \mu)\hat{c}^{\dagger}_{\uparrow-\kappa} - \Delta\hat{c}^{\dagger}_{\downarrow\kappa}\right\}, \tag{6.71}$$

$$\dot{\hat{c}}^{\dagger}_{\downarrow\kappa} = \frac{i}{\hbar}\left\{(\epsilon_\kappa - \mu)\hat{c}^{\dagger}_{\downarrow\kappa} - \Delta^*\hat{c}^{\dagger}_{\uparrow-\kappa}\right\}. \tag{6.72}$$

Here, the parameter $\Delta = -\frac{U}{M}\sum_q\langle\hat{c}_{\downarrow-\kappa-q}\hat{c}_{\uparrow\kappa+q}\rangle$, and $\Delta^* = -\frac{U}{M}\sum_q\langle\hat{c}^{\dagger}_{\uparrow\kappa+q}\hat{c}^{\dagger}_{\downarrow-\kappa-q}\rangle$. In obtaining these equations, we employed the Bardeen, Cooper, and Schrieffer approximation and assumed that pairing only occurs with net zero momentum, $\kappa' = 0$. Since the overall phase does not play any role, we take $\Delta = \Delta^*$. Hence, we get a set of coupled equations that are solved by a Bogoliubov transformation, which decouples the equations and diagonalizes the Hamiltonian. Thus, we obtain

$$\begin{pmatrix} \hat{c}_{\uparrow-\kappa} \\ \hat{c}^{\dagger}_{\downarrow\kappa} \end{pmatrix} = \begin{pmatrix} u^{(-)}_\kappa & v_\kappa \\ v_\kappa & u^{(+)}_\kappa \end{pmatrix}\begin{pmatrix} \hat{c}_{\uparrow-\kappa}(0) \\ \hat{c}^{\dagger}_{\downarrow\kappa}(0) \end{pmatrix}, \tag{6.73}$$

where

$$u_{\kappa}^{(\pm)}(t) = \cos\omega_{\kappa}t \pm i((\epsilon_{\kappa} - \mu)/\hbar\omega_{\kappa})\sin\omega_{\kappa}t, \tag{6.74}$$

$$v_{\kappa}(t) = i(\Delta/\hbar\omega_{\kappa})\sin\omega_{\kappa}t, \tag{6.75}$$

and

$$\hbar\omega_{\kappa} = \sqrt{\Delta^2 + (\epsilon_{\kappa} - \mu)^2} = E_{\kappa}. \tag{6.76}$$

The operators $\hat{c}_{\sigma,n}$ and $\hat{c}_{\sigma,n}^{\dagger}$ obey the fermionic anti-commutation rules, which lead us to the normalization condition $u_{\kappa}^2 + v_{\kappa}^2 = 1$. Hence, we can write $\langle\hat{c}_{\uparrow\kappa}^{\dagger}\hat{c}_{\uparrow\kappa}\rangle = f(E_{\kappa})$, where the Fermi function $f(E_{\kappa}) = (\exp(E_{\kappa}/k_{\mathrm{B}}T) + 1)^{-1}$ at finite temperature $T$. The number of particles can be written as

$$\begin{aligned} N &= \sum_{\kappa}\langle\hat{c}_{\downarrow\kappa}^{\dagger}\hat{c}_{\downarrow\kappa} + \hat{c}_{\uparrow\kappa}^{\dagger}\hat{c}_{\uparrow_{\kappa}}\rangle \\ &= 2\sum_{\kappa}\{u_{\kappa}^2 f(E_{\kappa}) + v_{\kappa}^2[1 - f(E_{\kappa})]\}. \end{aligned} \tag{6.77}$$

### 6.8.1 Fermionic coherent state and Schrödinger cat state

We define a coherent state for a fermion $|\alpha_{\sigma,\kappa_{\sigma}}\rangle$ with spin $\sigma$ and linear momentum $\kappa_{\sigma}$, such that

$$\hat{c}_{\sigma\kappa_{\sigma}}|\alpha_{\sigma,\kappa_{\sigma}}\rangle = \alpha_{\sigma\kappa_{\sigma}}|\alpha_{\sigma 1_{\sigma}}, \alpha_{\sigma 2_{\sigma}}, \dots, \alpha_{\sigma M_{\sigma}}\rangle,$$

where $\kappa_{\sigma} = -\kappa$ for $\sigma = \uparrow$ and $\kappa_{\sigma} = \kappa$ for $\sigma = \downarrow$. Moreover, $\alpha_{\sigma\kappa_{\sigma}}$ are Grassmann numbers that follow the anti-commutation relation, similar to fermionic creation and annihilation operators, that is, $\alpha_{\sigma\kappa_{\sigma}}^2 = 0$, $\{\alpha_{\sigma\kappa_{\sigma}}, \alpha_{\sigma\kappa_{\sigma}}^*\} = 0$, $\{\alpha_{\sigma\kappa_{\sigma}}, \hat{c}_{\sigma\kappa_{\sigma}}\} = 0$.

The coherent state for the fermions can be obtained from the vacuum state as

$$|\alpha_{\sigma,\kappa_{\sigma}}\rangle = \exp\left(-\sum_{\kappa=1}^{N}\alpha_{\sigma\kappa_{\sigma}}\hat{c}_{\sigma\kappa_{\sigma}}^{\dagger} - \frac{1}{2}\alpha_{\sigma\kappa_{\sigma}}^{*}\alpha_{\sigma\kappa_{\sigma}}\right)|0_{\sigma,\kappa_{\sigma}}\rangle, \tag{6.78}$$

where $|0_{\sigma,\kappa_{\sigma}}\rangle$ is the vacuum state. The operators $\hat{c}_{\sigma\kappa_{\sigma}}$ and $\hat{c}_{\sigma\kappa_{\sigma}}^{\dagger}$, respectively, describe the annihilation and creation operators for the fermions with $\hat{c}_{\sigma\kappa_{\sigma}}|0_{\sigma,\kappa_{\sigma}}\rangle = 0$ and $\hat{c}_{\sigma\kappa_{\sigma}}^{\dagger}|1_{\sigma,\kappa_{\sigma}}\rangle = 0$, whereas $\hat{c}_{\sigma\kappa_{\sigma}}|1_{\sigma,\kappa_{\sigma}}\rangle = |0_{\sigma,\kappa_{\sigma}}\rangle$ and $\hat{c}_{\sigma\kappa_{\sigma}}^{\dagger}|0_{\sigma,\kappa_{\sigma}}\rangle = |1_{\sigma,\kappa_{\sigma}}\rangle$.

A fermionic coherent state can be developed in a laboratory by taking a low density of Fermi gas. Fermions occupy the lowest states of each potential well in a periodic lattice, and their distribution over lattice sites follows that of a coherent state. In addition, as there can be two fermions with opposite spins in each site, the collective two-fermion state, following a Poisson distribution over lattice sites, is the Schrödinger cat state, that is,

$$|\psi\rangle = \frac{\mathcal{A}}{\sqrt{2}}(|\alpha_{\uparrow,-\kappa}\rangle + |\alpha_{\downarrow,+\kappa}\rangle). \tag{6.79}$$

Here, $\mathcal{A}$ is the normalization constant.

### 6.8.2 Bogoliubov oscillations of fermionic coherent state

The explicit time dependence of a fermionic coherent state is stored in the evolution of its auto-correlation function, $C(t)$. The function measures the overlap of the time-evolved coherent state of the fermions with the initial coherent state, such that $C(t) = \langle\alpha(0)|\alpha(t)\rangle$, where $|\alpha(0)\rangle$ describes the initial coherent state and $|\alpha(t)\rangle$ represents the time-evolved coherent state.

Using the operator definition, we may express the time evolution of the fermionic coherent state as

$$\begin{aligned}|\alpha_{\sigma,\kappa_\sigma}\rangle &= \exp\left(\sum_{\kappa=1}^{N}\hat{c}^{\dagger}_{\sigma\kappa_\sigma}(t)\alpha_{\sigma\kappa_\sigma} - \frac{1}{2}\alpha^{*}_{\sigma\kappa_\sigma}\alpha_{\sigma\kappa_\sigma}\right)|0_{\sigma,\kappa_\sigma}\rangle \\ &= \prod_{\kappa=1}^{N}\left\{|0_{\sigma\kappa_\sigma}\rangle - \alpha_{\kappa_\sigma}u^{(+)}_{\kappa_\sigma}|1_{\sigma\kappa_\sigma}\rangle - \frac{1}{2}\alpha^{*}_{\sigma\kappa_\sigma}\alpha_{\sigma\kappa_\sigma}|0_{\sigma\kappa_\sigma}\rangle\right\}.\end{aligned} \tag{6.80}$$

Since $u^{(+)}_{\kappa_\sigma}$ is periodic in nature, we deduce that the coherent state $|\alpha_{\sigma\kappa_\sigma}\rangle$ restructures itself at $t = t_{\rm R}$, where $t_{\rm R}$ describes the Bogoliubov oscillation period, that is, $t_{\rm R} = 2\pi/\omega_\kappa$.

Taking into account the definitions of a coherent state in equation (6.78) and the time-evolved coherent state in equation (6.80), we can readily write the auto-correlation function for a fermionic coherent state:

$$C(t) = \langle\alpha(0)|\alpha(t)\rangle = \exp\left\{\sum_{\kappa}\alpha^{*}_{\sigma\kappa_\sigma}\alpha_{\sigma\kappa_\sigma}f^{(\pm)}_{\kappa}(t)\right\}, \tag{6.81}$$

where the function $f^{(\pm)}_{\kappa}(t) = -(1 - u^{(\pm)}_{\kappa}(t))$ varies between 0 and $-2$. Hence, $C(t)$ takes a value between 1 and $\exp(-2\sum\alpha^{*}_{\sigma\kappa_\sigma}\alpha_{\sigma\kappa_\sigma})$. The maximum value for $C(t)$ is obtained at $f_\kappa(t) = 0$ at $t = 0$. Since

$$f^{(\pm)}_{\kappa}(t) = -1 + \cos\omega_\kappa t \pm i\sin\omega_\kappa t\left(1 + \frac{\Delta^2}{(\epsilon_\kappa - \mu)^2}\right)^{-1/2} \tag{6.82}$$

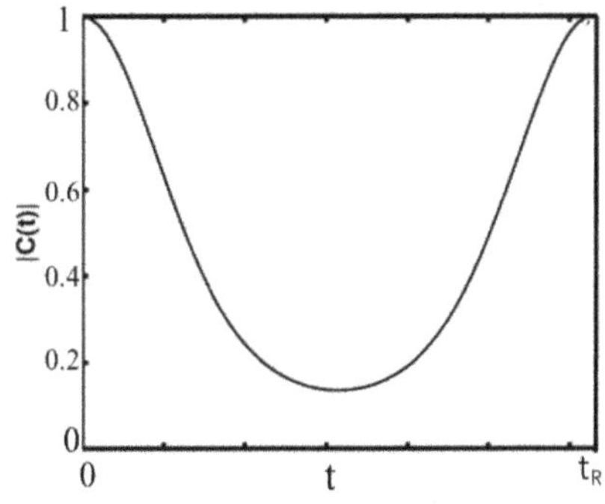


**Figure 6.7.** The absolute value of the auto-correlation function $C(t)$ vs time $t$ displays a complete Bogoliubov oscillation for the fermionic coherent state. The ratio of the energy gap at temperature $T = 0\,{\rm K}$ to $k_{\rm B}T_{\rm c}$ is considered to be $2\Delta/k_{\rm B}T_{\rm c} = 3.5$. The Fermi energy is $\varepsilon_{\rm F} = \mu$ and the single particle energy is $\epsilon_\kappa = (3/5)\varepsilon_{\rm F}$ [17]. Copyright © 2013, Springer Science Business Media New York. With permission of Springer.

is oscillatory, the fermionic coherent state repeats itself after a time $t = t_R$, where $t_R = 2\pi/\omega_\kappa$, which defines the Bogoliubov oscillation for the coherent state with linear momentum, $\kappa$.

Figure 6.7 shows the auto-correlation function, $C(t)$, for a fermionic coherent state with a complete Bogoliubov oscillation. The fermionic coherent state shows periodicity in time with the period $\omega_\kappa t_R = 2\pi$.

## Exercises

1. Solve the integrals in equations (6.6) and (6.7). Hint: recall that the WFs are related to Bloch's eigenstates $\varphi_{nk}$, i.e.

$$w_{n,j} = \frac{d}{2\pi}\int_{\mathrm{BZ}} dk \mathrm{e}^{-ijkd}\varphi_{nk}(x),$$

and follow the eigenvalue equation

$$\left\{\frac{p^2}{2m} + V_0(x)\right\}\varphi_{nk} = E_{nk}\varphi_{nk}.$$

2. Obtain equation (6.19) from equation (6.16) by using the WFs description in terms of Bloch's function, that is,

$$|n\rangle = \int dk|k\rangle\langle k|n\rangle = \frac{1}{\sqrt{2\pi d}}\int dk \mathrm{e}^{-indk}|k\rangle.$$

3. Show that an initial wave packet with a broad initial distribution in a one-dimensional optical lattice in the presence of a constant force follows a classical trajectory and performs non-dispersive evolution.
4. Calculate $m^*$ at the band edge, that is, around $k = \pm\pi/d$.
5. Obtain equation (B.9), which describes the evolution operator in Wannier space.
6. Obtain equation (B.4), which describes the evolution operator in quasi-momentum space.
7. Show $^{39}$K is a fermion, whereas $^{40}$K is a boson.

## References

[1] Pitaevskii L and Stringari S 2003 *Bose–Einstein Condensation* (Oxford: Oxford Science Publication)

[2] Yukalov V I 2009 Cold bosons in optical lattices *Laser Phys.* **19** 1–110

[3] Modugno G, de Mirandes E, Ferlaino F, Ott H, Roati G and Inguscio M 2004 Atom interferometry in a vertical optical lattice *Fortschritte der Physik* **52** 1173–9

[4] Pikovski I and Loeb A 2016 Quantum coherent oscillations in the early universe *Phys. Rev.* D **93** 101302(R)

[5] Hartmann T, Keck F, Korsch H J and Mossmann S 2004 Dynamics of Bloch oscillations *New J. Phys.* **6** 2

[6] Vutha A C 2010 A simple approach to the Landau-Zener formula *Eur. J. Phys.* **31** 389
[7] Quintana C M, Petersson K D, McFaul L W, Srinivasan S J, Houck A A and Petta J R 2013 Cavity-mediated entanglement generation via Landau–Zener interferometry *Phys. Rev. Lett.* **110** 173603
[8] Greiner M, Mandel O, Esslinger T, Hänsch T W and Bloch I 2002 Quantum phase transition from a superfluid to a Mott insulator in a gas of ultracold atoms *Nature* **415** 39–44
[9] Jaksch D, Bruder C, Cirac J I, Gardiner C W and Zoller P 1998 Cold bosonic atoms in optical lattices *Phys. Rev. Lett.* **81** 3108
[10] Greiner M, Mandel O and Esslinger T *et al* 2002 Quantum phase transition from a superfluid to a Mott insulator in a gas of ultracold atoms *Nature* **415** 39–44
[11] Fisher M P A, Weichman P B, Grinstein G and Fisher D S 1989 Boson localization and the superfluid-insulator transition *Phys. Rev.* B **40** 546
[12] Faccioli M and Salasnich L 2019 Gaussian quantum fluctuations in the superfluid-Mott-insulator phase transition *Phys. Rev.* A **99** 023614
[13] Saito H and Kato M 2018 Machine learning technique to find quantum many-body ground states of bosons on a lattice *J. Phys. Soc. Japan* **87** 014001
[14] Saito H 2017 Solving the Bose–Hubbard model with machine learning *J. Phys. Soc. Japan* **86** 093001
[15] Martikainen J P and Törmä P 2005 Quasi-two-dimensional superfluid fermionic gases *Phys. Rev. Lett.* **95** 170407
[16] Heikkinen M O J, Kim D H, Troyer M and Törmä P 2014 Nonlocal quantum fluctuations and Fermionic superfluidity in the imbalanced attractive Hubbard model *Phys. Rev. Lett.* **113** 185301
[17] Saif F 2013 Fermionic Coherent States in Optical Lattice *Journal of Russian Laser Research* **34** 496–502

# Chapter 7

# Forces on atoms in exponentially varying fields

*'After long reflection in solitude and meditation, I suddenly had the idea, during the year 1923, that the discovery made by Einstein in 1905 should be generalized by extending it to all material particles and notably to electrons.'*

—Prince Louis-Victor de Broglie

As discussed in previous chapters, a proper choice of spatial distribution in an optical field can create almost any desired optical potential. The potential provides a repulsive or attractive force to the interacting atoms as a function of the atom–field detuning. A gradually decaying optical field or an evanescent wave field may be obtained by the total internal reflection of the optical field at an interface. The optical field exerts an exponentially increasing repulsive or attractive force on an approaching atom, detuned to the blue or to the red, respectively. This is a crucial aspect for developing different kinds of optical systems.

## 7.1 Mirrors, cavities, and interferometers

*Atomic mirror.* How to generate a mirror for de Broglie waves is an interesting question. To answer this question, we first consider an electromagnetic field, $E(\mathbf{r}, t) = E(\mathbf{r})$, traveling in a dielectric medium with a dielectric constant, $n$, and undergoing total internal reflection. The electromagnetic field inside the dielectric medium reads

$$E(\mathbf{r}, t) = E_0 e^{i\mathbf{k}\cdot\mathbf{r} - i\omega_f t}\mathbf{e}_\mu, \tag{7.1}$$

where $\mathbf{e}_\mu$ is the polarization vector and $\mathbf{k}$ is the propagation vector.

The electromagnetic field, $E(\mathbf{r}, t)$, is incident on an interface between a medium with dielectric constant $n_\mathrm{i}$ and another medium with dielectric constant $n_\mathrm{r}$, such that

doi:10.1088/978-0-7503-2308-6ch7  

$n_r < n_i$. The angle of incidence of the field is $\theta_i$, normal to the interface. Recalling Snell's law, that is,

$$\frac{\sin\theta_i}{\sin\theta_r} = \frac{n_r}{n_i}, \tag{7.2}$$

since the index of refraction $n_r$ is smaller than $n_i$, the angle, $\theta_r$, at which the field refracts in the second medium will be larger than the angle of incidence $\theta_i$, that is, $\theta_r > \theta_i$. As we increase the angle of incidence $\theta_i$, we may reach a critical angle, $\theta_i = \theta_c$, for which $\theta_r = \pi/2$ and $\sin\theta_r = 1$. According to Snell's law, we define the critical incidence angle as

$$\theta_c \equiv \sin^{-1}\left(\frac{n_r}{n_i}\right). \tag{7.3}$$

Hence, for an electromagnetic wave with an angle of incidence larger than the critical angle, that is, $\theta_i > \theta_c$, we find the inequality $\sin\theta_r > 1$. As a result, we deduce that $\theta_r$ is imaginary. Equation (7.3) together with Snell's law lead us to the relation

$$\cos\theta_r = i\sqrt{\left(\frac{\sin\theta_i}{\sin\theta_c}\right)^2 - 1}\,. \tag{7.4}$$

Therefore, a field in a medium of smaller refractive index, $n_1$, reads

$$E(\mathbf{r}, t) = E_0 e^{ik_1x\sin\theta_r + ik_1z\cos\theta_r} e^{-i\omega_f t}\mathbf{e}_\mu = E_0 e^{-\kappa z} e^{i(\beta x - \omega_f t)}\mathbf{e}_\mu, \tag{7.5}$$

where

$$\kappa = k_1\sqrt{(\sin\theta_i/\sin\theta_c)^2 - 1}\,,$$

and

$$\beta = k_1\sin\theta_i/\sin\theta_c.$$

Here, $k_1$ defines the wave number in a medium with a refractive index $n_r$. This implies that, in the case of total internal reflection, the field decays along the normal to the interface and in the positive $z$-direction, that is, in the medium with the smaller refractive index, $n_r$.

*Force on an atom* due to the decaying field comes with the gradient of potential made by the electric field $E(\mathbf{r}, t)$. Following the discussion in section 3.1.1, the potential is obtained by the square of the spatial profile of the field as

$$V = \frac{\hbar\Omega_R^2}{\Delta} e^{-2\kappa z}.$$

Hence, the force on the atom appears as

$$F = -\nabla V = 2\kappa\frac{\hbar\Omega_R^2}{\Delta} e^{-2\kappa z}.$$

Hence, we write the electromagnetic field $E(\mathbf{r}, t)$ given in equation (7.5) in quantum mechanical form as

$$E(\mathbf{r}, t) = E_0\, e^{-\kappa z}\, (\hat{a}^\dagger e^{i(\beta x - \omega_f t)} - \text{c. c})\mathbf{e}_\mu, \tag{7.6}$$

where $\hat{a}^\dagger$ ($\hat{a}$) is the creation (annihilation) operator.

*Magnetic mirror.* We can also construct an atomic mirror by using magnetic fields instead of optical fields. Magnetic mirrors were first realized to study the reflection of neutrons [1]. In atom optics, the use of a magnetic mirror was suggested in [2], and later used to study the reflection of incident rubidium atoms perpendicular to the reflecting surface [3–6]. Recently, it has become possible to modulate a magnetic mirror by adding a time-dependent external field. We may also make controllable corrugations, which can be varied in a time span shorter than the time taken by the atoms to interact with the mirror [7, 8]. A possible mirror for atoms has also been achieved by means of surface plasmons [9–11]. Surface plasmons are electromagnetic charge density waves propagating along a metallic surface. Traveling light waves can excite surface plasmons. This technique provides a tremendous enhancement in the evanescent wave decay length [9].

### 7.1.1 Atomic trampoline as matter-wave cavity

A matter-wave cavity is a special arrangement where de Broglie waves move in a finite region. The atomic mirror is placed perpendicular to the gravitational field and, therefore, the de Broglie waves observe a normal incidence with the mirror and bounce back due to repulsive optical forcing. The bouncing atoms are confined from above due to gravitational pull toward the mirror. They exhaust their kinetic energy while moving against the gravitational field and return. As a consequence, the atoms undergo a bounded motion in this atomic trampoline or atomic gravitational cavity. Hence, an evanescent wave mirror together with a gravitational field constitutes a gravitational cavity for the atoms.

A slight change in the shape of the atomic mirror from flat to concave [12], or using hollow optical beams [13, 14], helps to make successful traps for atoms in a gravitational cavity. The dynamics of cold atoms in an atomic trampoline or atomic gravitational cavity attracted immense attention after the early experiments by the group of Chu *et al* at Stanford, California [15]. They used a cloud of cold sodium atoms stored in a magneto-optic trap and cooled to 25 $\mu$K. As the trap was switched off, the atoms approached the mirror under gravity and displayed a normal incidence. In their experiments two bounces of the atoms were reported. The major noise sources were fluctuations in the laser intensities and the number of initially stored atoms [16]. Successful control on these limitations made it possible to observe more than a thousand bounces in later experiments [13].

A mirror for the atomic de Broglie waves is a crucial component of atomic cavities. An atomic mirror is obtained by using an exponentially decaying optical field or an evanescent wave field [17]. Such an optical field exerts an exponentially increasing repulsive force on an approaching atom, detuned to the blue. The evanescent wave field, acting as a mirror, has led to the advent of various kinds

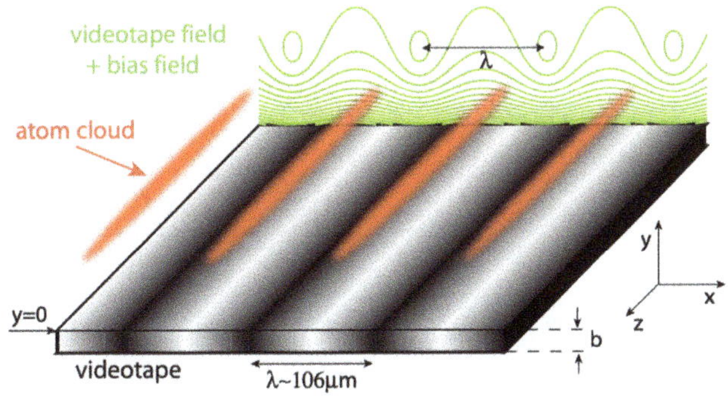


**Figure 7.1.** Magnetic fields used to create an array of videotape magnetic micro-traps. Contours of constant magnetic field strength are formed from the combined videotape field and bias field. As a result, atoms can be trapped in an array of elongated magnetic traps separated by a distance $\lambda$ along the $x$-axis. Reproduced from [21]. 

of atomic cavities, phase interferometers, [18], and amplitude interferometers [19]. In addition, the de Broglie mirror is the workhorse of recurrence tracking microscopes (RTMs) [20].

Such a gravitational cavity may also be obtained by generating exponentially decaying magnetic fields by magnetic films [21], as shown in figure 7.1. The atoms due to the magnetic moment experience exponentially repulsive force in the $y$-direction. Another kind of gravitational cavity can be realized by replacing the optical evanescent wave field by liquid helium, forming an atomic mirror for hydrogen atoms. In this setup hydrogen atoms are cooled below 0.5 K, and a specular reflection of 80% has been observed [22].

By an appropriate choice of attractive and/or repulsive evanescent waves, one can successfully trap or guide atoms in a particular system. In the presence of a blue detuned optical field and another red detuned optical field with a smaller decay length, a net potential is formed on or around a dielectric surface. This bi-dimensional trap can be used to store atoms. Based on the atomic mirror, various kinds of atomic cavities have been suggested.

A system of two atomic mirrors placed at a distance with their exponentially decaying fields in front of each other form a cavity or resonator for the de Broglie waves. The atomic cavity is regarded as an analog of the Fabry–Pérot cavity for radiation fields. By using more than two mirrors, other possible cavities can be developed, as well. For example, one can create a ring cavity for matter waves by combining three or four atomic mirrors.

### 7.1.2 Atom interferometers

In interferometry, the wave function describing the incoming wave is coherently separated, redirected, and then recombined again. Thus, any change along the separated paths of the diffracted waves produces observable variations in the interference pattern as they recombine. In atom interferometry, we split, redirect, and recombine the matter waves using optical potentials [23, 24]. Due to atom–field

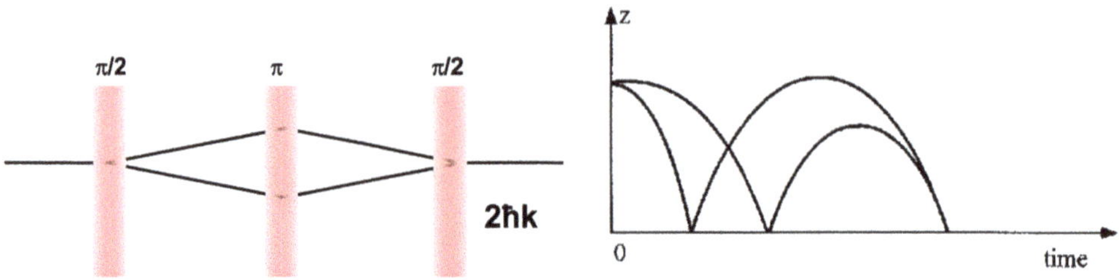


**Figure 7.2.** An atomic Mach–Zehnder interferometer is made for atoms. (a) Shown in momentum space, and (b) in the time domain.

interaction in the Bragg regime, for example, the atomic de Broglie wave splits into coherent beams. As discussed in section 4.2, the scattering along two paths with equal probability takes place as the scaled interaction time is $\pi/2$. Another optical lattice introduces a mirror effect as the scattering atoms interact with it for a scaled time equal to $\pi$. The de Broglie waves recombine and display interference, as shown in figure 7.2(a). The arrangement creates an atomic Mach–Zehnder interferometer for atoms [19].

The atom interferometer has been used to develop gyroscopes, atom clocks, inertial sensors, and gravimeters. The interferometric levitation of free-falling ultra-cold atoms is obtained by their being periodically submitted to multiple-order diffraction by a vertical one-dimensional standing wave. The various diffracted matter waves recombine coherently, resulting in high-contrast interference in the number of atoms detected at constant height, as shown in figure 7.3. A quantum trampoline has been developed based on periodically applying an imperfect Bragg mirror, which not only reflects upwards the falling atoms, but also acts as a beam splitter that separates and recombines the atomic wave packets. This results in multiple-wave atom interference, evidenced by an efficient suspension of the atoms. This suspension is obtained at a precise tuning of the trampoline period, whose value yields directly the local value of gravity, $g$ [25]. The gravimeter can also be based on perfect Bragg reflection [26, 27].

In experiments with an atomic Mach–Zehnder interferometer, ultra-cold atoms have been applied to observe the universality of free fall (UFF) in the quantum mechanical domain. The UFF asserts the equality of inertial and gravitational mass and implies that all objects freely falling in the same gravitational field experience the same acceleration. The UFF is an essential requirement for testing the accuracy of Einstein's theory of general relativity. In a typical experimental cycle, $8 \times 10^8$ atoms ($3 \times 10^7$ atoms) of rubidium ($^{87}$Rb) (potassium ($^{39}$K)) are loaded into a three-dimensional magneto-optical trap. The atoms are cooled to temperatures of 27 $\mu$K (32 $\mu$K). All the atoms are optically pumped into their respective ($|F = 1\rangle$) state. We release both ensembles simultaneously into free fall, which are then coherently split, redirected, and recombined [28, 29].

Efficient atomic diffraction at a normal incidence on an evanescent wave mirror is observed in the presence of a very weak spatial modulation. A reflection grating with modulation as small as 1.5% causes 66% of the atoms to be diffracted into orders $\pm 1$

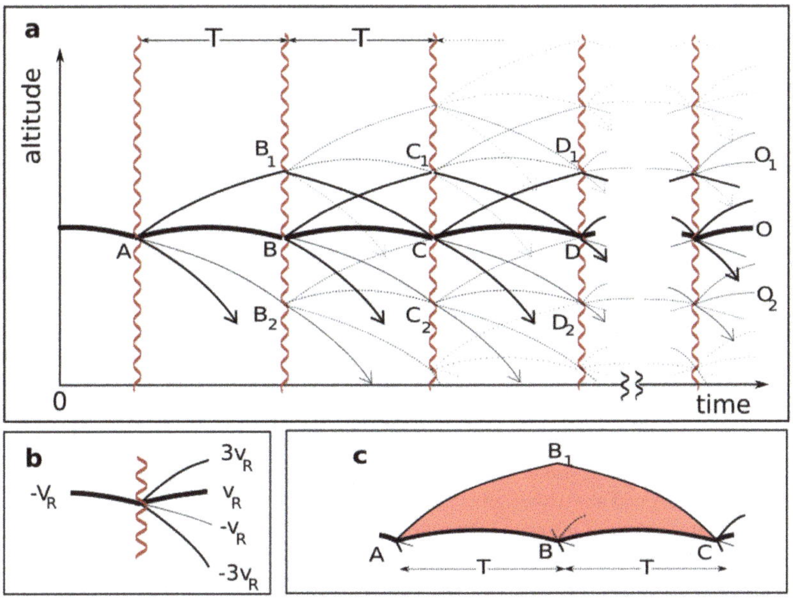


**Figure 7.3.** Atom trajectories in a quantum trampoline. (a) Atoms, diffracted by a periodically applied imperfect Bragg mirror, explore various paths that eventually recombine. The probability of a trajectory is represented by the line thickness. A thick line corresponds to a classical trampoline associated with perfect Bragg reflection. The arrows mark the loss channels at $-3V_R$. (b) An imperfect Bragg reflection: an incoming matter wave with vertical velocity $-V_R$ is mainly Bragg reflected to $+V_R$ (thick line). A small fraction is also diffracted to higher velocities ($+3V_R$ and $-3V_R$), and a smaller one transmitted without deviation. (c) Elementary interferometer: from a zero-order trajectory (thick line), an atom can be diffracted to $+3V_R$, at point A in the example shown. It is then Bragg reflected from $+V_R$ to $-V_R$ at point $B_1$ one period later, and finally recombines at point C with the zero-order trajectory. Reproduced from [25]. 

[30, 31]. Diffraction of ultra-cold atoms from an evanescent wave mirror at grazing incidence leads to an efficient large-angle atomic beam splitter. The phenomenon of beam splitting has applications in interferometric measurements of gravitational gradients or rotation measurements. The phase of the oscillations themselves amounts to an interferometric observation of the van der Waals interactions [32, 33]. The atomic phase interferometry is performed as an atomic de Broglie wave moving under gravity reflects back from two different positions of a modulated atomic mirror, recombines, and thus interferes in time, as shown in figure 7.2(b). The interferometer can be applied to analyze various characteristics, such as the van der Waals forces, surface roughness, and to develop a bi-model atomic trap.

A fascinating achievement of the gravitational cavity is the development of RTMs to study surface structures with nano- and sub-nanometer resolution [20]. The atomic de Broglie wave reflects back from the atomic mirror and recombines in time repeatedly. In its time evolution the gradual destructive self-interference is overcome by constructive interference at the quantum revival time. Hence, the initial

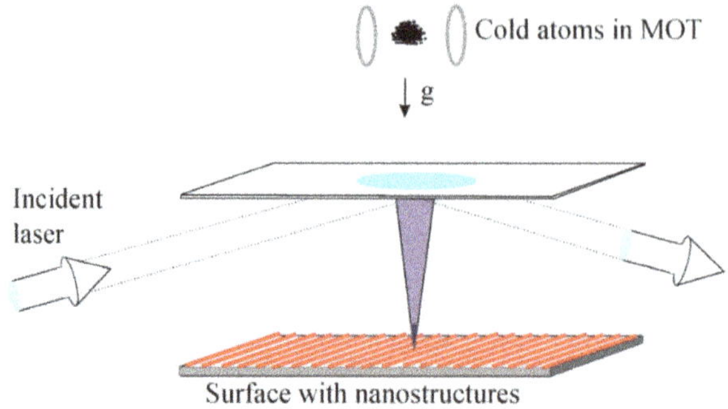


**Figure 7.4.** A model of a recurrence tracking microscope: a cloud of ultra-cold atoms is trapped and cooled in a magneto-optic trap (MOT) down to the microkelvin scale. The MOT is placed at a certain height above an evanescent wave atomic mirror. The mirror for the atoms results from the total internal reflection of the incident laser light field from the surface of the dielectric film. Thus, an evanescent wave field is produced on the surface that has polarization inside the plane of the reflection. The dielectric film is connected to a cantilever that has its other end above the surface under investigation. We consider the atomic dynamics along the $z$-axis, normal to the surface of the mirror. Reprinted figure with permission from [20], Copyright (2006) by the American Physical Society.

atomic wave packet rebuilds itself above the atomic mirror after successive reflections, as shown in figure 7.4. The mirror is joined to a cantilever that has its tip on the surface under investigation. As the cantilever varies its position following the surface structures, the atomic mirror changes its position in the upward or downward direction, thus varying the initial distance of bouncing atoms from the atomic mirror in the gravitational cavity. The time of a quantum revival depends upon the initial height of the atoms above the mirror, which, thus, varies as a result of the change in cantilever position. Hence, a change in the time of revival reveals the surface structures under investigation.

## 7.2 Bouncing atom on an atomic trampoline

A cold atom dropped from a certain initial height above an atomic mirror experiences a linear gravitational potential as it approaches the mirror, which is

$$V_{\text{gr}} = mgz. \tag{7.7}$$

Here, $m$ denotes the mass of the atom, $g$ expresses the constant gravitational acceleration, and $z$ describes the atomic position above the mirror. We write the atomic evolution inside the optical field by following the mathematical description developed in chapter 3. Therefore, by taking the optical field given in equation (7.5) and applying the dipole approximation and rotating-wave approximation, the atom–gravity–field system is expressed by the effective Hamiltonian, that is,

$$H = \frac{p^2}{2m} + mgz + \hbar\Omega_{\text{eff}}e^{-2\kappa z}. \tag{7.8}$$

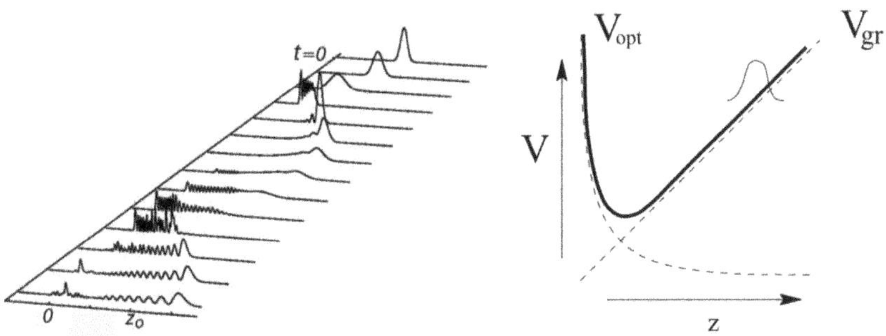


**Figure 7.5.** Left: As we switch off the magneto-optic trap (MOT) at time $t = 0$, the atomic wave packet evolves from an initial height $z_0$. Reprinted figure with permission from [20], Copyright (2006) by the American Physical Society. Right: It moves under the influence of the linear gravitational potential $V_{gr}$ (dashed line) toward the evanescent wave atomic mirror. Close to the surface of the mirror the effect of the evanescent light field is dominant and the atom experiences an exponential repulsive optical potential $V_{opt}$ (dashed line). Both the potentials together make the gravitational cavity for the atom (solid line). Reprinted from [34], Copyright (2005), with permission from Elsevier.

The effective Hamiltonian governs the center-of-mass motion of the atom in the presence of the gravity above the evanescent wave field. Here, $\Omega_{\text{eff}} = \Omega_{\text{R}}^2/\Delta$ describes the effective Rabi frequency, and $\kappa^{-1}$ describes the decay length of the atomic mirror.

The optical potential is dominant for smaller values of $z$ and decays exponentially as $z$ becomes larger. Thus, for larger positive values of $z$, as the optical potential approaches zero the gravitational potential takes over, as shown in figure 7.5 (left).

### 7.2.1 Mode structure

For simplicity, we may consider that an ultra-atom observes almost an instantaneous impact with an atomic mirror, when (i) the atom is initially placed away from the atomic mirror in the gravitational field, and (ii) the atomic mirror is made up of an exponentially decaying optical field with a very short decay length. Thus, we may take the gravitational cavity as a triangular potential, made up of a linear gravitational potential and an infinite potential, resulting in a bounded motion of the atom.

We express the corresponding effective Hamiltonian as

$$H = \frac{p^2}{2m} + V(z), \tag{7.9}$$

where

$$V(z) \equiv \begin{cases} mgz & z > 0, \\ \infty & z \leqslant 0. \end{cases} \tag{7.10}$$

The time-independent Schrödinger wave equation, $H\psi_n = E_n\psi_n$, is solved for the boundary conditions, which are

$$\psi_n(z = 0) = 0 \quad \text{and} \quad \psi_n(z = \infty) = 0.$$

Therefore, we obtain Airy functions as the eigenfunctions

$$\psi_n = \mathcal{N} Ai(s(z - z_n)), \tag{7.11}$$

where $s = (\hbar^2/2m^2g)^{-1/3}$, and the normalization constant $\mathcal{N}$ is calculated as

$$\mathcal{N} = \left( \int_0^\infty |Ai(s(z - z_n))|^2 \, dz \right)^{-1/2}. \tag{7.12}$$

Moreover, the displacement $z_n$ is obtained as $z_n = (\hbar^2/2m^2g)^{1/3} f_n$, such that

$$\psi_n(z = 0) = \mathcal{N} Ai(-sz_n) = \mathcal{N} Ai(-f_n) = 0. \tag{7.13}$$

Here, the function $f_n$ defines the $n$th negative zero of the Airy function, that is,

$$f_n = f(\zeta_n) = \zeta_n^{2/3}\left(1 + \frac{5}{48\zeta_n^2} - \frac{5}{36\zeta_n^4} + \cdots\right), \tag{7.14}$$

and

$$\zeta_n = 3\pi(n - 1/4)/2.$$

Hence, the integer $n$ describes the $n$th eigenstate $\psi_n$ with eigenenergy $E_n$, obtained as

$$E_n = mgz_n = \left(\frac{mg^2\hbar^2}{2}\right)^{1/3} f_n.$$

In the case of large $n$, we may express $f_n$ as $(3n\pi/2)^{2/3}$, which yields the eigenenergy as

$$E_n = \frac{1}{2} m^{1/3} (3n\pi\hbar g)^{2/3}. \tag{7.15}$$

### 7.2.2 Wigner function

The phase-space probability distributions of the eigenstates of the gravitational cavity are obtained using the Wigner function description, discussed in chapter 2, such that

$$W(z, p) = \frac{1}{2\pi\hbar} \int_{-\infty}^{\infty} \psi_n^*(z + y/2)\, \psi_n(z - y/2)\, e^{i\frac{p}{2\hbar}y}\, dy. \tag{7.16}$$

The first three eigenfunctions, that is, $n = 1, 2, 3$, in the scaled coordinates $z$ and $p$, are obtained as

$$W(z, p) = \frac{\mathcal{N}^2}{2\pi\hbar} \int_0^\infty Ai(z + y/2 - z_n)\, Ai(z - y/2 - z_n)\, e^{i\frac{p}{2\hbar}y}\, dy, \tag{7.17}$$

as shown in figure 7.6. The distributions are symmetric around the origin along the $p$-axis. Moreover, we note that the distributions extend in space along the $z$-axis as $n$

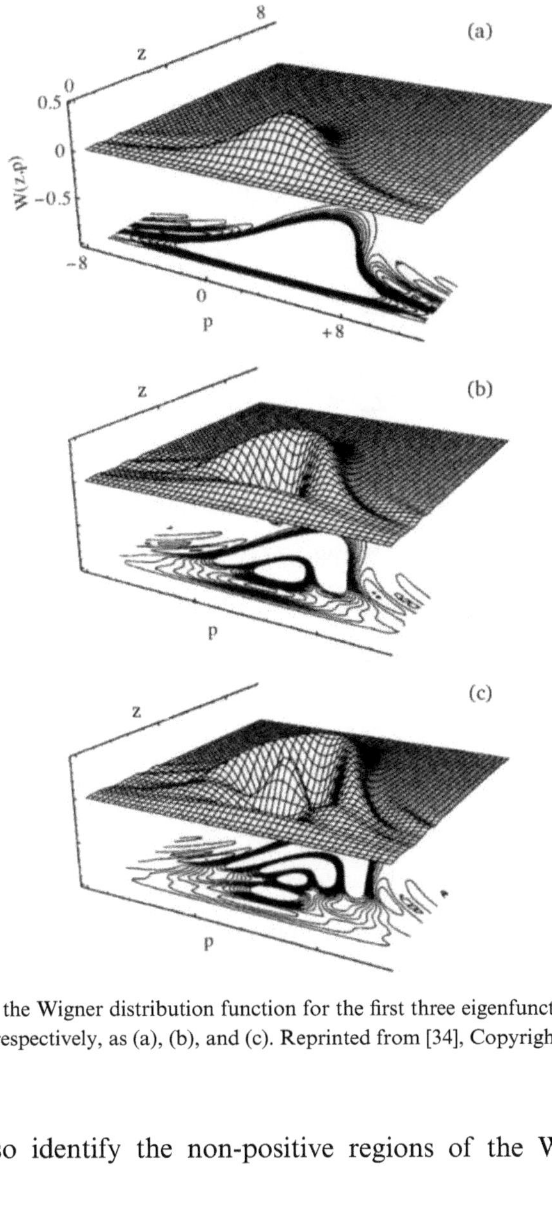


**Figure 7.6.** We express the Wigner distribution function for the first three eigenfunctions, defined in equation (7.11), for $n = 1, 2, 3$, respectively, as (a), (b), and (c). Reprinted from [34], Copyright (2005), with permission from Elsevier.

increases. We also identify the non-positive regions of the Wigner distribution functions.

## 7.3 Space-time evolution: quantum carpets

The space-time dynamics of a wave packet over a potential surface can explain several interesting dynamical features. In general, we express the atomic wave packet, $\Psi$, as a superposition of eigenstates, $\psi_n(z)$, that is,

$$\Psi(z) = \sum_n c_n \psi_n(z). \tag{7.18}$$

Here, $c_n$ is the dimensionless probability amplitude. The wave packet evolves in time following unitary transformation, such that

$$\Psi(z, t) = e^{-\frac{i\hat{H}t}{\hbar}}\Psi(z, t = 0), \tag{7.19}$$

$$= e^{-\frac{i\hat{H}t}{\hbar}}\sum_n c_n\psi_n(z), \tag{7.20}$$

$$= \sum_n c_n e^{-\frac{iE_nt}{\hbar}}\psi_n(z), \tag{7.21}$$

$$= \sum_n c_n e^{-i\omega_n t}\psi_n(z). \tag{7.22}$$

The probability of finding an atom in space at a particular event in time is described by the corresponding space-time probability distribution function, written as

$$\begin{aligned} |\Psi(z, t)|^2 &= \Psi^*(z, t)\Psi(z, t), \\ &= \sum_{n'} c_{n'}^* e^{i\omega_{n'}t}\psi_{n'}^*(z) \sum_n c_n e^{-i\omega_n t}\psi_n(z), \\ &= \sum_{n'}\sum_n c_{n'}^* c_n e^{-i(\omega_n - \omega_{n'})t}\psi_{n'}^*(z)\psi_n(z). \end{aligned} \tag{7.23}$$

The probability distribution function can easily be written in two parts:

$$|\Psi(z, t)|^2 = \sum_n |c_n|^2|\psi_n(z)|^2 + \sum_{n'}\sum_n c_{n'}^* c_n e^{-i(\omega_n - \omega_{n'})t}\psi_{n'}^*(z)\psi_n(z). \tag{7.24}$$

The first term in equation (7.24) is obtained by setting $n = n'$; therefore, it has no dependence on the time of evolution, $t$. The second term of the equation corresponds to $n \neq n'$, and it has time-dependent exponential functions with non-zero frequency differences, $\omega_n - \omega_{n'}$ in the exponent. For this reason, it manifests in constructive and destructive interference in time. The evolution of a matter-wave packet in time, therefore, weaves interesting quantum carpets.

We show quantum carpets for an atom in a gravitational cavity in figure 7.7. The dark gray regions express higher probability, whereas, the in-between light gray regions indicate negligibly small probabilities of finding the atom in its space-time evolution. Close to the surface of the atomic mirror, at $z = 0$, these structures appear as vertical canals sandwiched between two high-probability dark gray regions. These canals curve gradually away from the atomic mirror where the gravitational field is significant as a manifestation of non-linear frequency difference.

## 7.4 Quantum revivals

The quantum dynamics of a wave packet is a beautiful manifestation of classical mechanics, wave mechanics, and quantum laws. In order to explain these dynamics,

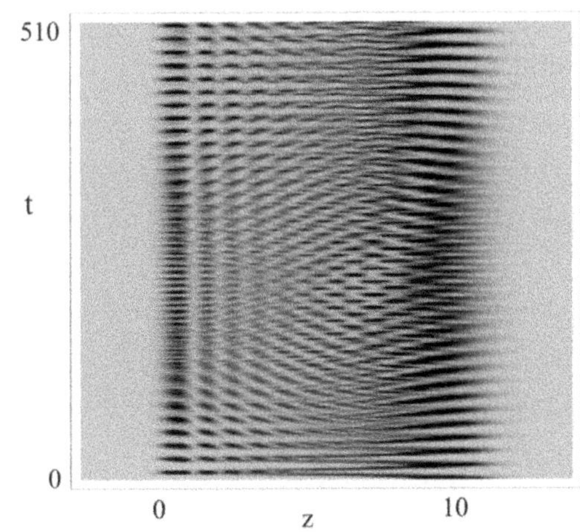


**Figure 7.7.** The space-time dynamics of an atomic wave packet, representing a quantum carpet. Reprinted from [34], Copyright (2005), with permission from Elsevier.

we calculate the auto-correlation function, $C(t)$, as it correlates the time-evolved wave packet with the initial wave packet, such that

$$C(t) = \langle \Psi(z, t = 0) | \Psi(z, t) \rangle, \tag{7.25}$$

$$= \sum_{n'} \sum_{n} c_{n'}^{*} c_n \mathrm{e}^{-i\omega_n t} \int dz \psi_{n'}^{*}(z) \psi_n(z). \tag{7.26}$$

The integration is performed over the position coordinate, where the property of ortho-normality of the eigenstates is at work. Hence, the remaining expression achieves significant simplicity as it displays only time dependency. We obtain

$$C(t) = \sum_{n} |c_n|^2 \mathrm{e}^{-i\frac{E_n t}{\hbar}}. \tag{7.27}$$

The time dependence of the wave packet appears in the exponential factor. We express the above equation by using the Taylor expansion of energy, $E_n$, around the expectation value or mean value of the quantum number, that is, $\langle n \rangle = n_0$. Hence, we obtain

$$C(t) = \sum_{n} |c_n|^2 \exp\left\{ -i \left( E_{n_0} + (n - n_0) \frac{\partial E_n}{\partial n} \bigg|_{n=n_0} + \frac{1}{2!}(n - n_0)^2 \frac{\partial^2 E_n}{\partial n^2} \bigg|_{n=n_0} + \cdots \right) \frac{t}{\hbar} \right\}, \tag{7.28}$$

$$= \sum_{n} |c_n|^2 \exp\{ -i(\omega_{n_0} + (n - n_0)\omega^{(1)} + (n - n_0)^2 \omega^{(2)} + \cdots )t \}, \tag{7.29}$$

where

$$\omega^{(j)} = \frac{1}{j! \hbar} \frac{\partial^j E_n}{\partial n^j} \bigg|_{n=n_0}. \tag{7.30}$$

Here, $j$ is an integer. Evidently, for every $j$th frequency, $\omega^{(j)}$, we write the corresponding time period as

$$T^{(j)} = \frac{2\pi}{\omega^{(j)}}. \tag{7.31}$$

Hence, the auto-correlation, $C(t)$, obtained in equation (7.29), becomes

$$C(t) = \sum_n |c_n|^2 \exp\left\{-2\pi i\left(\frac{1}{T^{(0)}} + (n - n_0)\frac{1}{T^{(1)}} + (n - n_0)^2\frac{1}{T^{(2)}} + \cdots.\right)t\right\}. \tag{7.32}$$

The frequency $\omega_{n_0} = E_{n_0}/\hbar$, and the corresponding time period, $T^{(0)} = 2\pi/\omega_{n_0}$, are independent of $n$.

*Classical periodicity.* For $j = 1$, we have the frequency $\omega^{(1)}$, written as

$$\omega^{(1)} = \frac{1}{\hbar}\frac{\partial E_n}{\partial n}\bigg|_{n=n_0}. \tag{7.33}$$

From the Bohr–Sommerfeld quantization rule, the classical action $I$ is defined as $I = \oint p dz = n\hbar$. Hence, the frequency $\omega^{(1)}$ becomes

$$\omega^{(1)} = \frac{\partial E_I}{\partial I}\bigg|_{I=I_0}, \tag{7.34}$$

which is independent of $\hbar$. The corresponding time period is

$$T^{(1)} = 2\pi\hbar\left(\frac{\partial E_n}{\partial n}\bigg|_{n=n_0}\right)^{-1} = 2\pi\left(\frac{\partial E_I}{\partial I}\bigg|_{I=I_0}\right)^{-1}, \tag{7.35}$$

In classical mechanics the time period defines the time taken by a classical particle in traversing a closed trajectory, or twice the time taken by a classical particle between two turning points. Hence, we call the time period $T^{(1)}$ the classical time period.

*Quantum revivals.* Keeping in view the property of the Taylor expansion, the next higher-order correction terms have relatively smaller contributions. For this reason, $\omega^{(2)}$ for $j = 2$ is smaller than $\omega^{(1)}$. For $j = 2$, we have

$$\omega^{(2)} = \frac{1}{\hbar}\frac{\partial^2 E_n}{\partial n^2}\bigg|_{n=n_0} = \hbar\frac{\partial^2 E_I}{\partial I^2}\bigg|_{I=I_0}. \tag{7.36}$$

The corresponding quantum revival time is obtained as

$$T^{(2)} = 2\pi\hbar\left(\frac{1}{2}\frac{\partial^2 E_n}{\partial n^2}\bigg|_{n=n_0}\right)^{-1} = \frac{2\pi}{\hbar}\left(\frac{1}{2}\frac{\partial^2 E_I}{\partial I^2}\bigg|_{I=I_0}\right)^{-1}, \tag{7.37}$$

which is the time at which the reconstruction of the wave packet takes place beyond the classical time period. The expression explicitly depends on $\hbar$; for this reason, the

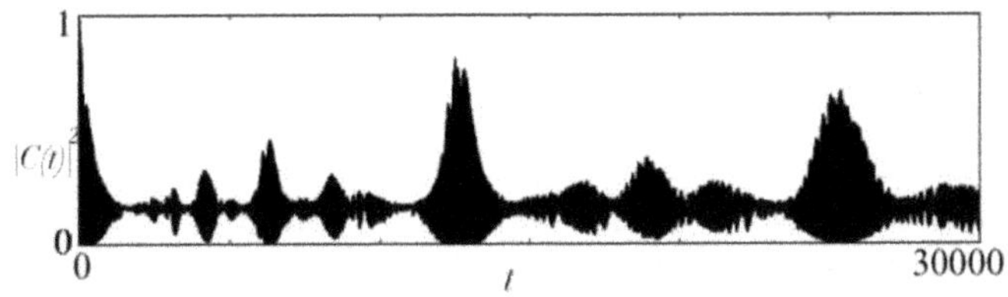


**Figure 7.8.** For an atom in a gravitational cavity square of the auto-correlation function $|C(t)|^2 = |\langle\Psi(0)|\Psi(t)\rangle|^2$ plotted vs time, $t$. The quantum revival of the atomic wave packet and fractional quantum revivals at fractions of the revival time manifest themselves as a function of the evolution time. Reprinted figure with permission from [20], Copyright (2006) by the American Physical Society.

revival phenomenon is purely a quantum mechanical effect, and the time it takes to occur is named as the quantum revival time, or simply as the revival time.

The atomic wave packet initially evolves over a short period of time in a potential, following classical mechanics. While moving along its classical trajectory, it displays inherent quantum dispersion and self-interference. The wave packet, therefore, tends to rebuild itself periodically at classical periods, $T^{(1)}$. However, due to the increasing phase difference between constituting wavelets, gradually the destructive interference dominates, which leads to a collapse of the wave packet. The occurrence of the collapse is a manifestation of the non-linear dependence of the eigenenergy $E_n$ on the quantum number $n$.

The collapse of the wave packet in its time evolution is comparable with the time evolution of a wave following wave mechanics. However, from equation (7.22), we find that the discreteness of the energy spectrum appears in the discrete phases $\varphi_n = \omega_n t$ associated with the constituting wavelets. This depicts the phase difference between the constituting wavelets, which results in constructive interference and leads to the restoration and restructuring of the wave packet at the quantum revival time. The quantum coherence also manifests at a fraction of the quantum revival time, that is, $\frac{p}{q}T^{(2)}$, where $p$ and $q$ are relative prime numbers. These time events are named *fractional revival times*. The phenomena of quantum revival and fractional revivals are shown in figure 7.8.

**Example 1.** With the help of equation (7.15), we calculate the quantum revival time for the gravitational cavity, that is,

$$T^{(2)} = \frac{16E_{n_0}^2}{m\pi\hbar g^2}, \tag{7.38}$$

where $E_{n_0} = mgz(0)$, and $z(0)$ is the initial height of the bouncing ultra-cold atoms above the atomic mirror. As explained in section 7.1, equation (7.38) describes the principle of a RTM. Following equation (7.38), we understand that a change in the position of the cantilever will change the initial height of the atoms above the mirror, $z(0)$. The change in the height shows itself by modifying the quantum revival time. Experimentally measured variations in the quantum revival time reveal the nature of the surface structures under study, with nanometer resolution.

## 7.5 Nano optical-fiber cavities

### 7.5.1 Atomic trapping

Let us consider an ultra-cold atom outside a dielectric nano-fiber that moves in a circular orbit near the surface of the nano-fiber. The atom displays a stable and localized evolution around the nano-fiber as a consequence of a balance between the attractive van der Waals force and the centrifugal force experienced by the atom. In the considered configuration, the total effective potential of the atom includes the repulsive centrifugal potential and the attractive van der Waals potential due to the interaction of atom A with a cylindrical surface. This can be written as

$$U_{\text{eff}}(r) = \frac{\hbar^2}{2mr^2}\left(\ell^2 - \frac{1}{4}\right) - \frac{C_0\mu}{(r-a)^3}, \tag{7.39}$$

where $m$ is the mass of the atom, $\ell$ is the quantum number defining the projection of the angular momentum on the fiber axis, and $\mu$ is an external surface factor. According to the above equation, the equilibrium radial position of the atom is defined by the equation

$$(r' - 1)^3 - \frac{4c}{4\ell^2 - 1}r'^2 = 0, \tag{7.40}$$

where $r' = r/a$, and the dimensionless parameter $c = 2mC_0\mu/\hbar^2 a$ defines the strength of the van der Waals interaction.

From equation (7.40), we can evaluate the basic parameters of the atomic trap. Consider, for example, cesium atoms located around a silica fiber of radius $a = 100$ nm. In this case, $C_0 = 1.6$ MHz $\mu\text{m}^3$, and the dimensionless parameter is evaluated as $c \approx 440$. Such a value of $c$ means that stable localization of the atom can be achieved at a distance of about $r_0 - a \approx a$, and at angular momentum quantum number $\ell \approx 20$. In frequency units, the depth of the effective potential in this case is about $U_{\text{eff}}(r_0)/h \approx 1$ MHz; in energy units the depth of the potential is $4 \times 10^{-9}$ eV [35] (see figure 7.9).

### 7.5.2 Bi-modal atomic trap

An evanescent wave field around an optical fiber with nano-scale diameter exerts force on cold atoms and may be used to store atoms and guide them by balancing the centrifugal force due to their rotation around the nano-fiber [36] The laser light propagating through the nano-fiber generates an evanescent wave around it. The field frequency and atomic transition frequency are far off-resonance, therefore, the atom may experience an attractive force or repulsive force. Hence, as the outward centrifugal force on the rotating atom is canceled by so generated attractive optical force, the atom finds itself trapped. We show a schematic diagram of such an atomic trap in figure 7.10.

In addition, we make an optical cylinder within a hollow optical fiber to trap and guide cold atoms [37, 38]. A laser light propagating in the glass makes an evanescent wave within the fiber. The optical field is detuned to the blue of atomic resonance,

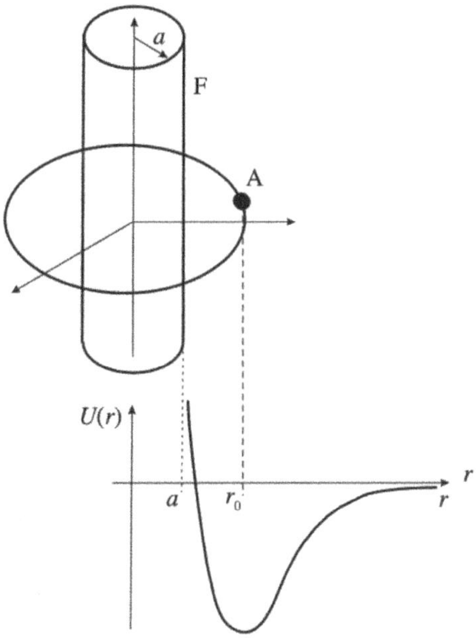


**Figure 7.9.** Trapping scheme of an atom A around an optical fiber F, including the effective atom potential as a function of distance $r$. Reproduced from [35]. 

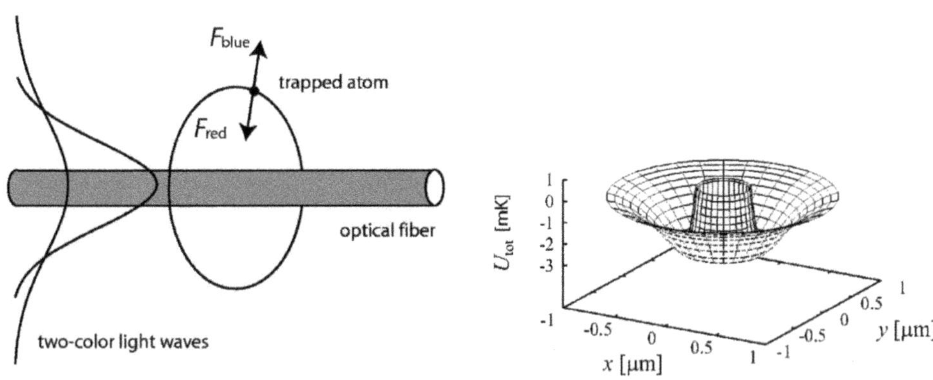


**Figure 7.10.** Left: Schematic diagram of a bi-dimensional atomic trap around an optical fiber. Right: Transverse plane profile of the total potential $U_{tot}$ produced by the net optical potential and the van der Waals potential. Reprinted from [34], Copyright (2005), with permission from Elsevier.

thus it exerts repulsive force on the atoms, pushing them toward the center of the fiber where the atoms get trapped. The system also serves as a useful waveguide for the atoms. Two light field modes, 1 and 2, with their frequencies, $\omega_1$ and $\omega_2$, respectively, develop a potential minima outside the nano-fiber radius that is used as a surface trap for the cold atoms. A schematic diagram of the arrangement is shown in figure 7.10. In addition, we consider that the two incoming fields are circularly polarized; in this case, the orbit rotates circularly in space and the spatial distribution of the field intensity is cylindrically symmetric. The optical potentials of the two fields, proportional to their intensities, add up to give the net optical potential, that is,

$$\begin{aligned} U &= U_1 + U_2, \\ &= -G_1K_0^2(q_1r) + G_2K_0^2(q_2r). \end{aligned} \tag{7.41}$$

Here, $K_0$ is the zeroth-order modified Bessel function of the second kind, whereas $G_i$ is associated with the intensity of the fields, $i = 1$ and $i = 2$ [39].

### 7.5.3 Atomic coherent tractor effect

The basic principle of the atomic coherent optical tractor effect is the superposition of two co-propagating electromagnetic fields [40]. Assuming for simplicity the two fields have the same polarization and amplitude, $E_0/2$, the general form of the superposed fields is

$$E(z, t) = E_0 \cos(\Delta^+kz - \Delta^+\omega t)\cos(\Delta^-kz - \Delta^-\omega t),$$

where $\Delta^{\pm}k = (k_2 \pm k_1)/2$, $\Delta^{\pm}\omega = (\omega_2 \pm \omega_1)/2$, and $k_i$ and $\omega_i$ are the wave number and angular frequency of the fields, $i = 1$ and $i = 2$. The field is a traveling wave with spatial and temporal frequencies given by the respective mean of the two incident fields' spatial and temporal frequencies. This traveling wave is amplitude modulated by the beat between the two fields.

Under the resonance condition, that is, $\omega_1 = \omega_2 = \omega$, the field is given by

$$E(z, t) = E_0 \cos(\Delta^+kz - \omega t)\cos(\Delta^-kz).$$

The field displays an important property with electromagnetic standing waves, namely, it has nodes and anti-nodes with fixed spatial positions decided by the time-independent beat envelope $\cos(\Delta^-kz)$. At this point, the advantage of using few-mode optical nano-fiber can be seen: atoms can interact with the evanescent region of the guided modes and a standing beat field can be generated using two co-propagating modes of the optical fiber. It is precisely this principle with which we engineer a tractor-beam effect.

The diameter of the optical nano-fiber is 430 nm and $\lambda = 588$ nm. The two-mode trap is created by the interference of the quasi-$y$-polarized HE11 mode of the nano-fiber with the TE01 mode, with optical powers of 3.47 mW and 0.47 mW, respectively. An experimentally realistic power limit of 50 mW for the total optical power in the optical nano-fiber sets the trapping position at least 200 nm from the optical nano-fiber surface (see figure 7.11).

## Exercises

1. Show that in a gravitational cavity the quantum revival time is

$$T^{(2)} = \frac{16E_{n_0}^2}{m\pi\hbar g^2}.$$

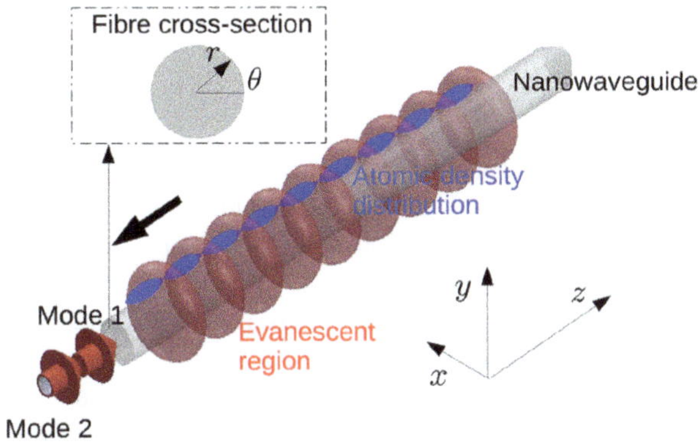


**Figure 7.11.** Schematic diagram of a bi-dimensional atomic trap around an optical fiber. Two guided modes of a nano-waveguide beat to form a periodic, evanescent field near its surface. Sodium atoms of temperature 10 $\mu$K are trapped in the field and moved toward the light source [40]. Copyright © 2016, The Author(s). With permission of Springer.

## References


[1] Vladimirskiĭ V V 1960 Magnetic mirrors, channels and bottles for cold neutrons *Zh. Éksp. Teor. Fiz.* **39** 1062
*Sov. Phys. JETP* **12** 740 1960

[2] Opat G I, Wark S J and Cimmino A 1992 Electric and magnetic mirrors and gratings for slowly moving neutral atoms and molecules *Appl. Phys.* B **54** 396

[3] Roach T M, Abele H, Boshier M G, Grossman H L, Zetie K P and Hinds E A 1995 Realization of a magnetic mirror for cold atoms *Phys. Rev. Lett.* **75** 629

[4] Hughes I G, Barton P A, Boshier M G and Hinds E A 1997 Atom optics with magnetic surfaces: I. Storage of cold atoms in a curved floppy disk *J. Phys. B: At. Mol. Opt. Phys.* **30** 647

[5] Hughes I G, Barton P A and Hinds E A 1997 Atom optics with magnetic surfaces: II. Microscopic analysis of the floppy disk mirror *J. Phys. B: At. Mol. Opt. Phys.* **30** 2119

[6] Saba C V, Barton P A, Boshier M G, Hughes I G, Rosenbusch P, Sauer B E and Hinds E A 1999 Reconstruction of a cold atom cloud by magnetic focusing *Phys. Rev. Lett.* **82** 468

[7] Rosenbusch P, Hall B V, Hughes I G, Saba C V and Hinds E A 2002 Manipulation of cold atoms using a corrugated magnetic reflector *Phys. Rev.* A **61** R031404

[8] Rosenbusch P, Hall B V, Hughes I G, Saba C V and Hinds E A 2000 Manipulation of cold atoms by an adaptable magnetic reflector *Appl. Phys.* B **70** 709

[9] Esslinger T, Weidemüller M, Hemmerich A and Hänsch T W 1993 Surface-plasmon mirror for atoms *Opt. Lett.* **18** 450

[10] Feron S *et al* 1993 Reflection of metastable neon atoms by a surface plasmon wave *Opt. Commun.* **102** 83

[11] Christ M, Scholz A, Schiffer M, Deutschmann R and Ertmer W 1994 Diffraction and reflection of a slow metastable neon beam by an evanescent light grating *Opt. Commun.* **107** 211

[12] Wallis H, Dalibard J and Cohen-Tannoudji C 1992 Trapping atoms in a gravitational cavity *Appl. Phys.* B **54** 407

[13] Ovchinnikov Yu B, Manek I and Grimm R 1997 Surface trap for Cs atoms based on evanescent-wave cooling *Phys. Rev. Lett.* **79** 2225

[14] Kulin S, Aubin S, Christe S, Peker B, Rolston S L and Orozco L A 2001 A single hollow-beam optical trap for cold atoms *J. Opt. B: Quantum Semiclass. Opt.* **3** 353

[15] Kasevich M A, Weiss D S and Chu S 1990 Normal-incidence reflection of slow atoms from an optical evanescent wave *Opt. Lett.* **15** 607–9

[16] Henkel C, Steane A M, Kaiser R and Dalibard J 1994 A modulated mirror for atomic interferometry *J. Phys. II* **4** 1877–96

[17] Gea-Banacloche J 1999 A quantum bouncing ball *Am. J. Phys.* **67** 776

[18] Steane A, Szriftgiser P, Desbiolles P and Dalibard J 1995 Phase modulation of atomic de Broglie waves *Phys. Rev. Lett.* **74** 4972

[19] Rasel E M, Oberthaler M K, Batelaan H, Schmiedmayer J and Zeilinger A 1995 Atom wave interferometry with diffraction gratings of light *Phys. Rev. Lett.* **75** 2633

[20] Saif F 2006 Recurrence tracking microscope *Phys. Rev.* A **73** 033618

[21] Llorente Garcia I, Darquie B, Curtis E A, Sinclair C D J and Hinds E A 2010 Experiments on a videotape atom chip: fragmentation and transport studies *New J. Phys.* **12** 093017

[22] Berkhout J J, Luiten O J, Setija I D, Hijmans T W, Mizusaki T and Walraven J T M 1989 Quantum reflection: focusing of hydrogen atoms with a concave mirror *Phys. Rev. Lett.* **63** 1689

[23] Roberts T D, Cronin A D, Tiberg M V and Pritchard D E 2004 Dispersion compensation for atom interferometry *Phys. Rev. Lett.* **92** 060405

[24] Perreault J D and Cronin A D 2005 Observation of atom wave phase shifts induced by van der Waals atom-surface interactions *Phys. Rev. Lett.* **95** 133201

[25] Robert-de-Saint-Vincent M and Brantut J P 2010 A quantum trampoline for ultra-cold atoms *Europhys. Lett.* **89** 10002

[26] Rasel E M, Oberthaler M K, Batelaan H, Schmiedmayer J and Zeilinger A 1995 Atom wave interferometry with diffraction gratings of light *Phys. Rev. Lett.* **75** 2633

[27] Gupta S, Dieckmann K, Hadzibabic Z and Pritchard D E 2002 Contrast interferometry using Bose–Einstein condensates to measure *h*/*m* and *α* *Phys. Rev. Lett.* **89** 140401

[28] Schlippert D, Hartwig J, Albers H, Richardson L L, Schubert C, Roura A, Schleich W P, Ertmer W and Rasel E M 2014 Quantum test of the Universality of Free Fall *Phys. Rev. Lett.* **112** 203002

[29] Albers H *et al* 2020 *Eur. Phys. J.* D **74** 145

[30] Landragin A, Cognet L, Horvath G-Zs K, Westbrook C-I, Westbrook N and Aspect A 1997 A reflection grating for atoms at normal incidence *Europhys. Lett.* **39** 485

[31] Estéve J, Stevens D, Savalli V, Westbrook N, Westbrook C I and Aspect A 2003 Resolved diffraction patterns from a reflection grating for atoms *J. Opt. B: Quantum Semiclass. Opt.* **5** S103

[32] Cognet L, Savalli V, Horvath G, Zs K, Holleville D, Marani R, Westbrook N, Westbrook C-I and Aspect A 1998 Atomic interference in grazing incidence diffraction from an evanescent wave mirror *Phys. Rev. Lett.* **81** 5044

[33] Westbrook N *et al* 1998 New physics with evanescent wave atomic mirrors: the van der Waals force and atomic diffraction *Phys. Scr.* **T78** 7–12

[34] Saif F 2005 Classical and quantum chaos in atom optics *Phys. Rep.* **419** 207–58

[35] Frawley M C, Chormaic S N and Minogin V G 2012 The van der Waals interaction of an atom with the convex surface of a nanocylinder *Phys. Scr.* **85** 058103

[36] Balykin V I, Hakuta K, Kien F L, Liang J Q and Morinaga M 2004 Atom trapping and guiding with a subwavelength-diameter optical fiber *Phys. Rev.* A **70** 011401

[37] Renn M J, Montgomery D, Vdovin O, Anderson D Z, Wieman C E and Cornell E A 1995 Laser-guided atoms in hollow-core optical fibers *Phys. Rev. Lett.* **75** 3253

[38] Ito H, Nakata T, Sakaki K, Ohtsu M, Lee K I and Jhe W 1996 Laser spectroscopy of atoms guided by evanescent waves in micron-sized hollow optical fibers *Phys. Rev. Lett.* **76** 4500

[39] Kien F L, Balykin V I and Hakuta K 2004 Atom trap and waveguide using a two-color evanescent light field around a subwavelength-diameter optical fiber *Phys. Rev.* A **70** 063403

[40] Sadgrove M, Wimberger S and Chormaic S N 2016 Quantum coherent tractor beam effect for atoms trapped near a nano-waveguide *Sci. Rep.* **6** 28905

**IOP** Publishing

Optical Forces on Atoms

**Farhan Saif and Shinichi Watanabe**

# Chapter 8

## Atoms in multi-dimensional systems

*'I think the next [21st] century will be the century of complexity. We have already discovered the basic laws that govern matter and understand all the normal situations. We don't know how the laws fit together, and what happens under extreme conditions. But I expect we will find a complete unified theory sometime this century. There is no limit to the complexity that we can build using those basic laws.'*

—Stephen W Hawking

In general, experimental systems are beyond one-dimensional ideal scenarios, where composite forces act and interactions take place between higher dimensions. In order to understand the associated subtleties in the quantum domain, we first must understand them in the classical domain.

Engineering coherence and synchronization present a great challenge in dynamical systems, with applications ranging from chemical reactions [1] to ultra-cold atomic dynamics in optical lattices [2, 3]. The issues are non-trivial in systems with coupled degrees of freedom. At the quantum level, coupling between the various subsystems can lead to entanglement, quantum scars, and quantum recurrences, which have no classical counterpart. By contrast, in the classical limit these systems may exhibit complexity or chaos. An understanding of the dynamical characteristics at the classical–quantum boundary explains the transition from the macroscopic to microscopic level. Moreover, a study of the classical and quantum dynamics of these systems reveals distinct dynamical characteristics [3]. These three regimes provide the understanding necessary to develop quantum devices and sensors. Further, they have importance in explaining key dynamical effects.

doi:10.1088/978-0-7503-2308-6ch8  

## 8.1 The state of a system

Independent coordinates $q_j$ that serve to uniquely determine the orientation and location of a system in physical space are called generalized or canonical or good coordinates. Here, $j = 1,...N$. The number, $N$, of generalized coordinates needed to describe the motion of the system under its constraints is the number of *degrees of freedom*.

To fully specify the *state of the system* at any instant, we need to express also the velocities, $\dot{q}_j$, or we may specify certain functions of the generalized coordinates and velocities, $p_j = p_j(q_1...q_N, \dot{q}_1...\dot{q}_N)$, called canonical momenta. The $2N$-dimensional space of generalized coordinates and their canonical momenta is called the *phase space*. Below, we will use $(p, q)$ to represent all of the momenta and coordinates $(p_1...p_N, q_1...q_N)$, so $(p, q)$ represents *a point* in phase space.

The evolution of continuous dynamical systems is described by differential equations:

$$dq/dt = f(p, q, t); \quad dp/dt = g(p, q, t).$$

A system is autonomous if the functions $f$ and $g$ depend only on $(p, q)$, and have no explicit time dependence. A system is Hamiltonian if the functions $f$ and $g$ are derivatives of a single function $H(q, p)$, that is,

$$f(p, q, t) = \partial H(p, q, t)/\partial p; \quad g(p, q, t) = -\partial H(p, q, t)/\partial q.$$

More explicitly, for systems with several degrees of freedom,

$$\frac{dq_j}{dt} = \frac{\partial H(p, q, t)}{\partial p_j}, \tag{8.1}$$

$$\frac{dp_j}{dt} = -\frac{\partial H(p, q, t)}{\partial q_j}. \tag{8.2}$$

For ordinary mechanical systems in Cartesian coordinates, the Hamiltonian is the sum of kinetic plus potential energies:

$$H = \sum_j \{p_j^2/2\,m + V(q_j)\}.$$

On writing out these equations, we find the velocity of each particle to be $p_j/m$, and the force is the rate of change of the momentum with time, $\dot{p}_j$. It was one of the glories of Issac Newton to have postulated the relation $\dot{p}$ as force, rather than $m\ddot{q}$ as force[1].

[1] As pointed out by John B. Delos, '$dp/dt = F$ can go over to generalized coordinates in Lagrangian formulation by redefinition of p and F, and a similar form carries over into Hamiltonian formulation. Furthermore it even carries over to relativistic formulations, where $F = ma$ is not true. So seemingly Newton had the insight that this would be the best way to formulate it.'

*Energy conservation.* If the Hamiltonian has no explicit time dependence, then energy (the value of the Hamiltonian) is conserved. We calculate the rate of change of energy as

$$\frac{dH}{dt} = \frac{\partial H}{\partial q}\frac{\partial q}{\partial t} + \frac{\partial H}{\partial p}\frac{\partial p}{\partial t} + \frac{\partial H}{\partial t}, \tag{8.3}$$

$$= \frac{\partial H}{\partial r}\frac{\partial H}{\partial p} + \frac{\partial H}{\partial p}\left(-\frac{\partial H}{\partial r}\right) + \frac{\partial H}{\partial t}, \tag{8.4}$$

$$= \frac{\partial H}{\partial t}. \tag{8.5}$$

Hence, in the absence of any explicit time dependence in the Hamiltonian, $\partial H/\partial t = 0$, as the system evolves along its phase-space path the energy of the system, $H(p(t), q(t))$, remains constant.

*Other constants of motion.* Just as the energy of an autonomous Hamiltonian system acts as a constant of motion, there may exist other constants of motion in a higher-dimensional system. A constant of motion, $F(q, p)$, is a function whose value remains fixed as the system evolves along its trajectory, that is, $F(q(t), p(t)) = f^0$, where $f^0$ is the initial value. The constant value of $F$ along a trajectory requires that

$$\frac{dF}{dt} = 0 = \frac{\partial F}{\partial p}\frac{\partial p}{\partial t} + \frac{\partial F}{\partial q}\frac{\partial q}{\partial t}, \tag{8.6}$$

$$= -\frac{\partial F}{\partial p}\frac{\partial H}{\partial q} + \frac{\partial F}{\partial q}\frac{\partial H}{\partial p}, \tag{8.7}$$

$$= \{F, H\}, \tag{8.8}$$

where $\{F, H\}$ is called the Poisson bracket. The vanishing of the Poisson bracket between $F$ and the Hamiltonian $H$ throughout the phase space makes $F$ a constant of motion. In its geometric interpretation, the vector field $(-\partial F/\partial q, \partial F/\partial p)$ is tangent to the energy surface $H(q, p) = E$. Conversely, the vector field $(-\partial H/\partial q, \partial H/\partial p)$ is tangent to the surface, $F(q, p) =$ constant.

Suppose for some system with $N$ degrees of freedom there are $N$ independent functions $\{F_i(p, q)|i = 1, \ldots, N\}$, one of which is the Hamiltonian for the system, and all of which have the property that their mutual Poisson brackets vanish everywhere:

$$\{F_i(p, q), F_j(p, q)\} = 0, \quad \text{for all } i, j, p, q.$$

Then, each such function $F_i(p, q)$ is a conserved quantity. (Independent means that their phase-space gradients $(\partial F_i/dp, \partial F_i/\partial q)$ are linearly independent vectors at every point in phase space. It is possible to prove that no more than $N$ independent functions can have their mutual Poisson brackets vanish everywhere.) Conservation of energy restricts the trajectory to the $(2N\text{-}1)$-dimensional energy shell in the

$2N$-dimensional phase space. Each additional conservation law $F_i(q(t), p(t)) = f_i^0$ restricts the trajectories to a lower-dimensional surface, so the full set of conservation laws restricts the trajectory to an $N$-dimensional surface in the $2N$-dimensional phase space. An important theorem in classical mechanics asserts that if the points on that surface form a closed, bounded and connected set, then that surface is topologically equivalent to an $N$-dimensional torus.

## 8.2 Integrable systems

Systems having such a full set of conservation laws are said to be 'integrable', and the trajectories are said to be 'regular', 'multiply-periodic', or 'quasi-periodic'. Physical systems are integrable if they are the systems for which solutions to the equations of motion can be expressed analytically. However, there are many systems for which the number of constants of motion is less than the degrees of freedom. Such systems are called non-integrable, and they may display complex or chaotic dynamics.

**Example 1:** Show that the ordered dynamics of two uncoupled harmonic oscillators follows torus dynamics.

A harmonic oscillator provides an example of a linear system, where we express the oscillator equation as

$$m\frac{d^2q}{dt^2} = -kq. \tag{8.9}$$

The corresponding first-order differential equations are

$$m\frac{dq}{dt} = p, \tag{8.10}$$

$$\frac{dp}{dt} = -kq. \tag{8.11}$$

The Hamiltonian for the harmonic oscillator is

$$H(p, q) = p^2/2\,m + kq^2/2,$$

which describes an ellipse in phase space $(p, q)$. It can be converted into a circle by the transformation

$$p' = (mk)^{-1/4}p, \quad q' = (mk)^{1/4}q,$$

which is

$$H = \omega(p'^2 + q'^2)/2. \tag{8.12}$$

Here, $\omega = (k/m)^{1/2}$. The phase point representing the instantaneous position and momentum of the oscillator moves at constant angular velocity, $\omega$, around this circle:

$$r = (2H/\omega)^{1/2}, \quad \theta = \arctan(p'/q') = \theta_0 + \omega t.$$

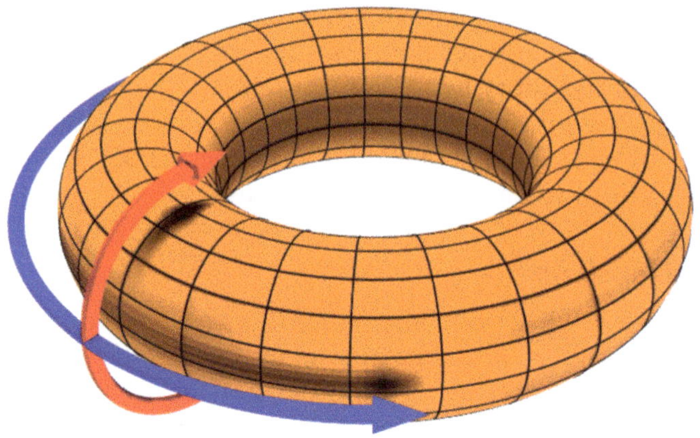

**Figure 8.1.** The motion in a system with two degrees of freedom and two independent conservation laws must lie on a two-dimensional surface in four-dimensional phase space. Typically, for bounded motion, that surface is topologically equivalent to a torus. A torus is said to be the topological product of two circles. Motion in the sense of the red curve is called poloidal, and that in the sense of the blue curve is called toroidal. Reproduced from [23]. © IOP Publishing Ltd. CC BY 3.0.

By a similar transformation, the Hamiltonian for a two-dimensional harmonic oscillator can be written in the form

$$H = H_1 + H_2 = \omega_1(p_1^2 + q_1^2)/2 + \omega_2(p_2^2 + q_2^2)/2.$$

Both $H_1$ and $H_2$ are conserved and

$$\{H_1, H_2\} = 0.$$

The phase-space point moves in a circle at frequency $\omega_1$ in the $(p_1, q_1)$ plane, and in a circle at frequency $\omega_2$ in the $(p_2, q_2)$ plane. Hence, the system has the same numbers of degrees of freedom as that of constants of motion, therefore this is an integrable system and the motion is bounded. Thus, the combined motion lies around a torus in four-dimensional phase space, as shown in figure 8.1. In case the ratio $\omega_1/\omega_2$ is an irrational number, then the trajectory 'fills' the torus by approaching arbitrarily close to every point on the torus. However, if that ratio is a rational number, then the trajectory is closed and periodic, forming Lissajous figures in the configuration space $(q_1, q_2)$.

**Example 2:** Show that a Van der Pol oscillator displays convergent and divergent behavior in its time evolution.

A Van der Pol oscillator provides an example of a *non-linear oscillatory system*, and is written as

$$\frac{d^2x}{dt^2} - \mu(1 - x^2)\frac{dx}{dt} + x = 0. \tag{8.13}$$

Here, the second term acts like a dissipation term. A Van der Pol oscillator can be expressed as a set of two first-order differential equations as

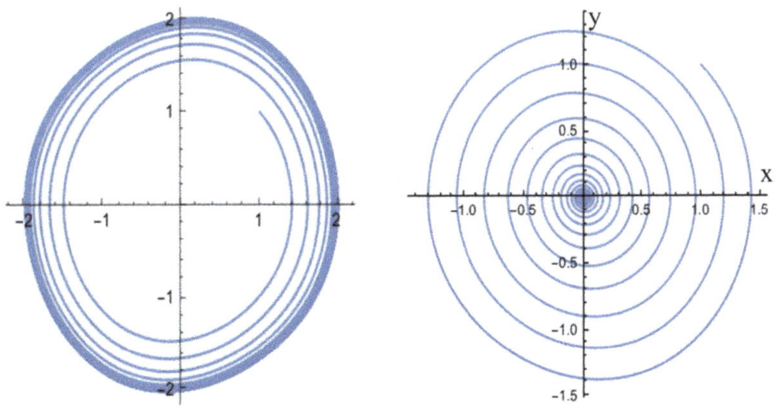


**Figure 8.2.** Evolution in a Van der Pol oscillator for (a) $\mu = 0.1$, and (b) $\mu = -0.1$.

$$\frac{dy}{dt} = \mu(1 - x^2)y - x,$$
$$\frac{dx}{dt} = y. \tag{8.14}$$

Hence, it is a two-dimensional system. It should be noted that for $\mu = 0$ the system behaves as a harmonic oscillator. In this case, we have regular evolution for all the initial conditions. However, if $\mu$ is a small positive number, that is, $\mu \ll 1$, the evolution is independent of the initial conditions, as all the initial phase points evolve to a *limit cycle*. The situation is different if $\mu$ is a negative small number: depending on the initial conditions, the solution may approach zero (i.e. be *convergent*), or escape to infinity (i.e. be *divergent*). The two situations are shown in figure 8.2.

*Action and angle variables.* Another important theorem in classical mechanics asserts that if a system is integrable and the motion lies on a torus, then there is a set of action and angle variables with similar properties: the actions are conserved and the angles increase linearly with time. Furthermore, the actions and angles are canonical variables, meaning that we may think of the angles as coordinates and the actions as their conjugate momenta. The Hamiltonian can be written as a function of actions and angles so the equations of motion can be written as

$$dJ_i/dt = -\partial H/\partial\theta_i = 0, \quad d\theta_i/dt = \partial H/\partial J_i = \omega_i(J_1 \ldots J_N). \tag{8.15}$$

Hence, the actions are constants of motion, and the angles increase linearly with time in an integrable system.

The general definition of the action variables is as follows. An $N$-dimensional torus has $N$ distinct fundamental loops that cannot be shrunk to zero. In figure 8.1, one loop would be toroidal and one would be poloidal. Each action variable is an integral around one of these fundamental loops:

$$J_k = \frac{1}{2\pi} \oint \sum_i p_i dq_i. \tag{8.16}$$

## 8.3 Nearly integrable systems

Action and angle variables are especially useful for classical perturbation theory. For example, if the system is an unperturbed harmonic oscillator, then we write the Hamiltonian simply as a function of the action variable. For nearly integrable dynamical systems Hamiltonian is a function of action and angle variables. Typically, in the full Hamiltonian, the angle-dependent terms will be small, so the action variables will change slowly.

For example, if the system is an unperturbed harmonic oscillator, then we write the Hamiltonian simply as a function of the action variable. For nearly integrable dynamical systems Hamiltonain is a function of action and angle variables.

More generally, we want to know what happens to the motion in integrable systems subjected to small perturbations. This question was studied rigorously by Andrey Kolmogorov, Vladimir Arnold, and Jürgen Moser, and their pioneering research led to what is known today as the KAM theorem. This theorem states, 'When a small perturbative (symmetry breaking) term is added to the Hamiltonian almost all of the phase space behaves as if the symmetries still exist. However, in regions where the perturbative or symmetry-breaking terms allow resonances to occur between otherwise uncoupled degrees of freedom the dynamics begins to change its character'. Thus, given a torus of an initially unperturbed system, perturbation distorts the torus, but in most cases does not destroy it: for weak perturbations, most tori survive, each as a continuous deformation of a corresponding unperturbed torus. This raises two questions: first, what does 'almost all' mean; and, second, what happens when tori are destroyed?

*Resonances.* The answers to these questions come from understanding resonant and off-resonant dynamics. In a Hamiltonian system with two coupled degrees of freedom, each torus can be labeled by the values of its action variables, and the frequencies are continuous functions of the actions, as given in equation (8.17). (A harmonic oscillator is exceptional in that the frequencies are fixed, independent of the actions.) A torus is resonant when its two frequencies satisfy the equality

$$r\omega_1 - s\omega_2 = 0. \tag{8.17}$$

Here, $r$ and $s$ are relatively prime integers and their ratio, $r/s$, is called the winding number or magic number. Kolmogorov, Arnold, and Moser proved that those tori with a winding number that is irrational, and not too close to a rational number with small $r$ and $s$, are distorted but preserved. Surprisingly, this means that for weak perturbations almost all tori are preserved.

In contrast, 'resonant tori' of unperturbed systems, that is, those for which $r/s$ is rational, are annihilated. Also, the tori that are close to those resonant tori are annihilated. George D Birkhoff proved that what survives is an alternating chain of stable and unstable periodic orbits. The stable orbits are surrounded by small 'islands of stability' in which the trajectories are regular, forming a new family of tori surrounding a central periodic orbit. The unstable orbits in this chain are surrounded by zones of chaotic behavior. We will show pictures of this behavior later.

**Example 3:** Show that a periodically forced Van der Pol oscillator satisfies the minimum criteria to display chaotic dynamics.

In the presence of a forcing term, the Van der Pol oscillator equation becomes

$$\frac{d^2x}{dt^2} - \mu(1 - x^2)\frac{dx}{dt} + x = \nu \sin \omega t, \tag{8.18}$$

where $\nu$ expresses the amplitude of the driving force, with $\omega$ being its frequency. Hence, the system comprises three dimensions. In addition, it has two degrees of freedom in the presence of two frequencies: one the natural frequency of the system, and the other the driving frequency $\omega$, competing with each other. Equation (8.20), used by Van der Pol to study the range of stability of heart dynamics in the presence of an external driving signal for an example pacemaker, can equivalently be written as

$$\begin{aligned} \frac{dp}{dt} &= \mu(1 - x^2)p - x + \nu \sin \omega t, \\ \frac{dx}{dt} &= p. \end{aligned} \tag{8.19}$$

The dependence of future evolution on initial conditions is a characteristic of non-linear systems, and is regarded as a signature of classical chaos. A forced Van der Pol oscillator system satisfies *minimum criteria* to display chaotic dynamics, in that it has *two degrees of freedom* and *a coupling* between them. In addition, energy is no more a conserved quantity, indicating one missing constant of motion. Hence, the system given in equation (8.21) displays both stable and chaotic dynamics.

## 8.4 Kicked-rotator model

We may express a particle in circular motion and subject to external forcing after equal intervals of time by the Hamiltonian

$$H = \frac{\mathcal{P}^2}{2I} + V(\theta)\sum_n \delta\left(\frac{t}{T} - n\right), \tag{8.20}$$

where $T$ expresses the time period. The Hamiltonian expresses a matter wave in a phase-modulated optical lattice or, in an equivalent system, a diatomic molecule moving in an intense electric field or simply a rotor subject to periodic kicks [4]. Here, $\mathcal{P}$ is the angular momentum of the particle rotating with a moment of inertia $I$, whereas $V(\theta) = \mu\varepsilon_0 \cos\theta$, which may take any value in between +1 and −1 depending on $\theta$.

Classical dynamics follows the Hamiltonian equations

$$\begin{aligned} \dot{\mathcal{P}} &= -\frac{\partial H}{\partial \theta} = +\mu\varepsilon_0 T \sin\theta \sum_n \delta(t - nT), \\ \dot{\theta} &= \frac{\partial H}{\partial \mathcal{P}} = \frac{\mathcal{P}}{I}. \end{aligned} \tag{8.21}$$

The change in action at time $t = mT$ can be calculated by considering an incremental time step, $\epsilon$, before and after the $m$th impact as

$$\triangle \mathcal{P} = \int_{m-\epsilon}^{m+\epsilon} \left(\frac{\partial L}{\partial t}\right) dt = \mu \varepsilon_0 T \sin\theta \sum_n \int_{m-\epsilon}^{m+\epsilon} \delta(t - nT) dt, \\ = \mu\varepsilon_0 T \sin\theta_m, \tag{8.22}$$

whereas the change in angle at the $m$th impact is calculated as

$$\triangle\theta = \int_{m-\epsilon}^{m+\epsilon} \frac{\partial\theta}{\partial t} dt = \frac{2L}{I}\epsilon = 0 \tag{8.23}$$

as $\epsilon \longrightarrow 0$. Hence, we conclude that at the $m$th impact the action and angle are modified such that

$$\mathcal{P}_m - \mathcal{P}'_m = \mu\varepsilon_0 \sin\theta_m T, \\ \theta_m - \theta'_m = 0. \tag{8.24}$$

Here, $\mathcal{P}_m$ and $\theta_m$ are the angular momentum and angle just before the $m$th kick, and $\mathcal{P}'_m$ and $\theta'_m$ are the angular momentum and angle just after the $m$th kick. After the $m$th impact and just before the $(m + 1)$th impact the rotator undergoes free motion, which means $\mathcal{P}_{m+1} - \mathcal{P}'_m = 0$ and $\theta_{m+1} - \theta_m = \frac{\mathcal{P}_{m+1}T}{I}$.

Hence, we can write the evolution of the rotator over one period through a two-dimensional map:

$$\mathcal{P}_{m+1} = \mathcal{P}_m - \mu\varepsilon_0 T \sin\theta_m, \\ \theta_{m+1} = \theta_m - \mathcal{P}_{m+1}\frac{T}{I}. \tag{8.25}$$

Using the scaled angular momentum $l = -\frac{T}{I}\mathcal{P}$, we may express this map as

$$l_{m+1} = l_m + K \sin\theta_m, \\ \theta_{m+1} = \theta_m + l_{m+1}, \tag{8.26}$$

where $K = \frac{\mu\varepsilon_0 T^2}{I}$ is called the chaos parameter. This map is known as a standard map or Chirikov–Taylor map. For non-zero $K$, equations (8.22) and (8.28) describe a kicked-rotor model with two degrees of freedom in the presence of coupling. For $K = 0$, the same map is known as a circle map and corresponds to integrable dynamics.

## 8.5 Poincaré surface of sections

In order to analyze a dynamical system, Henri Poincaré expressed the flow of a dynamical system with two degrees of freedom as a one degree-of-freedom system taking the other degree as a constant of motion. Therefore, the phase space is plotted by taking into account all those phase points where the other degree of freedom is constant. Therefore, the dynamical system in the reduced phase space appears as a stroboscopic map of the phase space. For example, a one degree-of-freedom system driven periodically with a frequency $\omega = 2\pi/T$ is described by a Poincaré map, given as

$$\{(q(t), p(t)) \ t = t_0 + mT, \ m = 1, 2, 3, \ldots\}. \tag{8.27}$$

Since the Hamiltonian is periodic over the time period $T$, that is, $H(t + T) = H(t)$, the Poincaré surface of section describes an equi-energy plane, where energy is conserved on the plane.

The Poincaré surface of section is a very useful tool to understand the integrability of a system and its detailed dynamical response. Using the surface of section, we can determine the local stability and transition from ordered to stochastic evolution, resonances, resonance overlap, and other interesting dynamical characteristics. As a consequence, the plotting of Poincaré surface of sections (or, for simplicity, Poincaré sections), is essential for numerical studies. Furthermore, it plays a central role in checking the consistency of analytical results obtained by using perturbative methods. In a $2N$-dimensional system, a Poincaré section has $2N - 2$ dimensionality. In a higher-dimensional system, more dynamical subtleties occur such as *Arnold diffusion*.

**Example 4:** The classical dynamics of a driven kicked-rotor model depends upon a single control parameter, $K$. Here, the natural question arises, can we observe a transition to chaos as a function of $K$ using Poincaré surface of sections?

For a small value of $K$, for example $K = 0.5$, we find regularity and periodicity. The phase-space structures are periodic in $l$ with a period of $2\pi$, as shown in figure 8.3. A monotonic increase of $l$ is impossible without crossing sealing curves (also named in the literature as invariant curves or KAM surfaces). The angular momentum and the kinetic energy of the rotor are bounded. Therefore, the sealing curves (or invariant curves) stretching across the phase space from $\theta = 0$ to $\theta = 2\pi$ are dynamically disconnected, as shown in figure 8.3. This is because each of them

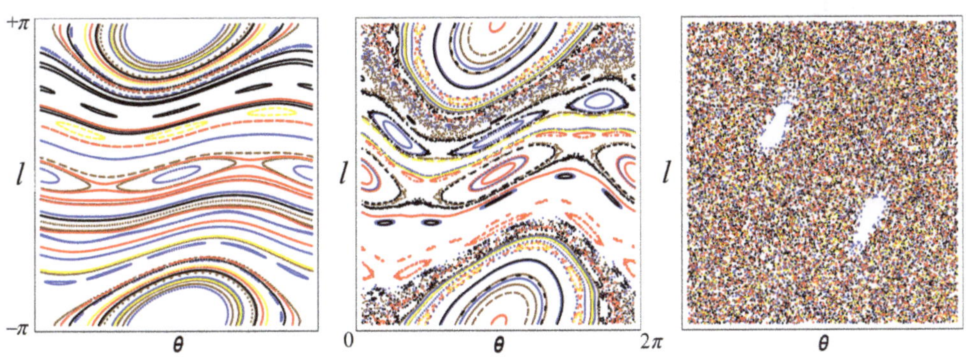


**Figure 8.3.** Poincaré surface of sections are plotted for various values of chaos parameter $K = 0.5$ (left), $K = 0.961$ (middle), and $K = 5$ (right). The angle variable $\theta$ ranges from from 0 to $2\pi$, whereas the momentum $l$ is from $-\pi$ to $+\pi$.

divides the phase space into segments, such that the dynamics in one segment is completely independent of the other.

With increasing value of $K$ more and more sealing curves are destroyed. At a critical value of $K$, calculated numerically by Greene as $K_c = 0.961$, the last sealing-invariant spanning curve or KAM surface is broken, and the angular momentum is no longer confined. The behavior of the phase space for $K = 5$ is shown in figure 8.3. It is found that for still larger values of $K$, the angular momentum is no longer confined and the rotor is able to absorb energy unbounded from the external field. This means energy grows diffusively following Arnold diffusion.

## 8.6 Arnold diffusion

We can understand Arnold diffusion by recalling that, in a system with two degrees of freedom, where $N = 2$, a weak modulation or coupling strength causes the diffusion to occur only between isolating KAM (invariant) surfaces or tori. As a result, the whole Poincaré surface of section is divided into regions bounded by the KAM surfaces. However, in a system with $2N$ dimensions and $N$ degrees of freedom, the KAM surfaces no longer separate the stochastic regions from each other, as we find in a two degrees-of-freedom system. Therefore, in an $N$ degrees-of-freedom system there occurs a global diffusion for a very weak perturbation or coupling strength, which is known as Arnold diffusion. Due to this Arnold diffusion, the whole phase space is connected for a system with more than two degrees of freedom, that is, $N > 2$.

## 8.7 Lyapunov exponent

A sensitive dependence on initial conditions is a hallmark of complex dynamics. The complex or chaotic dynamics can be *measured* quantitatively with the help of a Lyapunov exponent, which is a measure of distance between two nearby trajectories over a long time. If in the long-time dynamics the distance does not change appreciably or remains constant, the exponent approaches zero and the dynamical system displays ordered dynamics. However, if the distance effectively grows as a function of time, the exponent becomes positive, which defines complex or chaotic dynamics in the system. A negative exponent corresponds to a dissipative system. Mathematically, we may express the situation as

$$d(t) = d(0)\mathrm{e}^{\Lambda(t-t_0)},$$

which reveals the maximum Lyapunov exponent as

$$\Lambda = \lim_{t\to\infty} \frac{1}{t - t_0} \ln\left(\frac{d(t)}{d(0)}\right), \tag{8.28}$$

where $d(0)$ is the distance between two neighboring trajectories at initial time $t_0$, and $d(t)$ is the distance after an evolution time $t$.

The numerical calculation of the Lyapunov exponent for a chaotic system requires renormalization of the distance as two nearby trajectories diverge

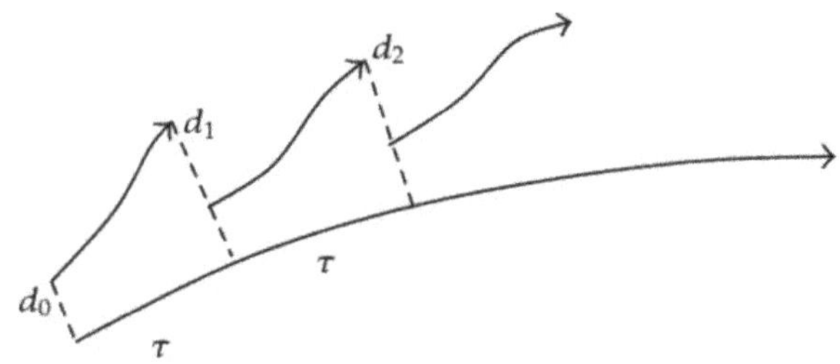


**Figure 8.4.** Schematic diagram showing the renormalization process of successive threshold crossings and rescaling.

exponentially. We may consider a test trajectory and a reference trajectory so that they originate with a very small distance, $d_0$, in phase space. Keeping in mind an exponential growth of the distance in case of chaotic trajectory, a rescaling of the distance between the test trajectory and reference trajectory is performed as it goes beyond a certain fixed value of distance $\mathcal{D}$. It is important to remember that $\mathcal{D}$ is small enough so that linearized equations of motion are a comfortable description of the system. At a time $t_1$ when the test trajectory goes beyond the distance $\mathcal{D}$ from the reference trajectory, that is, $d(t_1) \geqslant \mathcal{D}$, we calculate the exponent as

$$\Lambda = \frac{1}{t_1 - t_0} \ln \frac{d(t_1)}{d(0)} = \frac{1}{t_1 - t_0} \ln a_1, \tag{8.29}$$

where $a_1$ is the rescaling parameter:

$$a_1 = \frac{d(t_1)}{d(0)} = \frac{d_1}{d_0}. \tag{8.30}$$

Following the same procedure at successive threshold crossings and rescaling (figure 8.4), we obtain

$$\Lambda = \frac{1}{t_2 - t_0} \ln \frac{d_2 \cdot a_1}{d_0} = \frac{1}{t_2 - t_0} \ln (a_1 \cdot a_2), \tag{8.31}$$

$$\Lambda = \frac{1}{t_3 - t_0} \ln \frac{d_3 \cdot a_1 \cdot a_2}{d_0} = \frac{1}{t_3 - t_0} \ln (a_1 \cdot a_2 \cdot a_3), \tag{8.32}$$

and so on. Hence, if we repeat the process many times, the implication is that, in the long-time limit, we obtain

$$\Lambda = \frac{1}{t_n - t_0} \ln (a_1 \cdot a_2 \cdot a_3 \cdots a_n) = \frac{1}{t_n - t_0} \sum_{j=1}^{n} \ln a_n. \tag{8.33}$$

Evidently, the time steps, $t_j$, have no periodicity. Here,

$$a_j = \frac{d(t_j)}{d(0)} = \frac{d_j}{d_0}. \tag{8.34}$$

A trajectory with a zero Lyapunov exponent is regarded as a stable trajectory, whereas one with a non-zero Lyapunov exponent is an unstable trajectory.

## 8.8 Secular theory for non-linear resonances

We develop a general approach to understand a dynamical system that fulfils the minimum criteria to display complex dynamics. For this purpose, we write the corresponding Hamiltonian in action-angle coordinates, such that

$$H = H_0(J_1, J_2) + \varepsilon H_1(J_1, J_2, \theta_1, \theta_2), \tag{8.35}$$

where $H_0$ describes the interaction-free Hamiltonian, and $H_1$ expresses the interaction Hamiltonian. The symbols $J_1$, $J_2$ express the action coordinates, $\theta_1$, $\theta_2$ are the corresponding angle coordinates, and $\varepsilon$ describes the coupling strength between them. Keeping in view the angular periodicity, the interaction Hamiltonian is expressed as a power series, such that

$$H_1(J, \theta) = \sum_{l,m} H_{l,m} \mathrm{e}^{-i(l\theta_1 + m\theta_2)}. \tag{8.36}$$

A two degree-of freedom system possesses two frequencies, $\omega_1$ and $\omega_2$, defined as

$$\omega_1 = \dot{\theta}_1 = \frac{\partial H}{\partial J_1}, \quad \text{and} \quad \omega_2 = \dot{\theta}_2 = \frac{\partial H}{\partial J_2}. \tag{8.37}$$

The dynamical system displays a resonance whenever the two frequencies satisfy the resonance criterion, that is,

$$r\omega_1 - s\omega_2 = 0 \implies \frac{\omega_2}{\omega_1} = \frac{r}{s}, \tag{8.38}$$

where $r$ and $s$ are relatively prime integers.

We introduce a generating function $F = (r\theta_1 - s\theta_2)I_1 + \theta_2 I_2$, which provides a canonical transformation from coordinates $(J, \theta)$ to $(I, \phi)$, that is,

$$J_1 = \frac{\partial F}{\partial \theta_1} = rI_1, \tag{8.39}$$

$$J_2 = \frac{\partial F}{\partial \theta_2} = I_2 - sI_1, \tag{8.40}$$

$$\phi_1 = \frac{\partial F}{\partial I_1} = r\theta_1 - s\theta_2, \tag{8.41}$$

$$\phi_2 = \frac{\partial F}{\partial I_2} = \theta_2. \tag{8.42}$$

With the choice of the generating function, we move to a rotating frame, where

$$\dot{\phi}_1 = r\dot{\theta}_1 - s\dot{\theta}_2 = r\omega_1 - s\omega_2$$

determines detuning from the resonance, and $\dot{\phi}_2 = \dot{\theta}_2 = \omega_2$ is the higher frequency. With the help of the canonical transformation, obtained from the generating function, we express the Hamiltonian $H$ in equation (8.38), as

$$H(I, \phi) = H_0(I) + \varepsilon \sum_{l,m} H_{l,m}(I) \mathrm{e}^{-\frac{i}{r}(l\phi_1 + (mr+sl)\phi_2)}. \tag{8.43}$$

We consider a special case as we set

$$mr + sl = 0 \implies \frac{l}{m} = -\frac{r}{s},$$

which is possible for $\{l, m\} = \{-r, s\}, \{r, -s\}$, or $\{0, 0\}$. The choice of $l$ and $m$ eliminates $\phi_2$ dependence in the Hamiltonian, suppresses one degree of freedom, and leads us to the effective Hamiltonian in the $(I_1, \phi_1)$ space defining a resonance. This step can also be understood as a rotating-wave approximation, where we average over the rapidly changing angle $\phi_2$, and obtain a cycle-averaged Hamiltonian. It is valid only when $\dot{\phi}_2 = \omega_2 \gg r\omega_1 - s\omega_2 = \dot{\phi}_1$.

The disappearance of $\phi_2$ makes the corresponding action, $I_2$, a constant of motion, that is, $\{I_2, H\} = 0 = \frac{dI_2}{dt}$. Hence, the dynamical system effectively evolves in a plane, $(I_1, \phi_1)$, and is integrable. We write the cycle-averaged Hamiltonian as

$$H(I_1, \phi_1) = H_0(I_1) + \varepsilon H_{00} + 2\varepsilon H_{-r,s} \cos \phi_1, \tag{8.44}$$

where we have considered the symmetry due to resonance, that is, $H_{r,-s} = H_{-r,s}$.

In general, the fixed points present in the dynamical system in the plane $(I_1, \phi_1)$ are obtained from the relations, expressed as

$$\dot{\phi}_1 = \frac{\partial H}{\partial I_1} = 0 \text{ and } \dot{I}_1 = -\frac{\partial H_1}{\partial \phi_1} = 0, \tag{8.45}$$

which implies that the rates of change of angle and action are, respectively, zero at the fixed points. For simplicity, hereafter we write $(I_1, \phi_1)$ as $(I, \phi)$.

In order to study the dynamics in the region of a resonance, we expand energy up to second order around the fixed point $(I_0, \phi_0)$, that is,

$$H_0 \cong H_0(I_0) + C \triangle I + \frac{1}{2} G \triangle I^2. \tag{8.46}$$

Here, $C = \left.\frac{\partial H}{\partial I}\right|_{I=I_0}$, $G = \left.\frac{\partial^2 H}{\partial I^2}\right|_{I=I_0}$, and $\triangle I = I - I_0$. The first term in equation (8.48) is constant, whereas the second term is zero from the definition of a fixed point. Hence, on substituting equation (8.48) in equation (8.46), we find the effective Hamiltonian governing the dynamics around resonance as

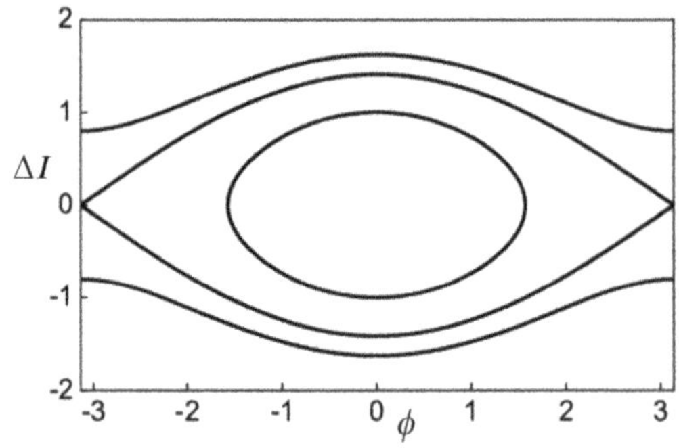


**Figure 8.5.** Phase-space of a resonance. Within the resonance the dynamics is oscillatory, whereas it changes to rotational dynamics outside.

$$H = \frac{1}{2}G\,\triangle I^2 - V\cos\phi, \tag{8.47}$$

where $V = -2\varepsilon H_{-r,s}(I_0)$. This explains that the evolution around resonance follows *rigid pendulum dynamics*, as shown in figure 8.5.

In order to calculate the *maximum excursion*, we consider that $\triangle I^{\max}$ is the maximum excursion given by half of the separatrix width (at $\phi = 0$) as

$$\frac{1}{2}G\,(\triangle I^{\max})^2 - V = 0, \tag{8.48}$$

which implies that

$$\triangle I^{\max} = \sqrt{\frac{2V}{G}}. \tag{8.49}$$

**Example 5:** Find the fixed points for the rigid pendulum given in equation (8.49). Discuss the dynamics in the neighborhood of these fixed points.

By applying equation (8.47), we find the fixed points of the reduced Hamiltonian, given in equation (8.49), as (0, 0) and (0, $\pm\pi$). In the neighborhood of the *stable fixed point* (0, 0), the phase space is elliptical, with the ratio of lengths of the semi-axis of the ellipse given by

$$H = \frac{1}{2}G\triangle I^2 + \frac{1}{2}V\triangle\phi^2. \tag{8.50}$$

Here, we have ignored the constant terms. It is interesting to note that in the neighborhood of the *unstable fixed points* (0, $\pm\pi$) the phase space is hyperbolic, given as

$$H = \frac{1}{2}G\triangle I^2 - \frac{1}{2}V\triangle\phi^2. \tag{8.51}$$

## 8.9 Quantum scars

As was explained earlier, stable periodic trajectories are surrounded in phase space by trajectories that form continuous families of tori, which can be labeled by action variables. A discrete set of those tori, the ones with appropriately quantized action variables, correspond to quantum states. Chaotic systems may have no stable periodic orbits, and no tori. Therefore, this type of connection between classical trajectories and quantum states does not exist.

Chaotic systems have unstable periodic orbits, and they manifest themselves in quantum mechanics in several ways. They produce oscillations in the density of states as a function of energy, in the absorption spectrum of atoms or molecules in applied fields, and they produce 'scars' in wave functions, that is, the wave functions for some eigenstates are found to be concentrated near the classical unstable periodic orbits. Scarred eigenfunctions display anomalous enhancement in regions of the phase space that are traversed by one of the periodic orbits in the classical limit when $\hbar \to 0$, as shown in figure 8.6.

The first theory of scars was proposed by Eric Heller [5, 6], and was later modified [7–13]. Heller's theory studies the overlap integral

$$C(t) = \langle \psi(0, x) | \psi(t, x) \rangle \tag{8.52}$$

for a propagating wave packet that, at an initial time, has a Gaussian shape, and initial conditions in phase space $(x_0, p_0)$ that correspond to an unstable periodic orbit. Expanding the wave function $\psi(0, x)$ in energy eigenstates,

$$\psi(0, x) = \sum_n c_n \varphi_n(x), \tag{8.53}$$

one sees that the Fourier transform,$S(E)$, of the overlap, $C(t)$, is the spectral density weighted by the probabilities$|c_n|^2$, that is,

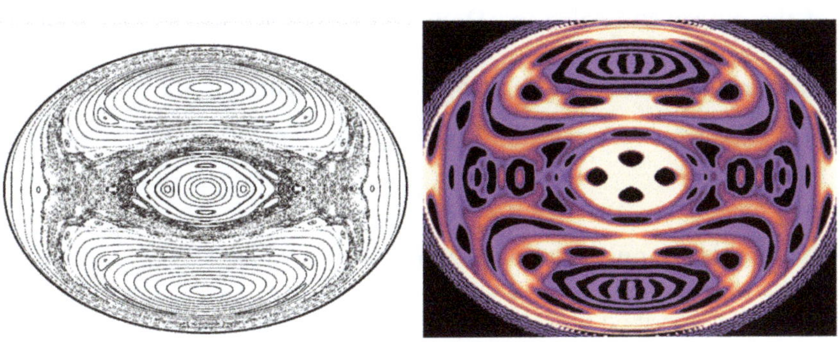

**Figure 8.6.** The motion of an electron of hydrogen in a magnetic field in a regime between regularity and chaos. The classical picture (left) shows structures on infinitely fine scales, including many small islands of stability resulting from bifurcations. In the quantum case (right), only the larger structures appear, since the Heisenberg uncertainty principle rules out details on scales smaller than Planck's constant [12]. However, we note features due to quantum interferences. Reproduced from [12]. 

$$S(E) = \sum_n |c_n|^2 \delta(E - E_n). \tag{8.54}$$

Now, if the period, $\tau$, of the classical periodic orbit and the largest positive Lyapunov exponent, $\lambda$, are such that $e^{-\tau\lambda/2}$ is not very small, the overlap $C(t)$ will display peaks at times $n\tau$. As the wave packet spreads, the amplitude of the peaks decreases after each orbit traversal at the rate $e^{-\tau\lambda/2}$. The Fourier transform of $C(t)$ will therefore have peaks of width $\lambda$ with spacing $\omega = 2\pi/\tau$. Referring to equation (8.54), one concludes that only the eigenstates that lie under the peaks contribute to the expansion of the wave packet. Since the wave packet has an enhanced intensity along the region of the periodic orbit, this is expected to carry over to the contributing energy eigenstates. The stronger the overlap resurgences are, the stronger the effect is expected to be. Therefore, the intensity of the effect varies as $1/\tau\lambda$ [14]. Recent works on quantum simulators [15–17] have shown that quantum many-body scars can occur in strongly interacting quantum systems. A quantum simulator can be developed by using a one-dimensional Rydberg atom platform in the regime of a Rydberg blockade [18–20], where nearest-neighbor excitations of the atoms are energetically prohibited. Periodically driven optical lattices present a great opportunity for exploring the quantum characteristics [21, 22].

## Exercises

1. A particle moves along a circle of constant radius. How many degrees of freedom are associated with it?
2. A diatomic molecule is trapped and bounded to move on a circular trajectory. How many degrees of freedom are associated with its motion?
3. A single-mode electromagnetic field exists between two end-mirrors, provided that the length of the cavity is an integer multiple of half of the wavelength of the field. One of the two mirrors is movable and follows harmonic motion due to radiation pressure force. Express the Hamiltonian of the mirror and solve the Hamilton equations to explain its evolution in phase space.
4. Consider an oscillator, following a linear potential in the positive half plane, defined as

$$H = \frac{p^2}{2m} + kz, \qquad z > 0.$$

Express the Hamiltonian in action-angle coordinates.

5. Show that the energy remains conserved for a one-dimensional harmonic oscillator described by the following set of Hamilton's equations,

$$\frac{\partial x_1}{\partial t} = x_2,$$
$$\frac{\partial x_2}{\partial t} = -\,\omega^2 x_1,$$

where $\omega$ is the constant frequency. The oscillator has initial values of $x_1 = x_0$ and $x_2 = v_0$.

6. Prove that energy is not a conserved quantity in an explicitly time-dependent system.
7. Using the definition of the Lyapunov exponent, show that at time intervals $t_2$ and $t_3$ we, respectively, obtain the expressions given in equations (8.33) and (8.34).

## References


[1] Rabitz H 2000 Whither the future of controlling quantum phenomena? *Science* **288** 824–8

[2] Mikhailov E E, Sautenkov V A, Rostovtsev Y V, Zhang A, Zubairy M S, Scully M O and Welch G R 2006 Spectral narrowing via quantum coherence *Phys. Rev.* A **74** 013807

[3] Tanabe S, Watanabe S, Saif F and Matsuzawa M 2002 Survival probability of a truncated radial oscillator subject to periodic kicks *Phys. Rev.* A **65** 033420

[4] Miyagi H, Morishita T and Watanabe S 2012 Electron scattering and photoionization of one-electron diatomic molecules *Phys. Rev.* A **85** 022708

[5] Heller E J 1984 Bound-state eigenfunctions of classically chaotic Hamiltonian systems: scars of periodic orbits *Phys. Rev. Lett.* **53** 1515

[6] Heller E J 1991 Wavepacket dynamics and quantum chaology *Chaos and Quantum Physics* 52 *(Les Houches, France)*M J Giannoni, A Voros and J Zinn-Justin pp 547–663

[7] Wilkinson P B, Fromhold T M, Eaves L, Sheard F W, Miura N and Takamasu T 1996 Observation of 'scarred' wavefunctions in a quantum well with chaotic electron dynamics *Nature* **380** 608

[8] Bogomolny E B 1988 Smoothed wave functions of chaotic quantum systems *Phys. D: Nonlinear Phenom.* **31** 169

[9] Berry M V 1989 Quantum scars of classical closed orbits in phase space *Proc. R. Soc. Lond. A: Math. Phys. Sci.* **423** 219–31

[10] Feingold M, Littlejohn R G, Solina B, Pehling J S and Piro O 1990 Scars in billiards: the phase space approach *Phys. Lett.* A **146** 199–203

[11] Feingold M 1994 Phase space scars and quantum billiards *Z. Phys. B: Condens. Matter* **95** 121–40

[12] Monteiro T S 1992 Quantum phase-space behaviour for the $H_2$ molecule in a magnetic field *J. Phys. B: At. Mol. Opt. Phys.* **25** L621

[13] Bies W E, Kaplan L and Heller E J 2001 Scarring effects on tunneling in chaotic double-well potentials *Phys. Rev.* E **64** 016204

[14] Mendes R V 1998 Saddle scars: existence and applications *Phys. Lett.* A **239** 223–7

[15] Bernien H *et al* 2017 Probing many-body dynamics on a 51-atom quantum simulator *Nature* **551** 579–84

[16] Turner C J, Michailidis A A, Abanin D A, Serbyn M and Papić Z 2018 Weak ergodicity breaking from quantum many-body scars *Nat. Phys.* **14** 745–9

[17] Ho W W, Choi S, Pichler H and Lukin M D 2019 Periodic orbits, entanglement, and quantum many-body scars in constrained models: matrix product state approach *Phys. Rev. Lett.* **122** 040603

[18] Schaus P, Cheneau M, Endres M, Fukuhara T, Hild S, Omran A, Pohl T, Gross C, Kuhr S and Bloch I 2012 Observation of spatially ordered structures in a two-dimensional Rydberg gas *Nature* **491** 87–91

[19] Hudomal A, Vasić I, Regnault N and Papić Z 2020 Quantum scars of bosons with correlated hopping *Commun. Phys.* **3** 99
[20] Labuhn H, Barredo D, Ravets S, de Léséleuc S, Macrí T, Lahaye T and Browaeys A 2016 Tunable two-dimensional arrays of single Rydberg atoms for realizing quantum Ising models *Nature* **534** 667–70
[21] Görg F, Sandholzer K, Minguzzi J, Desbuquois R, Messer M and Esslinger T 2019 Realization of density-dependent Peierls phases to engineer quantized gauge fields coupled to ultracold matter *Nat. Phys.* **15** 1161–7
[22] Schweizer C, Grusdt F, Berngruber M, Barbiero L, Demler E, Goldman N, Bloch I and Aidelsburger M 2019 Floquet approach to Z2 lattice gauge theories with ultracold atoms in optical lattices *Nat. Phys.* **15** 1168–73
[23] Das P, Pramanik S and Ghosh S 2016 Particle on a torus knot: constrained dynamics and semi-classical quantization in a magnetic field *Ann. Phys.* **374** 67–83

**IOP** Publishing

Optical Forces on Atoms

**Farhan Saif and Shinichi Watanabe**

# Chapter 9

# Time-periodic force on atoms

*'When you can measure what you are speaking about, and express it in numbers, you know something about it.'*

—Lord Kelvin

In general, higher-dimensional systems display complex or chaotic dynamics. The study of the quantum characteristics of a classically chaotic system received considerable attention after the work of James E Bayfield and Peter M Koch on microwave ionization of hydrogen [1, 2]. In such a system, the suppression of ionization due to the microwave field was attributed to dynamical localization [3–5]. Later, the phenomenon was observed experimentally [6–10]. The existence of dynamical localization in a system is regarded as a signature of quantum chaos [11].

The presence of time-periodic external modulation makes the potential energy of the system vary with time. As a result, it exerts an explicitly time-dependent force on the atom. We write the general Hamiltonian of the driven systems as

$$H = H_0(r, p) - V(t)\, u(r + \varphi(t)). \tag{9.1}$$

Here, $H_0$ controls the atomic dynamics in the absence of a driving force. Furthermore, the amplitude, $V(t)$, and/or the phase, $\varphi(t)$, can be periodic in time. In addition, the function $u$ expresses the dependence of the potential on spatial coordinates.

Classically, the modulated driven system expressed in equation (9.1) leads to a set of first-order differential equations:

$$\begin{aligned} \dot{r} &= \partial_p H_0, \\ \dot{p} &= -\partial_r H_0 - V \partial_r u, \\ \dot{\theta} &= \omega, \end{aligned} \tag{9.2}$$

 

where $\partial_r$ and $\partial_p$ denote partial differentiation with respect to position $r$ and momentum $p$, respectively. Moreover, $\theta = \omega t$ is the scaled time, namely a cyclic variable conjugate to the frequency of the modulation, $\omega$. This reveals that the presence of explicit time dependence leads to at least three-dimensional phase space in a dynamical system, $(r, p, \theta)$. Hence, in the presence of coupling, these systems fulfil the minimum criteria to display complex dynamics.

The dynamics of a one-dimensional oscillator experiencing external force can be written as

$$m\ddot{x} + f(x) = g(x, t).$$

The oscillator equation is obtained from the set of first-order coupled differential equations (9.2), following the discussion developed in chapter (8). Here, $f(x) = \partial_x H_0$ and the function $g(x, t) = -V(t)\partial_x u(x, t)$. In a periodically modulated optical lattice, we may have both symmetries, in the coordinate space and in the time domain, therefore

$$g(x - x') = g(-x - x'), \quad g(t - t') = g(-t - t'), \tag{9.3}$$

This may be related to the periodicity of the function, written as

$$g(x, t) = g(x, t + T) = g(x + d, t). \tag{9.4}$$

Here, $T$ and $d$ define the periodicity in time and space, respectively. The presence of double periodicity in a modulated optical lattice makes them special in the study of dynamical systems. Optical lattices in the presence of modulating force are categorized as phase-modulated optical lattices and amplitude-modulated optical lattices, respectively. Ions in a Paul trap in the presence of an optical lattice and ultra-cold atoms in a time-periodic quadropole potential are examples of the kind where we have no spatial periodicity [12, 13].

## 9.1 Floquet analysis

The presence of an explicit time-periodic dependence modifies the system dynamics and introduces one additional degree of freedom. The corresponding mathematical description of the experimental system therefore shows periodicity in time, that is,

$$H(t) = H(t + T),$$

where

$$T = 2\pi/\omega.$$

The time dependence in the Hamiltonian makes energy an unconserved quantity, and thus no more a constant of motion. However, the solution to the time-dependent Schrödinger wave equation [14, 15], that is,

$$i\hbar\frac{\partial}{\partial t}|\psi(t)\rangle = \hat{H}(t)|\psi(t)\rangle, \tag{9.5}$$

can be expressed in terms of the evolution operator $\hat{U}(t, t_i)$, as

$$|\psi(t)\rangle = \hat{U}(t, t_i)|\psi(t_i)\rangle. \tag{9.6}$$

The operator $\hat{U}(t, t_i)$ takes the system from an initial time $t = t_i$ to a later time $t$, and ensures quantum reversibility. As is well known, say from its integral representation, $\hat{U}(t, t_i)$ is a function of the time difference $t - t_i$, that is, $\hat{U}(t, t_i) = \hat{U}(t - t_i, 0)$. We will use this fact later.

Now, in parallel to the Bloch theorem, which appeared in the context of a spatially periodic system, the theorem applied to a temporally periodic system is referred to as the *Floquet theorem*. This states that there exist solutions of the form

$$|\psi(t)\rangle = e^{-i\epsilon_n t/\hbar}|u_n(t)\rangle, \tag{9.7}$$

such that $u_n(t)$ is periodic, namely $u_n(t + T) = u_n(t)$, and $\epsilon_n$ is an associated constant called the quasi-energy. We may rewrite equation (9.5) as

$$\left\{ i\hbar\frac{\partial}{\partial t} - \hat{H}(t) \right\} |\psi_n(t)\rangle = 0.$$

On substituting the Floquet form given in equation (9.7), we obtain the eigenvalue equation as[1]

$$\hat{H}_F|u_n(t)\rangle = \epsilon_n|u_n(t)\rangle,$$

where

$$\hat{H}_F = i\hbar\frac{\partial}{\partial t} - \hat{H}(t)$$

is stroboscopically reproduced after every period of time $T$.

Thus, it holds that $U(t + T, t)$ acts on $|\psi_n(t)\rangle$ as

$$\begin{aligned} \hat{U}(t + T, t)|\psi_n(t)\rangle &= |\psi_n(t + T)\rangle, \\ &= e^{-i\epsilon_n(t+T)/\hbar}|u_n(t + T)\rangle, \\ &= e^{-i\epsilon_n T/\hbar}e^{-i\epsilon_n t/\hbar}|u_n(t)\rangle, \\ &= e^{-i\epsilon_n T/\hbar}|\psi_n(t)\rangle. \end{aligned} \tag{9.8}$$

Since $\hat{U}(t + T, t) = \hat{U}(T, 0)$ from its translational property,

$$\hat{U}(T, 0)|\psi_n(t)\rangle = e^{-i\epsilon_n T/\hbar}|\psi_n(t)\rangle,$$

which is an alternative expression defining the Floquet states[2]. We may express the dynamics in time by incurring another operator that describes micro-motion over

[1] This may be solved by diagonalizing the matrix of $\hat{H}_F$ represented in a time-wise periodic basis.

[2] This gives another way of obtaining the quasi-energy eigenvalues, namely, associating the eigenvalues $\lambda_n$ of $\hat{U}(T, 0)$ with $e^{-i\epsilon_n T/\hbar}$. The time-translation operator $\hat{U}(T, 0)$ may be evaluated by various numerical schemes.

one period, say a *micro-motion operator*, $\hat{U}_{\mathrm{F}}(t)$, which is unitary, $\hat{U}_{\mathrm{F}}^{\dagger}(t)\hat{U}_{\mathrm{F}}(t) = 1$, and periodic, that is, $\hat{U}_{\mathrm{F}}(t+T) = \hat{U}_{\mathrm{F}}(t)$. Explicitly, we define it by

$$\hat{U}_{\mathrm{F}}(t) = \hat{U}(t, 0)\, \mathrm{e}^{iH_{\mathrm{F}}(t)t/\hbar}.$$

Indeed,

$$\begin{aligned}\hat{U}_{\mathrm{F}}(t+T) &= \hat{U}(t+T, 0)\mathrm{e}^{i\{H_{\mathrm{F}}(t+T)/\hbar\}(t+T)},\\ &= \hat{U}(t+T, T)[\hat{U}(T, 0)\mathrm{e}^{i[H_{\mathrm{F}}(t)/\hbar]T}]\mathrm{e}^{i\{\hat{H}_{\mathrm{F}}(t)/\hbar\}t},\\ &= \hat{U}(t, 0)\mathrm{e}^{i\{H_{\mathrm{F}}(t)/\hbar\}t} = \hat{U}_{\mathrm{F}}(t),\end{aligned}$$

since

$$\hat{U}(T, 0)\mathrm{e}^{i\{H_{\mathrm{F}}(t)/\hbar\}T} = 1.$$

It also readily follows from the defining equation of $\hat{U}_{\mathrm{F}}(t)$ and that of $\hat{U}(t, 0)$ that

$$H_{\mathrm{F}} = i\hbar\hat{U}_{\mathrm{F}}(t)\frac{\partial\hat{U}_{\mathrm{F}}^{\dagger}(t)}{\partial t} - \hat{U}_{\mathrm{F}}(t)H(t)\,\hat{U}_{\mathrm{F}}^{\dagger}(t).$$

In terms of the operators $\hat{H}_{\mathrm{F}}$ and $\hat{U}_{\mathrm{F}}$, the time evolution operator takes the form

$$\hat{U}(t, t_{\mathrm{i}}) = \hat{U}_{\mathrm{F}}^{\dagger}(t)\exp\left\{i\frac{(t-t_{\mathrm{i}})\hat{H}_{\mathrm{F}}(t)}{\hbar}\right\}\hat{U}_{\mathrm{F}}(t_{\mathrm{i}}). \tag{9.9}$$

Thus, the discussion illustrates that the time dynamics evolves as an interplay of two operators. The first is the micro-motion operator, $\hat{U}_{\mathrm{F}}$, which denotes the evolution within one period, or the micro-motion, and the time-periodic component of the dynamics. The second component is $H_{\mathrm{F}}$, which describes the linear phase evolution, and determines the time evolution in a similar way as the time-independent Hamiltonian.

*Characteristics.* (1) The eigenstate $|u_n(0)\rangle$ gives rise to a generalized static state $|\psi_n\rangle$ of the time-dependent Schrödinger equation, called the Floquet state $|\psi_n\rangle$.

(2) The time-periodic states related to the micro-operator $\hat{U}_{\mathrm{F}}$ are defined as $|u_n(t)\rangle = \hat{U}_{\mathrm{F}}(t)|u_n(0)\rangle$, such that $|u_n(t+T)\rangle = |u_n(t)\rangle$.

(3) The Floquet states are the eigenstates of the time evolution operator $\hat{U}(t, t_{\mathrm{i}})$ over a period

$$|\psi_n(t+T)\rangle = \hat{U}(t+T, t)|\psi_n(t)\rangle = \mathrm{e}^{-i\epsilon_n T/\hbar}|\psi_n(t)\rangle.$$

(4) The Floquet states are ortho-normal and span a complete Hilbert space. For this reason, we can express any wave function on their basis. Therefore, we write

$$
\begin{aligned}
\psi(t) &= U(t,\,0)|\psi(0)\rangle,\\
&= U(t,\,0)\sum_n a_n|\psi_n(0)\rangle,\\
&= U(t,\,0)\sum_n a_n|u_n(0)\rangle,\\
&= \sum_n a_n U_{\mathrm{F}}^{\dagger}(t)\mathrm{e}^{-iH_{\mathrm{F}}t/\hbar}|u_n(0)\rangle,\\
&= \sum_n a_n \mathrm{e}^{-i\epsilon_n t/\hbar}U_{\mathrm{F}}^{\dagger}(t)|u_n(0)\rangle,\\
&= \sum_n a_n \mathrm{e}^{-i\epsilon_n t/\hbar}|u_n(t)\rangle,\\
&= \sum_n a_n|\psi_n(t)\rangle,
\end{aligned}
$$

where the periodic part of the Floquet state function evolves as

$$|u_n(t)\rangle = \hat{U}_{\mathrm{F}}(t)|u_n(0)\rangle.$$

## 9.2 Floquet–Bloch solution

Following the discussion in the previous section, $\hat{H}_{\mathrm{F}}$ has a complete set of basis $\{|u_\alpha(0)\rangle\}$, such that

$$\hat{H}_{\mathrm{F}}|u_\alpha(0)\rangle = \epsilon_\alpha|u_\alpha(0)\rangle. \tag{9.10}$$

Here, we have changed the suffix of the Floquet states from $n$ to $\alpha$ for a reason we will explain shortly.

The quasi-energy periodic in time operator, $H_{\mathrm{F}}$, has the dimension of energy, and defines the Floquet operator with quasi-eigenfunction $|u_\alpha(0)\rangle$ and quasi-eigenenergy $\epsilon_\alpha$. In the eigenvalue equation (9.10), the eigenvalue, $\epsilon_\alpha$, acts as time-independent energy and is therefore known as quasi-energy. Moreover, the function $u_\alpha(t)$ at an arbitrary time $t$ is related to $u_\alpha(0)$ by

$$\hat{U}_{\mathrm{F}}^{\dagger}|u_\alpha(t)\rangle = |u_\alpha(0)\rangle. \tag{9.11}$$

It is of central importance to note that the solution of the eigenvalue equation falls into *equivalence classes*. This means that for any fixed integer value $n$, and similarly for any $m$ value, the expressions of eigenstate and eigenenergy can be expressed as

$$|u_\alpha(t)\rangle = |u_n(t)\rangle\mathrm{e}^{im\omega t}, \tag{9.12}$$

$$\epsilon_\alpha = \epsilon_n + m\hbar\omega. \tag{9.13}$$

Therefore, we use $\alpha$ as a double index, that is, $\alpha = (n,\, m)$. Hence, the energy axis is stratified into $n$ Brillouin zones, where each Brillouin zone corresponds to one Floquet state specified by $n$.

**Example 1:** In the presence of a negligible acting external force, explain the Floquet solution and the corresponding Brillouin zones.

The role of Floquet theory is to find the eigenfunction $|u_\alpha\rangle$ and eigenvalue $\epsilon_\alpha$ on the basis of invariant quantities. However, periodically driven time-dependent systems impose some mathematical subtlety. For example, let us consider a system expressed by the Hamiltonian $H(t) = H_0 + \lambda H_i(t)$, where $H_i$ contains all the time dependence of the system. The Hamiltonian $H_0$ describes a time-independent system that has the eigenfunction $\phi_n$ and eigenvalue $E_n$. In the case of $\lambda \approx 0$, we may write $H(t) \approx H_0$, and the eigenfunction of the time-independent system evolves as

$$|\phi_n(t)\rangle = \mathrm{e}^{-iE_n\, t/\hbar}|\phi_n(0)\rangle, \tag{9.14}$$

whereas, according to the Floquet theory, $|\phi_n(t)\rangle = \mathrm{e}^{-i\epsilon_\alpha t/\hbar}|u_n(t)\rangle$ is the solution of a weakly driven system. In the Floquet representation, the time-independent solution depends upon two quantum numbers, $n$, $m$, such that

$$|\phi_n(t)\rangle = \mathrm{e}^{-i(E_n+m\hbar\omega)\, t/\hbar}\left(|\phi_n(0)\rangle \mathrm{e}^{im\omega t}\right). \tag{9.15}$$

If $m$ is an integer then a comparison with the Floquet state reveals that $(|\phi_n(0)\rangle \mathrm{e}^{im\omega t})$ is equivalent to $|u_n(t)\rangle$ and periodic in time as required, that is,

$$|u_n(t+T)\rangle = |\phi_n(0)\rangle\, \mathrm{e}^{im\omega(t+T)} = |\phi_n(0)\rangle \mathrm{e}^{im(\omega t+2\pi)} = |u_n(t)\rangle, \tag{9.16}$$

where $T = \frac{2\pi}{\omega}$. The quasi-energy, $\epsilon_\alpha = E_n + m\hbar\omega$, is calculated up to an accuracy of $\hbar\omega$ and is given by energy eigenvalues modulo $2\pi$. Therefore, it can be divided into Brillioun zones of width $\hbar\omega$.

**Example 2:** Explain the interaction of a two-level atom in a single-mode field using the Floquet representation.

To develop a basic understanding of the Floquet picture, we solve a simple driven system in which a two-level atom interacts with a circularly polarized classical electromagnetic field of frequency $\omega$. We express the semi-classical governing Hamiltonian as

$$H = \frac{\hbar\omega_0}{2}\sigma_z + \frac{\hbar\Omega}{2}(\sigma_x \cos \omega t + \sigma_y \sin \omega t) = \frac{\hbar\omega_0}{2}\sigma_z + \frac{\hbar\Omega}{2}(\sigma_+ \mathrm{e}^{-i\omega t} + \sigma_- \mathrm{e}^{i\omega t}). \tag{9.17}$$

Here, the first term describes the atomic Hamiltonian and the second term presents the atom–field interaction. In addition, $\Omega$ describes the Rabi frequency. It is also the coupling strength between atom and field, being a product of the atomic dipole moment $\mu$ and field amplitude $\varepsilon$. The parameter $\omega$ describes the field frequency.

The Schrödinger equation reads

$$i\frac{\partial}{\partial t}|\psi\rangle = \frac{1}{2}\omega_0\sigma_z|\psi\rangle + \frac{1}{2}\Omega(\sigma_+ \mathrm{e}^{-i\omega t} + \sigma_- \mathrm{e}^{i\omega t})|\psi\rangle.$$

Writing

$$|\psi\rangle = \begin{pmatrix} \psi_1(t) \\ \psi_2(t) \end{pmatrix},$$

and going to the rotating frame by

$$\begin{pmatrix} \psi_1(t) \\ \psi_2(t) \end{pmatrix} = \begin{pmatrix} e^{-i\omega t/2} V_1(t) \\ e^{i\omega t/2} V_2(t) \end{pmatrix},$$

and then substituting this into the Schrödinger equation, we obtain

$$i\frac{\partial}{\partial t}\begin{pmatrix} V_1(t) \\ V_2(t) \end{pmatrix} = \begin{pmatrix} \frac{\omega_0 - \omega}{2} & \frac{\Omega}{2} \\ \frac{\Omega}{2} & -\frac{\omega_0 - \omega}{2} \end{pmatrix}\begin{pmatrix} V_1(t) \\ V_2(t) \end{pmatrix}.$$

We cast the solution into the following form:

$$\begin{pmatrix} V_1(t) \\ V_2(t) \end{pmatrix} = e^{-i\lambda t}\begin{pmatrix} a \\ b \end{pmatrix},$$

assuming $a$ and $b$ are constants, which indeed they are. We get the eigenvalue equation:

$$\left(\frac{\omega_0 - \omega}{2}\sigma_z + \frac{\Omega}{2}\sigma_x\right)\begin{pmatrix} a \\ b \end{pmatrix} = \lambda\begin{pmatrix} a \\ b \end{pmatrix}.$$

The eigenvalues are

$$\lambda_\pm = \pm\frac{1}{2}\sqrt{(\omega_0 - \omega)^2 + \Omega^2},$$

so that

$$\begin{pmatrix} \psi_1(t) \\ \psi_2(t) \end{pmatrix}_\pm = e^{-i\lambda_\pm t}\begin{pmatrix} e^{-i\omega t/2} a \\ e^{i\omega t/2} b \end{pmatrix}.$$

The solution is of the Floquet form as we identify

$$\frac{\epsilon_{(\pm,n)}}{\hbar} = \lambda_\pm + \omega\left(n + \frac{1}{2}\right),$$

then

$$\begin{pmatrix} \psi_1(t) \\ \psi_2(t) \end{pmatrix}_{(\pm,n)} = e^{-i\epsilon_{(\pm,n)}t}\begin{pmatrix} e^{-(n+1)i\omega t} a \\ e^{-in\omega t} b \end{pmatrix},$$

where integer $n$ is the band index and $\alpha = (j,\, n)$ is the composite index, where $j = \pm$. The labelling is valid in the limit of $\Omega \to 0$.

The preceding discussion is equivalent to having found the Floquet operator to be

$$\hat{H}_F = \frac{\hbar\omega}{2}1 + \frac{\hbar\Delta}{2}\sigma_z + \frac{\hbar\Omega}{2}\sigma_x, \tag{9.18}$$

where $\Delta = \omega_0 - \omega$ for $n = 0$. For a detailed discussion, see reference [15].

## 9.3 Einstein–Brillouin–Keller quantization

In 1917, Einstein presented his seminal paper at the German Physical Society meeting on the quantization of energy for mechanical systems. The work extended the Bohr–Sommerfeld quantization to higher-dimensional coupled systems. Later, the quantization conditions were independently discovered by Joseph Keller. The work was initially applied to some simple systems. However, these semi-classical quantization rules were not fully appreciated until the 1970s, when the quantization of non-integrable Hamiltonian systems was needed. This led to the new subject of quantum chaos.

One interesting advantage of the approach is that we obtain the quantization without involving heavy machinery based on the Schrödinger equation. In the Einstein–Brillouin–Keller (EBK) quantization scheme, a system containing $N$ degrees of freedom and $2N$-dimensional phase-space is quantized as

$$\oint_{\gamma_i} pdx = 2\pi\hbar\left(n_i + \frac{ind(\gamma_i)}{4}\right), \tag{9.19}$$

where $i = 1, 2, 3, \ldots, N$. Moreover, $\gamma_i$ describes the $i$th closed path in the phase space (figure 9.1), and $ind(\gamma_i)$ is the Maslov index. The index describes the number of turning points in a closed path.

For simplicity, we consider a one-dimensional periodically driven system that has two degrees of freedom and four-dimensional phase space in generalized coordinates. We consider time, $t$, as a coordinate and define the corresponding momentum as $p_t \equiv \frac{\partial}{\partial t}$, thus writing the generalized Hamiltonian as

$$\tilde{H} = H(x, p, t) + p_t, \tag{9.20}$$

which provides Hamilton's equations of motion for time variable $s$, written as

$$\frac{\partial p}{\partial s} = -\frac{\partial \tilde{H}}{\partial x} = -\frac{\partial H}{\partial x}, \tag{9.21}$$

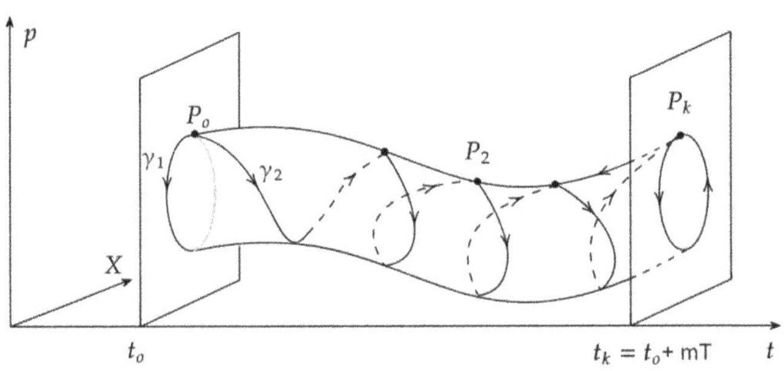


**Figure 9.1.** The trajectories in extended phase space. Torus quantization is performed following the trajectories. Reproduced from [16]. 

$$\frac{\partial x}{\partial s} = \frac{\partial \tilde{H}}{\partial p} = \frac{\partial H}{\partial p}, \tag{9.22}$$

$$\frac{\partial t}{\partial s} = \frac{\partial \tilde{H}}{\partial p_t} = 1, \tag{9.23}$$

$$\frac{\partial p_t}{\partial s} = -\frac{\partial \tilde{H}}{\partial t} = -\frac{\partial H}{\partial t}. \tag{9.24}$$

In a four-dimensional system there are two competing frequencies, $\omega_1$ and $\omega_2$. As discussed in chapter 8, the system exhibits non-linear resonance whenever $r\omega_1 = s\omega_2$; therefore, we obtain a *magic number*:

$$\frac{\omega_2}{\omega_1} = \frac{r}{s},$$

which can be rational or irrational. For a rational magic number, we find $r$ and $s$ as the relative prime numbers. EBK quantization provides the quantization rules for the systems that exhibit resonances, such as weakly coupled systems or systems with stable islands. Hence, we write the quantization condition for the periodically driven system as

$$\int_{\gamma_i} (pdx + p_t dt) = 2\pi\hbar\left(n_i + \frac{ind(\gamma_i)}{4}\right), \tag{9.25}$$

where $i = 1, 2$.

For a two degrees-of-freedom system, we can have two basic quantum numbers coming from two conditions [16]:

(i) Choosing a path into the $t$ = constant plane provides one quantization condition:

$$\int_{\gamma_1} pdx = 2\pi\hbar\left(n_1 + \frac{ind(\gamma_1)}{4}\right). \tag{9.26}$$

(ii) Choosing a second path for $x$ = constant, which provides

$$\int_{\gamma_2} pdx + p_t dt = 2\pi\hbar\left(n_2 + \frac{ind(\gamma_2)}{4}\right) = 2\pi n_2 \hbar, \tag{9.27}$$

where $ind(\gamma_2) = 0$ on the basis that in the time domain or $t$-direction there are no turning points for classical trajectories. These trajectories in extended phase space can be viewed as a path, $\gamma_1$, winding around the torus once, whereas the path $\gamma_2$ stretches along the $t$-direction so that it can periodically be followed again, as shown in figure 9.2. The EBK quantization is useful in simple systems and those higher-dimensional systems that have stable islands, however it breaks in systems with no stable islands such as in a stadium billiard.

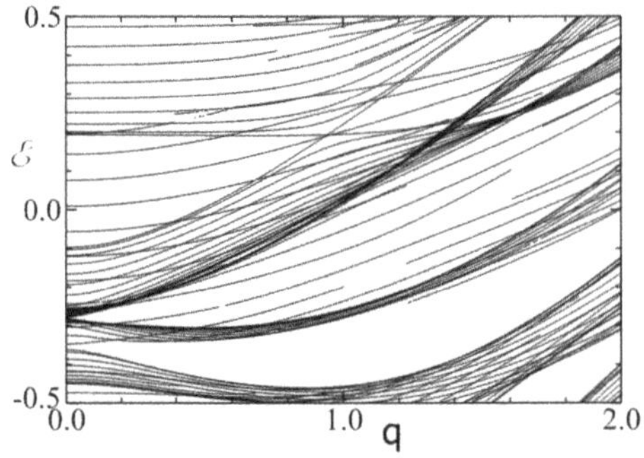


**Figure 9.2.** The lowest three quasi-energy bands are plotted. Each band is represented by 21 scaled wave numbers $k/k_L$ ranging from 0 to 1 in equal steps of 0.05. There appear to be more than three quasi-energy bands in the strong driving regime. For further details, see reference [9]. Reproduced from [15]. 

## 9.4 Quantization near non-linear resonances

In this section, we extend the Floquet formalism discussed above to non-linear resonances of periodically driven systems. We provide semi-classical and quantum mechanical treatments to the dynamics of a wave packet that is experiencing an external modulation. As a result, we find the quasi-energy eigenstates and the quasi-eigenenergies of the dynamical system.

### 9.4.1 The semi-classical approach

A primary resonance occurs in a dynamical system when a classical particle has an action, $I_N$, such that the cycle time of the unperturbed motion coincides with an integer multiple of the time period of the driving field. We write it mathematically as $T_1 = NT$, where $N$ is the integer and $T$ is the time period of the driving field, related to the driving field frequency $\omega$ as $T = 2\pi/\omega$. This implies $\omega_1 = \frac{1}{N}\omega$. Therefore, we calculate the action $I_N$ from the relation

$$\omega_1(I_N) = \left.\frac{\partial H_0}{\partial I}\right|_{I=I_N} = \frac{1}{N}\omega.$$

Following the discussion in section 8.5, we write the dynamics of a particle in a driven system around a $N$th primary resonance as

$$H = H_0(I) - \lambda V(I)\cos(N\phi - \omega t). \tag{9.28}$$

Here, $H_0$ is the Hamiltonian of the system in the absence of the modulation, $\lambda$ describes the strength of the modulation, and $V(I) = -2H_{-N,1}(I)$ is a system-dependent quantity.

*Effective resonance Hamiltonian.* In order to understand the dynamics around a resonance, we expand the Hamiltonian $H_0$ up to the second order around $I_N$, that is,

$$H = H_0(I_N) + H'(I - I_N) + \frac{1}{2}H''(I - I_N)^2 - \lambda V \cos(N\phi - \omega t). \tag{9.29}$$

Here, $H' = \frac{\partial H_0}{\partial I}\bigg|_{I=I_N}$ and $H'' = \frac{\partial^2 H_0}{\partial I^2}\bigg|_{I=I_N}$. Since $I - I_N$ and $\phi$ constitute a pair of canonical conjugates, the approximate pendulum Hamiltonian, in equation (9.29), can be quantized by introducing the operator representation of the action, $I - I_N$, such that

$$I - I_N = \frac{\hbar}{i}\frac{\partial}{\partial \phi}. \tag{9.30}$$

On substituting equation (9.30) in equation (9.29), we obtain the quantum mechanical description of the resonance Hamiltonian, which is

$$H = H_0(I_N) + \frac{H'\hbar}{i}\frac{\partial}{\partial \phi} - \frac{H''\hbar^2}{2}\frac{\partial^2}{\partial \phi^2} - \lambda V \cos(N\phi - \omega t). \tag{9.31}$$

Therefore, the corresponding time-dependent Schrödinger equation becomes

$$i\hbar\frac{\partial \psi}{\partial t} = \left[H_0(I_N) + \frac{H'\hbar}{i}\frac{\partial}{\partial \phi} - \frac{H''\hbar^2}{2}\frac{\partial^2}{\partial \phi^2} - \lambda V \cos(N\phi - \omega t)\right]\psi. \tag{9.32}$$

Further simplicity is introduced by writing $N\phi - \omega t = 2\varphi$ as the new angle variable. This is the Kramers–Henneberger transformation, discussed in detail in the next chapter. Hence, we obtain the simplified Schrödinger equation:

$$i\hbar\frac{\partial \psi}{\partial t} = \left[-\frac{N^2 H''\hbar^2}{8}\frac{\partial^2}{\partial \varphi^2} + H_0(I_N) + \lambda V \cos 2\varphi\right]\psi. \tag{9.33}$$

The Hamiltonian in equation (9.33) is independent of time $t$, therefore we factorize the wave function into time-dependent and time-independent parts, that is, $\psi(\varphi, t) = \psi_1(\varphi)\psi_2(t)$. The time-dependent solution is obtained as

$$\psi_2(t) = e^{-i\frac{\mathcal{E}t}{\hbar}},$$

whereas, for $\psi_1$, we get the effective time-independent Schrödinger equation,

$$\left[-\frac{N^2 H''\hbar^2}{8}\frac{\partial^2}{\partial \varphi^2} + H_0(I_N) + \lambda V \cos 2\varphi\right]\psi_1 = \mathcal{E}\psi_1. \tag{9.34}$$

Here, the quantity $\mathcal{E}$ is the quasi-energy. Equation (9.34) can easily be rewritten in the form of a standard Mathieu equation:

$$\left[\frac{\partial^2}{\partial \varphi^2} + a - 2q \cos 2\varphi\right]\psi_1 = 0, \tag{9.35}$$

where

$$a = \frac{8}{N^2 H'' \hbar^2}[\mathcal{E} - H_0(I_N)], \tag{9.36}$$

$$q = \frac{4\lambda V}{N^2 H'' \hbar^2}. \tag{9.37}$$

The general solution of the Mathieu equation has a Floquet form, that is, $\psi_1(\varphi) = \mathrm{e}^{i\nu\varphi} P_\nu(\varphi)$ with a characteristic exponent $\nu$. The Mathieu function, $P_\nu$, is $\pi$-periodic, and so is the function $\psi_1$ in $\varphi$. On substituting the expression for $\varphi$, we obtain

$$\psi_1(\phi) = \exp\left\{i\nu\left(\frac{N\phi - \omega t}{2}\right)\right\} P_\nu\left(\frac{N\phi - \omega t}{2}\right).$$

The angle variable, $\phi$, has $2\pi$ periodicity; for this reason, $\psi_1$ must be $2\pi$ periodic in $\phi$. Hence, we get a restriction on the values for $\nu$, that is,

$$\nu = \nu(j) = \frac{2j}{N}, \quad \text{where } j = 0, 1, 2, \ldots N - 1. \tag{9.38}$$

The second-order Mathieu equation has a discrete set of such solutions, labelled by $m$. Here, $m$ has integer values and acts as a new quantum number.

*Quasi-energy eigenstates and quasi-energy spectrum.* Hence, we write the complete solution as

$$\psi_{\nu(j),m}(\phi, t) = \mathrm{e}^{ij\phi} \exp\left\{-i\left(\mathcal{E} + \frac{j\hbar\omega}{N}\right)\frac{t}{\hbar}\right\} P_{\nu(j),m}\left(\frac{N\phi - \omega t}{2}\right) \tag{9.39}$$

$$= u_{\nu(j),m}(\phi, t)\mathrm{e}^{-i\mathcal{E}_{\nu(j),m}t/\hbar}, \tag{9.40}$$

where $m$ is an integer. Here, $u_{\nu(j),m}$ is

$$u_{\nu(j),m}(\phi, t) = \mathrm{e}^{ij\phi}\, P_{\nu(j),m}\left(\frac{N\phi - \omega t}{2}\right), \tag{9.41}$$

which is $2\pi$ periodic, such that

$$u_{\nu(j),m}(\phi + 2\pi, t) = u_{\nu(j),m}(\phi, t), \tag{9.42}$$

$$u_{\nu(j),m}(\phi, t + 2\pi/\omega) = u_{\nu(j),m}(\phi, t), \tag{9.43}$$

and the corresponding quasi-energy of the system is, therefore, expressed as

$$\mathcal{E}_{\nu(j),m} = \left[\mathcal{E} + \frac{j\hbar\omega}{N}\right] \bmod \hbar\omega \tag{9.44}$$

$$= \left[\frac{N^2 H'' \hbar^2}{8} a_{\nu(j),m}(q) + H_0(I_N) + \frac{j\hbar\omega}{N}\right] \bmod \hbar\omega. \tag{9.45}$$

This Floquet band spectrum has a simple interpretation. The Mathieu equation can be regarded as a stationary Schrödinger equation for a fictitious particle moving in a cosine lattice, and the solution for the $N$th resonance requires periodic boundary conditions, given in equations (9.42) and (9.43), after $N$ cosine wells. Thus, $N = 1$ leads to a single-well potential, and $N = 2$ to a double-well potential, and so on. For $N = 1$, we find $j = 0$, and $a_{\nu(0),m}$ is the characteristic parameter that corresponds to the $\pi$-periodic Mathieu function. We find

$$a_{\nu(0),m} = \begin{cases} a_m(q) & m = 0, 2, 4, \ldots \\ b_{m+1}(q) & m = 1, 3, 5, \ldots \end{cases}. \tag{9.46}$$

For $N = 2$, there are two group of quasi-energy eigenstates, labelled by $j = 0$ and $j = 1$. The states with $j = 1$ are the $2\pi$ periodic Mathieu functions, and the corresponding Mathieu characteristic parameter values are

$$a_{\nu(1),m} = \begin{cases} b_{m+1}(q) & m = 0, 2, 4, \ldots \\ a_m(q) & m = 1, 3, 5, \ldots \end{cases}. \tag{9.47}$$

### 9.4.2 The quantum mechanical approach

We study the quantum dynamics of a wave packet in a system experiencing an external periodic force to calculate the quasi-energy eigenstates and the quasi-energy spectrum. We write the Hamiltonian of the driven system as

$$H = H_0(x, p) + \lambda V(x)\sin(\omega t). \tag{9.48}$$

Here, $x$ and $p$ are the position and momentum operators, which satisfy the commutation relation $[x, p] = i\hbar$. Moreover, $H_0$ describes the Hamiltonian of the unmodulated system, such that $H_0|n\rangle = E_n|n\rangle$. Here, $E_n$ and $|n\rangle$ describe the $n$th eigenenergy and the eigenstate of the undriven system. In addition, $\omega$ and $\lambda$ are the frequency and strength of the external modulation, respectively. As discussed above, we find primary resonances in the driven system whenever the cycle time of a fictitious classical particle in the undriven system becomes equal to an integer multiple $N$ of the time period of the driving field.

We conjecture that the solution of the time-dependent Schrödinger equation in the vicinity of the $N^{th}$ resonance is written as an anstaz, i.e.

$$|\psi(\tau)\rangle = \sum_n C_n(\tau)|n\rangle \exp\left\{-i\left[E_{\bar{n}} + (n - \bar{n})\frac{\hbar}{N}\right]\frac{\tau}{\hbar}\right\}. \tag{9.49}$$

Here, $\tau = \omega t$ is the scaled time, $\bar{n}$ is the mean quantum number, and $E_{\bar{n}}$ is the mean energy. In addition, $C_n(\tau)$ is the time-dependent probability amplitude. In quantum mechanical solutions another scaled parameter appears, that is, the scaled Planck's constant, $\hbar$. The parameter $\lambda$ controls the degree of non-integrability,while the scaled Planck's constant, $\bar{k}$, controls the scale at which its quantum mechanical counterpart can resolve phase-space structures.

On substituting equation (9.49) in the time-dependent Schrödinger equation, we find that the probability amplitude, $C_n(\tau)$, changes with time following the equation

$$i\bar{k}\dot{C}_n(\tau) = \left[E_n - E_{\bar{n}} - (n - \bar{n})\frac{\bar{k}}{N}\right]C_n(\tau) + \frac{\lambda}{2i}(V_{n,n+N}C_{n+N} - V_{n,n-N}C_{n-N}).$$

By performing here a cycle average, the fast-oscillating terms are averaged out. Moreover, $V_{n,n\pm N} = \langle n|V(x)|n \pm N\rangle$.

*Effective resonance Hamiltonian.* We consider that a wave packet is narrowly peaked around the mean value, $\bar{n}$, as it evolves in a non-linear resonance. For this reason, we take slow variations in the energy, $E_n$, around the $\bar{n}$ in a non-linear resonance, and expand $E_n$ up to the second order using a Taylor expansion. Thus, the equation for the probability amplitudes, $C_n(\tau)$, becomes

$$\begin{aligned} i\bar{k}\dot{C}_n = \bar{k}(n - \bar{n})\left(\omega_1 - \frac{1}{N}\right)C_n(\tau) + \frac{1}{2}\bar{k}^2(n - \bar{n})^2\zeta \\ C_n(\tau) + \frac{\lambda V}{2i}(C_{n+N} - C_{n-N}). \end{aligned} \tag{9.50}$$

Bearing in mind the symmetry around a resonance, we consider $V_{n,n+N} = V_{n,n-N} = V$. Equation (9.50) for $C_n$ is based on

$$\omega_1 = \frac{1}{\bar{k}}\frac{\partial E_n}{\partial n}\bigg|_{n=\bar{n}},$$

and

$$\zeta = \frac{1}{\bar{k}^2}\frac{\partial^2 E_n}{\partial n^2}\bigg|_{n=\bar{n}},$$

which correspond to the frequency and the non-linearity of the time-independent system, respectively. We introduce the Fourier coefficient of $C_n$ as a $2\pi$ periodic function, $g(\theta)$, such that the index $n = \bar{n}$ corresponds to the zero mode. We write

$$C_n = \frac{1}{2\pi}\int_0^{2\pi} g(\theta)e^{-i(n-\bar{n})\theta}d\theta, \tag{9.51}$$

$$= \frac{1}{2N\pi}\int_0^{2N\pi} g(\phi)e^{-i(n-\bar{n})\phi/N}d\phi. \tag{9.52}$$

Here, $g(\phi, \tau)$ has $2N\pi$ periodicity in the $\phi$ coordinate. Furthermore, we write

$$(n - \bar{n})C_n = \frac{iN}{2N\pi}\int_0^{2N\pi} g(\phi)\frac{\partial}{\partial\phi}e^{-i(n-\bar{n})\phi/N}d\phi \tag{9.53}$$

$$= \frac{-iN}{2N\pi} \int_0^{2N\pi} \frac{\partial g}{\partial \phi} e^{-i(n-\bar{n})\phi/N} d\phi, \tag{9.54}$$

and

$$(n - \bar{n})^2 C_n = \frac{-N^2}{2N\pi} \int_0^{2N\pi} g(\phi) \frac{\partial^2}{\partial \phi^2} e^{-i(n-\bar{n})\phi/N} d\phi \tag{9.55}$$

$$= \frac{-N^2}{2N\pi} \int_0^{2N\pi} \frac{\partial^2 g}{\partial \phi^2} e^{-i(n-\bar{n})\phi/N} d\phi. \tag{9.56}$$

Hence, equation (9.50) becomes $i\hbar \dot{g}(\phi) = H(\phi)g(\phi)$, where the Hamiltonian $H(\phi)$ is given as

$$H(\phi) = -\frac{N^2 \hbar^2 \zeta}{2} \frac{\partial^2}{\partial \phi^2} - iN\hbar \left( \omega_1 - \frac{1}{N} \right) \frac{\partial}{\partial \phi} - \lambda V \sin \phi. \tag{9.57}$$

Due to the time-independent behaviour of the Hamiltonian, $H(\phi)$, we write

$$g(\phi) = \chi(\varphi) \, e^{-\frac{i\mathcal{E}\tau}{\hbar}} \exp\left( -2i \frac{(N\omega_1 - 1)\varphi}{N^2 \zeta \hbar} \right),$$

and substitute $\phi = 2\varphi + \pi/2$. Hence, the Schrödinger equation for $g(\phi, \tau)$ reduces to the Mathieu equation:

$$\left[ \frac{\partial^2}{\partial \phi^2} + a - 2q \cos 2\varphi \right] \chi(\varphi) = 0. \tag{9.58}$$

The Mathieu characteristic parameters are

$$a = \frac{8}{N^2 \hbar^2 \zeta} \left[ \frac{(N\omega_1 - 1)^2}{2N\zeta} + \mathcal{E} \right], \tag{9.59}$$

and

$$q = \frac{4\lambda V}{N^2 \hbar^2 \zeta}. \tag{9.60}$$

*Quasi-energy eigenstates and quasi-energy spectrum.* The $\pi$-periodic solutions to equation (9.58) correspond to even functions of the Mathieu equation with corresponding eigenvalues as real. These solutions are defined by Floquet states, that is, $\chi(\varphi) = e^{i\nu\varphi} P_\nu(\phi)$, where the Floquet function has $\pi$ periodicity, that is, $P_\nu(\varphi) = P_\nu(\varphi + \pi)$, and $\nu$ is the characteristic exponent. In order to have $2\pi$-periodic solutions in the $\phi$ coordinate, we require $\nu$ to be $\nu = \nu(j) = 2j/N$, where $j = 0, 1, 2, \ldots, N - 1$.

The allowed values of $\nu(j)$ can exist as a characteristic exponent of the solution to the Mathieu equation for discrete $m$, which is an integer for a certain value $a_{\nu(j),m}(q)$

when $q$ is fixed. Hence, with the help of equation (9.59), we obtain the values of the unknown $\mathcal{E}$. Therefore, we may express $|\psi(\tau)\rangle$ as

$$|\psi\rangle = e^{i\mathcal{E}_{\nu(j),m}\tau/k} \; |u_{\nu(j),m}\rangle, \tag{9.61}$$

where

$$|u_{\nu(j),m}\rangle = \frac{1}{2\pi}\sum_{n}\int_{0}^{2\pi} g_{\nu(j),\,m}^{(n)}(\theta)d\theta \; |n\rangle, \tag{9.62}$$

and

$$g_{\nu(j),\,m}^{(n)} = \exp\left\{i\left(\frac{j}{N} - \frac{N\omega_1 - 1}{N^2\xi k}\right)\left(N\theta - \frac{\pi}{2}\right)\right\}e^{-i\frac{(n-\bar{n})(N\theta+\tau)}{N}}P_{\nu(j),m}\left(\frac{N\theta - \pi/2}{2}\right). \tag{9.63}$$

The quasi-energy of the system becomes

$$\mathcal{E}_{\nu(j),m} = \left[\frac{N^2 k^2 \zeta}{8}a_{\nu(j),m}(q) + \frac{N\omega_1 - 1}{N^2\xi k} + E_{\bar{n}} + k\frac{j}{N}\right] \text{ mod } k. \tag{9.64}$$

In the case of exact resonance, the term $N\omega_1 - 1 = 0$. The structure of the resonant quasi-energy spectrum obtained in this way is equal to the structure of equation (9.45).

Interpretation of the quantum numbers $j$ and $m$ can be done using EBK quantization rules. The quantum number $m$ corresponds to the $t$ = constant plane, identified as the $\gamma_1$ path, which winds around the torus once. However, $j$ corresponds to the $x$ = constant plane and stretches along the $t$-axis. Therefore, a fictitious particle cuts the plane $N - 1$ times and winds around the torus following the $\gamma_2$ path, as discussed in section 9.3.

## Exercises

1. Consider a two-level atom interacting with a linearly polarized classical electromagnetic field of frequency $\omega$, expressed as

$$H = \frac{\hbar\omega_0}{2}\sigma_z + \frac{\hbar\Omega}{2}\sigma_x \cos\omega t, \tag{9.65}$$

   where $\Omega = \frac{\mu\varepsilon}{\hbar}$. Show that the Floquet operator for the system is

$$\hat{F} = \frac{\hbar\omega}{2}1 + \frac{\hbar\Delta}{2}\sigma_z + \frac{\hbar\Omega}{2}(\sigma_x(1 + \cos 2\omega t) - \sigma_y \sin\omega t), \tag{9.66}$$

   where $\Delta = \omega_0 - \omega$.
2. Apply EBK quantization to obtain the quantization of a periodically modulated harmonic oscillator.
3. Apply EBK quantization to obtain the quantization of a periodically modulated truncated linear potential.

## References


[1] Bayfield J E and Koch P M 1974 Multiphoton ionization of highly excited hydrogen atoms *Phys. Rev. Lett.* **33** 258

[2] Yoshida S, Reinhold C O, Kristöfel P, Burgdörfer J, Watanabe S and Dunning F B 1999 Floquet analysis of the dynamical stabilization of the kicked hydrogen atom *Phys. Rev.* A **59** R4121(R)

[3] Casati G, Chirikov B V and Shepelyansky D L 1984 Quantum limitations for chaotic excitation of hydrogen atom in monochromatic field *Phys. Rev. Lett.* **53** 2525

[4] Bayfield J E, Casati G, Guarneri I and Sokol D W 1989 Localization of classically chaotic diffusion for hydrogen atoms in microwave fields *Phys. Rev. Lett.* **63** 364

[5] Koch P M and van Leeuwen K A H 1995 *Phys. Rep.* **255** 289

[6] Galvez E J, Sauer B E, Moorman L, Koch P M and Richards D 1988 *Phys. Rev. Lett.* **61** 2011

[7] Blümel R, Kappler C, Quint W and Walther H 1989 *Phys. Rev.* A **4** 808

[8] Bayfield J E, Casati G, Guarneri I and Sokol D W 1989 *Phys. Rev. Lett.* **63** 364

[9] Arndt M, Buchleitner A, Mantegna R N and Walther H 1991 *Phys. Rev. Lett.* **67** 2435

[10] Segev B, Côté R and Raizen M G 1997 *Phys. Rev.* A **56** R3350

[11] Haake F 2001 *Quantum Signatures of Chaos* (Berlin: Springer)

[12] El Ghafar M, Törmä P, Savichev V, Mayr E, Zeiler A and Schleich W P 1997 Dynamical localization in the Paul trap *Phys. Rev. Lett.* **78** 4181

[13] Riedel K, Törmä P, Savichev V and Schleich W P 1999 Control of dynamical localization by an additional quantum degree of freedom *Phys. Rev.* A **59** 797

[14] Eckardt A 2017 Colloquim: atomic quantum gases in periodically driven optical lattices *Rev. Mod. Phys.* **89** 011004

[15] Holtlhaus M 2016 Floquet engineering with quasi-energy bands of periodically driven optical lattice *J. Opt. B: At. Mol. Opt. Phys.* **49** 013001

[16] Bensch F, Korsch H J, Mirbach B and Ben-Tal N 1992 *J. Phys. A: Math. Gen.* **25** 6761–77

# Chapter 10

## Atoms in modulated optical lattice—I

> *'The atoms become like a moth, seeking out the region of higher laser intensity.'*
>
> —Steven Chu

The dynamics of an atom in an optical lattice in the presence of a time-dependent external force is intricate, and different from the case of no such force. The presence of a modulating force introduces another degree of freedom, leading to non-linear dynamical characteristics. In this book, we consider examples of an external periodic force on atoms, for instance, by considering a periodic modulation of the phase or amplitude of the two counter-propagating laser beams that constitute the optical lattice. The amplitude modulation or phase modulation is introduced, for example, by an acousto-optic modulator. The dynamics of ultra-cold atoms in amplitude-modulated and phase-modulated one-dimensional, two-dimensional, and three-dimensional optical lattices simulates electron dynamics in a semi-conductor crystal under the influence of an AC electric field. Optical lattice systems in the presence of phase modulation have been the subject of interest in the observation of quantum chaos, photon-assisted tunneling, the control of Mott insulator transitions, quantum transport, and Wannier–Stark resonance under the effect of a constant force.

As discussed in section 3.2, the atoms in a one-dimensional optical lattice experience a periodic potential, that is,

$$V(x) = \frac{V_0}{2} \cos 2k_{\mathrm{L}}x,$$

where $V_0$ is directly proportional to the intensity of the optical field. The wave number is related to the wavelength, such that $k_{\mathrm{L}} = 2\pi/\lambda$. The lattice constant or

doi:10.1088/978-0-7503-2308-6ch10

optical lattice spacing of the optical lattice is $d = \frac{\lambda}{2}$, which may take a value of 421 nm for a wavelength of $\lambda = 842$ nm.

The single photon recoil energy, which may be used to scale the energy corresponding to the atom–field interaction, is

$$E_R = \frac{\hbar^2 k_L^2}{2m}.$$

For a rubidium atom ($^{87}$Rb) of mass $m = 1.443 \times 10^{-25}$ kg the recoil energy is $E_R = 1.34 \times 10^{-11}$ eV. A typical lattice depth is of the order of $10^{-10}$ eV, which is five to ten recoil energies. This implies that the various phenomena relating to ultra-cold atoms in optical lattices take place at around ten orders of magnitude higher in energy scale in comparison with electrons in solid-state physics.

The density of the particles at the center of an ultra-cold atomic sample is of the order of $10^{12}$ cm$^{-3}$ to $10^{14}$ cm$^{-3}$, much lower than the density of molecules in air at room temperature, that is, ($10^{19}$ cm$^{-3}$), the density of atoms in liquids ($10^{22}$ cm$^{-3}$), and the density of nucleons in atomic nuclei ($10^{38}$ cm$^{-3}$).

The temperature of the atoms is directly proportional to the kinetic energy and the average of the square of the velocity, $\overline{v^2}$, that is,

$$k_B T = \frac{1}{2} M \overline{v^2} = \frac{\overline{p^2}}{2M}.$$

Therefore, the de Broglie wavelength is

$$\Lambda_{dB} = \frac{h}{\sqrt{\overline{p^2}}} = \frac{h}{\sqrt{2mk_B T}}.$$

Hence, we note that the de Broglie wavelength is inversely proportional to the square root of the temperature, $T$. This implies that as the temperature reduces, the de Brogile wavelength increases, which is essential to observe quantum effects. For example, at room temperature, that is, around 300 K, the de Brogile wavelength of a rubidium atom is 0.2 pm, much smaller than the size of the atom; at 1 K temperature, the corresponding de Broglie wave associated with the rubidium atom has a wavelength near 0.3 nm; whereas at an eight orders of magnitude smaller temperature it becomes 3 $\mu$m, that is, around ten thousand times bigger than the size of the atom.

An ultra-cold atom can be understood by considering its recoil energy, $E_R$, in comparison with the ensemble average energy, $k_B T$. For this reason, we find the de Broglie wavelength to be

$$\Lambda_{dB} = \frac{h}{\sqrt{2mk_B T}} = \frac{h}{\sqrt{2mE_R}} = \lambda,$$

when $E_R$ is taken to be the same as $k_B T$. This implies that the de Broglie wavelength becomes comparable to the optical wavelength under this condition. However, quantum characteristics only become visible if the de Broglie wavelength has a higher value and covers few optical wavelengths. This creates the requirement that $k_BT$ should be smaller than $E_R$.

## 10.1 Phase-modulated optical lattice

External periodic forcing on ultra-cold atoms in optical lattices is introduced by modulating the amplitude, the phase, or both. An acousto-optic modulator is capable of introducing such a modulation. Further, the optical lattices display inherent periodicity in space as well. Therefore, an optical lattice in the presence of a phase modulation exhibits a double-periodic system (figure 10.1), expressed by the governing Hamiltonian as

$$H = \frac{p^2}{2m} - \frac{V_0}{2}\cos[2k_L\{x - \Delta L\cos(\omega_m t)\}]. \tag{10.1}$$

Here, $\Delta L$ and $\omega_m$ are the amplitude and frequency of the external phase modulation. For a modulation frequency of 1 MHz acting on a rubidium atom, and displacing it by a distance 100 nm, the force applied on it is around $10^{-20}$ Newton. In comparison, the gravitational force on a healthy person standing on Earth is about $10^3$ Newton.

As stated above, the Hamiltonian given in equation (10.1) is periodic in space and time. The spatial periodicity of the Hamiltonian

$$H(x + d, t) = H(x, t)$$

implies the application of Bloch theory, whereas temporal periodicity due to external periodical forcing shakes the optical potential such that

$$H(x, t + T) = H(x, t),$$

which enables the use of Floquet theory, as discussed previously. Here, $k_L = \pi/d$, where $d$ and $T$ are, respectively, the lattice spacing and time period. Hence, a Hamiltonian with twofold translational symmetry possesses a complete set of spatio-temporal Floquet–Bloch solutions.

Cold atoms in modulated optical lattices are a prototype of an electron moving in a crystal with an impurity. Such a comparison can extend the applications developed in condensed-matter physics and solid-state physics to atomic molecular and optical

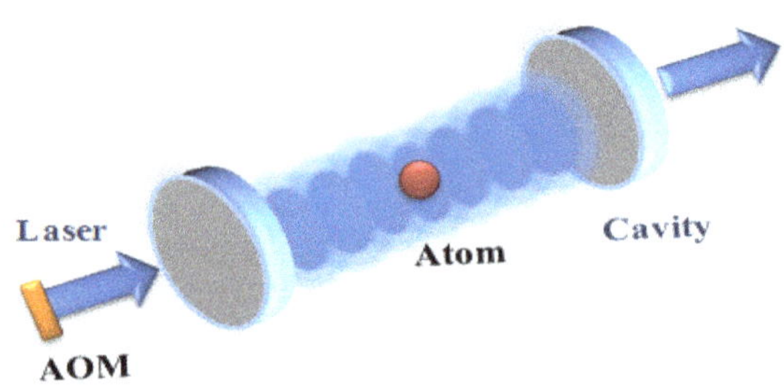


**Figure 10.1.** Schematic diagram showing ultra-cold atom dynamics in a one-dimensional phase-modulated optical lattice. The optical field is modulated by means of an acousto-optic modulator (AOM).

sciences, and has led to technological applications that range from superlattices to computers based on a few atoms.

*Kramers–Henneberger transformation.*

We may rewrite the above Hamiltonian using the Kramers–Henneberger transformation, that is,

$$\begin{aligned} \tilde{x} &= x - L(t), \\ \tilde{t} &= t. \end{aligned} \tag{10.2}$$

Here, $L(t) = \Delta L \cos(\omega_m t)$. This implies

$$\frac{\partial}{\partial x} = \frac{\partial}{\partial \tilde{x}} \frac{\partial \tilde{x}}{\partial x} + \frac{\partial}{\partial \tilde{t}} \frac{\partial \tilde{t}}{\partial x} = \frac{\partial}{\partial \tilde{x}},$$
$$\frac{\partial}{\partial t} = \frac{\partial}{\partial \tilde{x}} \frac{\partial \tilde{x}}{\partial t} + \frac{\partial}{\partial \tilde{t}} \frac{\partial \tilde{t}}{\partial t} = -\dot{L}(\tilde{t}) \frac{\partial}{\partial \tilde{x}} + \frac{\partial}{\partial \tilde{t}}.$$

Here, $\dot{L}(\tilde{t}) = \partial L(\tilde{t})/\partial \tilde{t}$, which becomes $-\Delta L \omega_m \sin \omega_m \tilde{t}$. On substituting the transformation in the corresponding time-dependent Schrödinger wave equation and rearranging the terms, we obtain the transformed Hamiltonian as

$$\tilde{H} = \frac{\tilde{p}^2}{2m} + V(\tilde{x}) + m\tilde{x}\ddot{L}(\tilde{t}), \tag{10.3}$$

$$\tilde{H} = \frac{\tilde{p}^2}{2m} + V(\tilde{x}) + F(\tilde{t})\, \tilde{x}. \tag{10.4}$$

Here,

$$F(\tilde{t}) = m\ddot{L}(\tilde{t}) = -m\omega_m^2 \Delta L \cos(\omega_m \tilde{t}) = -F_1 \sin(\omega_m \tilde{t}). \tag{10.5}$$

For simplicity, we remove the 'tilde' sign (~) in the next discussion.

*Generalized force.* The Hamiltonian in equation (10.4) has periodicity in space and time. The periodicity in the space coordinate is restricted by adding a linearly changing term, as discussed in chapter 6. The inclusion of the term in the Hamiltonian in equation (10.4) introduces a generalization. Here, the net force is composed of DC and AC parts, such that

$$F_0 - F_1 \cos(\omega_m t). \tag{10.6}$$

This is a prototype of the situation in which an electron is subject to a constant electric field (or DC field) that is exerting a constant force $-eE_0$ on the electron, and a time periodic field (or AC field) applying a periodic force $-eE_1 \sin \omega_m t$; here, $e$ can be considered as the charge of the electron.

In the case of $F_0 = 0$ and $F_1 = 0$, we have Wannier functions as the eigenstates of the system. In the case in which we have $F_0 \neq 0$ only, that is, the time-independent case with a constant DC field, this supports a Hilbert space spanned by Wannier–Stark states with Wannier–Stark energy levels, as discussed in section 6.2. The general case can be solved using a vector potential.

## 10.2 Time evolution

In the Kramers–Henneberger transformation, we write the Schrödinger wave equation, which controls the evolution of a cold atom in a phase-modulated optical lattice, that is,

$$i\hbar\frac{\partial|\Psi(x,\,t)\rangle}{\partial t} = \left[\frac{\hat{p}^2}{2m} - \frac{V_0}{2}\cos(2k_{\rm L}\hat{x}) + \hat{x}\,F(t)\right]|\Psi(x,\,t)\rangle. \tag{10.7}$$

A gauge transformation,

$$\hat{U}(\hat{x},\,t) = \exp\left\{-\frac{i}{\hbar}\hat{x}\int_0^t F(t')dt'\right\}, \tag{10.8}$$

re-expresses the transformed system in a co-moving frame or laboratory frame as

$$\begin{aligned}|\Psi(x,\,t)\rangle &= \hat{U}(\hat{x},\,t)|\tilde{\Psi}(x,\,t)\rangle \\ &= \exp\left\{-\frac{i}{\hbar}\hat{x}\int_0^t F(t')dt'\right\}|\tilde{\Psi}(x,\,t)\rangle.\end{aligned} \tag{10.9}$$

Moreover, the transformation shifts the left-hand side of the Schrödinger wave equation (10.7) as

$$\hat{U}^{\dagger}(\hat{x},\,t)\left(i\hbar\frac{\partial}{\partial t}\right)\hat{U}(\hat{x},\,t) = i\hbar\frac{\partial}{\partial t} + \hat{x}F(t), \tag{10.10}$$

whereas, on the right-hand side, the potential energy term commutes with the operator $\hat{U}$. We get the transformed momentum operator

$$\hat{U}^{\dagger}(\hat{x},\,t)\,\hat{p}\,\hat{U}(\hat{x},\,t) = \hat{p} - \int_0^t F(t')dt'. \tag{10.11}$$

Here, we have applied the commutation relation $[\hat{x},\,\hat{p}] = i\hbar$.

Hence, in the gauge transformation, the Schrödinger wave equation becomes

$$i\hbar\frac{\partial|\tilde{\Psi}(x,\,t)\rangle}{\partial t} = \hat{\tilde{H}}|\tilde{\Psi}(x,\,t)\rangle, \tag{10.12}$$

where the corresponding Hamiltonian is expressed as

$$\hat{\tilde{H}} = \frac{\left(\hat{p} - \int_0^t F(t')dt'\right)^2}{2m} + V_1(\hat{x}). \tag{10.13}$$

The Hamiltonian describes an ultra-cold atom in a periodic optical lattice in the presence of an external time-dependent period force in the laboratory frame. Here, the external force appears the same as the coupling of a gauge-invariant, space-independent vector potential, $A(t)$, with a particle of charge, $e$, that is,

$$eA(t) = \int_0^t dt'\,F(t'). \tag{10.14}$$

This implies that the presence of the modulation in the otherwise time-independent system introduces a vector potential in the dynamical system. Due to the temporal periodicity of the transformed Hamiltonian, the vector potential is zero over the period $T$, that is,

$$\int_0^T dt' \, F(t') = 0. \tag{10.15}$$

Following equation (10.14), the vector potential for the phase-modulated periodic optical lattice, as defined in equation (10.5), is

$$eA(t) = -m\omega_m \Delta L \sin \omega_m t. \tag{10.16}$$

From equation (10.13), we note that the external periodic forcing modifies the momentum of the ultra-cold atom as

$$p(t) = p - \int_0^t F(t')dt'. \tag{10.17}$$

This helps to write the corresponding time-dependent wave vector, $k(t)$, so that

$$k(t) = k - \frac{1}{\hbar}\int_0^t F(t')dt'. \tag{10.18}$$

With the help of the *Housten function*, we write the unknown wave function as

$$|\psi_k(t)\rangle = \exp\left\{-\frac{i}{\hbar}\int_0^t \mathcal{E}(k(t'))dt'\right\}|\psi_{k(t)}\rangle. \tag{10.19}$$

The time periodicity of the system, as defined in equation (10.16), brings a constraint on the periodicity of the wave function, that is, $|\psi_k(t)\rangle = |\psi_k(t+T)\rangle$. Hence, the solution with periodicity in time is written as

$$|\psi_k(t+T)\rangle = \exp\left\{-\frac{i}{\hbar}\mathcal{E}(k)T\right\}|\psi_k(t)\rangle. \tag{10.20}$$

Here, $T = 2\pi/\omega$ is the time period of the periodic external force.

## 10.3 Floquet–Bloch solution

Keeping in view the Houston function given in equation (10.19), we express the state function of the system as

$$|\psi_k(t)\rangle = \exp\left\{-\frac{i}{\hbar}\int_0^t \mathcal{E}(k(t'))dt'\right\}|\psi_{k(t)}\rangle, \tag{10.21}$$

$$=\exp\left\{-\frac{i}{\hbar}\int_0^t \mathcal{E}(k(t'))dt'\right\}\sum_l |l\rangle \exp(ilk(t)d), \tag{10.22}$$

where we have expressed $|\psi_{k(t)}\rangle$ in the Hilbert space spanned by the Wannier states, $\{|l\rangle\}$. We further simplify it by recognizing that we can write

$$\exp\left(-\frac{i}{\hbar}\int_0^t \mathcal{E}(k(t'))dt'\right) = \exp\left(-\frac{i}{\hbar}\int_0^t \{\mathcal{E}(k(t')) - \mathcal{E}(k)\}dt'\right) \times \exp\left(-\frac{i}{\hbar}\mathcal{E}(k)t\right), \tag{10.23}$$

so that we can express equation (10.22) as

$$|\psi_k(t)\rangle = |u_k(t)\rangle e^{-i\mathcal{E}(k)t/\hbar}, \tag{10.24}$$

where

$$|u_k(t)\rangle = \sum_l |l\rangle \exp(ilk(t)d) \exp\left(-\frac{i}{\hbar}\int_0^t (\mathcal{E}(k(t')) - \mathcal{E}(k))dt'\right), \tag{10.25}$$

which satisfies the requirement $|u_k(t)\rangle = |u_k(t+T)\rangle$. Interestingly, $\mathcal{E}(k)$ corresponds to the time-independent energy from Bloch theory, and $\int_0^t \mathcal{E}(k(t'))dt'$, being time periodic, has its origin in Floquet analysis.

In a more general way, we may express the time-dependent wave function of the one-dimensional phase-modulated optical potential as

$$|\psi_{nk}(x, t)\rangle = e^{-i(kx - \varepsilon_{nk}t/\hbar)}|w_{nk}\rangle, \tag{10.26}$$

where

$$|w_{nk}(x, t)\rangle = \exp\left\{-\frac{i}{\hbar}\int_0^t dt'\{\mathcal{E}_n[k - eA(t')/\hbar] - \varepsilon_{nk}\}\right\}|u_{nk(t)}(x)\rangle. \tag{10.27}$$

Here, $|u_{nk}\rangle$ is the solution of the time-independent Schrödinger equation (10.7) for $F = 0$, and $\varepsilon_{nk}$ is the corresponding eigen energy depending upon the quantum number $n$.

## 10.4 Dispersion relation

From equations (10.19) and (10.20), the quasi-energy of the system, $\mathcal{E}(k)$, is obtained as

$$\begin{aligned}\mathcal{E}(k) &= \frac{1}{T}\int_0^T \mathcal{E}(k(t))dt,\\ &= \frac{1}{T}\int_0^T \mathcal{E}\left(k - \frac{1}{\hbar}\int_0^t F(t')dt'\right)dt.\end{aligned} \tag{10.28}$$

The dispersion relation for an ultra-cold atom in an unmodulated optical lattice is derived in chapter 6, that is,

$$\mathcal{E}(k) = -J\cos kd,$$

and we recall $J$ as the time-independent tunneling matrix element. We substitute the expression in equation (10.28); therefore,

$$\mathcal{E}(k) = -J\frac{1}{T}\int_0^T \cos\left(kd - \frac{d}{\hbar}\int_0^t F(t')dt'\right)dt. \tag{10.29}$$

In order to solve equation (10.29), we consider the general expression for the force written in equation (10.6), such that $F(t) = F_0 - F_1 \cos \omega_m t$. The constant force, $F_0$, may originate as the cold atom experiences gravity during its interaction with the optical lattice. This constant force helps in manipulating the atom, and contributes in an interesting way as we note that the periodicity condition may lead to integer multiples of transfers of 'photons'. It is a condition for the existence of bands, that is,

$$\int_0^T F(t')dt' = 2n_p\hbar k_L, \tag{10.30}$$

where $n_p$ is an integer. The presence of a constant force limits the periodicity in space and leads to Wannier–Stark states as eigenfunctions. Thus, a tilted potential has equal spacings in its energy spectrum, as the energy becomes $n_p\hbar\omega$, as discussed in chapter 6. As a conclusion, we note that the presence of constant force introduces transitions for $n_p \neq 0$ in the presence of phase modulation. Equation (10.29) hence can be written as

$$\mathcal{E}(k) = -J\frac{1}{T}\int_0^T \cos\left(kd - \frac{F_0 d}{\hbar}t' + \frac{F_1 d}{\hbar\omega_m}\sin\omega_m t'\right)dt'.$$

With the help of the condition given in equation (10.30), the above equation takes the shape

$$\mathcal{E}(k) = -J\frac{1}{T}\int_0^T \cos\left(kd - n_p\omega_B t' + \frac{F_1 d}{\hbar\omega_m}\sin\omega_m t'\right)dt'. \tag{10.31}$$

Here, $\omega_B\left(=\frac{F_0 d}{\hbar}\right)$ is the Bloch oscillation frequency.

We solve equation (10.31) in two steps:

(i) We apply the following trigonometric identity:

$$\cos(A - B) = \cos A \cos B + \sin A \sin B,$$

where $A = kd$ and $B = n_p\omega_B t + \frac{F_1 d}{\hbar\omega_m}\sin\omega_m t$, and obtain

$$\begin{aligned}\mathcal{E}(k) = &- J\cos(kd)\frac{1}{T}\int_0^T \cos\left(n_p\omega_B t' - \frac{F_1 d}{\hbar\omega_m}\sin\omega_m t'\right)dt' \\ &- J\sin(kd)\frac{1}{T}\int_0^T \sin\left(n_p\omega_B t' - \frac{F_1 d}{\hbar\omega_m}\sin\omega_m t'\right)dt'.\end{aligned} \tag{10.32}$$

(ii) We use the Bessel function identities

$$\frac{1}{2\pi}\int_0^{2\pi} d\theta \cos\left(\nu_p\theta - z\sin\theta\right) = J_\nu(z), \tag{10.33}$$

$$\frac{1}{2\pi}\int_0^{2\pi} d\theta \sin\left(\nu_p\theta - z\sin\theta\right) = 0. \tag{10.34}$$

Here, we put $\theta = \omega_m t$, $\nu_p = n_p\frac{\omega_B}{\omega_m}$, and $z = \frac{F_1 d}{\hbar\omega_m}$, and substitute equations (10.33) and (10.34) in equation (10.32). Therefore, the simplified dispersion relation for a phase-modulated optical lattice in the presence of a linear potential becomes

$$\mathcal{E}(k) = -JJ_{\nu_p}\left(\frac{F_1 d}{\hbar\omega_m}\right)\cos kd = -J_{\text{eff}}\cos kd. \tag{10.35}$$

(a) In the case of a resonance condition between $\omega_B$ and $\omega_m$ (i.e. $\omega_B = \omega_m$), we note that $\nu_p$ becomes an integer (i.e. $\nu_p = n_p$). This leads to

$$J_{\text{eff}} = JJ_{n_p}\left(\frac{F_1 d}{\hbar\omega_m}\right). \tag{10.36}$$

Here, $J_{n_p}$ is the Bessel function of the order $n_p$. Therefore, we obtain the dispersion relation

$$\mathcal{E}(k) = -JJ_{n_p}\left(\frac{F_1 d}{\hbar\omega_m}\right)\cos kd. \tag{10.37}$$

(b) The values of $\nu_p$ and $n_p$ are related to the linear potential. In the absence of a linear potential, which is responsible for the constant force $F_0$, we find $n_p = 0$. In this case, we obtain

$$J_{\text{eff}} = -JJ_0\left(\frac{F_1 d}{\hbar\omega_m}\right), \tag{10.38}$$

leading to

$$\mathcal{E}(k) = -JJ_0\left(\frac{F_1 d}{\hbar\omega_m}\right)\cos kd, \tag{10.39}$$

where $J_0$ is the zeroth-order Bessel function. Equation (10.39) defines the dispersion relation for an ultra-cold atom in a phase-modulated optical lattice that has both space and time periodicity.

(c) We note that, for $F_1 = 0$,

$$J_{\text{eff}} = JJ_0\left(\frac{F_1 d}{\hbar\omega_m}\right) = J.$$

This makes equation (10.39) the dispersion relation for an ultra-cold atom in an optical lattice, as obtained in chapter 6, that is,

$$\mathcal{E}(k) = -J \cos kd.$$

From equation (10.35), we note that $\mathcal{E}(k)$ shows oscillatory behavior with respect to quasi-momentum $k$. Furthermore, equations (10.36) and (10.38) show that the effective tunneling exhibits variation as a function of scaled modulation amplitude, $\frac{F_1 d}{\hbar\omega_m}$. In addition, at the zeros of the Bessel function the dispersion relation attains a zero value for all $k$, displaying a *band-narrowing* phenomenon.

## 10.5 Bloch acceleration and group velocity

Let us consider an initial wave packet centered around a certain momentum, that is, $p = \hbar k(0)$. Its center evolves in time as

$$\hbar \dot{k}(t) = F(t), \tag{10.40}$$

which is the Bloch acceleration theorem. In phase-modulated optical lattices, the external forcing is expressed as $F_1 \cos \omega_m t$. Therefore, from equation (10.18), we find that

$$k(t) = k(0) + \left(\frac{F_1}{\hbar\omega_m}\right) \sin \omega_m t. \tag{10.41}$$

The material wave packet evolving in the time-dependent system as a superposition of wavelets is expressed in equation (10.25). The wave packet has a group velocity, which is obtained as the derivative of the dispersion relation $\mathcal{E}_k$. Therefore, equation (10.39) leads to

$$\begin{aligned} v_g &= \left.\frac{\partial \mathcal{E}(k)}{\hbar \partial k}\right|_{k(0)}, \\ &= J_{\text{eff}} d \sin k(0) d. \end{aligned} \tag{10.42}$$

Here, the effective hopping matrix element, $J_{\text{eff}}$, for the atom in the phase-modulated optical lattice is

$$J_{\text{eff}} = J J_0\left(\frac{F_1 d}{\hbar\omega_m}\right).$$

## 10.6 Effective mass

Following the discussion in section 5.3, we calculate the effective mass corresponding to the driven system. We expand the single-band energy using Taylor expansion, that is,

$$\mathcal{E}(k) = \mathcal{E}(0) + \left.\frac{\partial \mathcal{E}(k)}{\partial k}\right|_{k(0)} (k - k(0)) + \frac{1}{2!}\left.\frac{\partial^2 \mathcal{E}(k)}{\partial k^2}\right|_{k(0)} (k - k(0))^2, \tag{10.43}$$

and ignore the higher powers. On the right-hand side, the first term describes the constant energy value corresponding to the center of the wave packet, while the second term expresses the group velocity. The third term, when expressed as

$$\frac{\hbar^2(k - k(0))^2}{2m^*},$$

expresses the effective mass, $m^*$, that is,

$$m^* = \hbar^2 \left( \frac{\partial^2 \mathcal{E}(k)}{\partial k^2} \bigg|_{k(0)} \right)^{-1}. \tag{10.44}$$

Hence, we write

$$m^* = \hbar^2 J_{\text{eff}} d^2 \cos k(0)d. \tag{10.45}$$

## 10.7 Dynamical localization

The Floquet–Bloch solution, given in equation (10.20), has three interesting features with regard to the dynamics in the time periodic system. First, the state function $|\psi_k(t)\rangle$ for the system in the presence of an external forcing is expressed in the same way as that of a time-independent system. Second, following the Floquet states characteristic, the state function $|u_k(t)\rangle$ returns to its initial state after a complete time period $T$. Third, we find that there exist values of the amplitude of the external force, $F_1$, which correspond to the zeros of the Bessel function, $J_\nu$. At these values the state function modifies itself drastically. Using the dispersion relation, we calculate the width of the lowest band as the difference of energy at $k = 0$ and $k = \frac{\pi}{d}$, that is,

$$\begin{aligned} \Delta\mathcal{E} &= \mathcal{E}(k = 0) - \mathcal{E}(k = \pi/d) = 2J_{\text{eff}}, \\ &= 2JJ_\nu\left(\frac{F_1 d}{\hbar\omega_m}\right). \end{aligned} \tag{10.46}$$

Therefore, the width of the lowest band approaches zero at particular values of scaled modulation amplitude, $F_1 d/\hbar\omega_m$. This is *band narrowing*, as shown in figure 10.2, and is a manifestation of the fact that the phasers, $e^{-i\mathcal{E}(k)t/\hbar}$, corresponding to each $|\psi_k(t)\rangle$ approach unity as the effective tunneling $J_{\text{eff}}$ becomes zero. As a result, the material wave packet displays restricted dynamics and freezes for certain scaled modulation amplitudes $F_1 d/\hbar\omega_m$, which is known as *dynamical localization*.

*Group velocity and effective mass.* We find that for any initial momentum $\hbar k(0)$ the group velocity is zero when $J_{\text{eff}} = 0$, which is possible when $F_1 d/\hbar\omega_m$ corresponds to zeros of the Bessel function. Hence, the wave packet dynamics freezes and is restricted by the external forcing for some windows of modulation $F_1$, in contrast to time-independent dynamics. This defines dynamical localization. We find that as the group velocity $v_g$ approaches zero, the effective mass $m^*$ reaches an infinite value, thus limiting the dynamics in phase space and suppressing the dispersion.

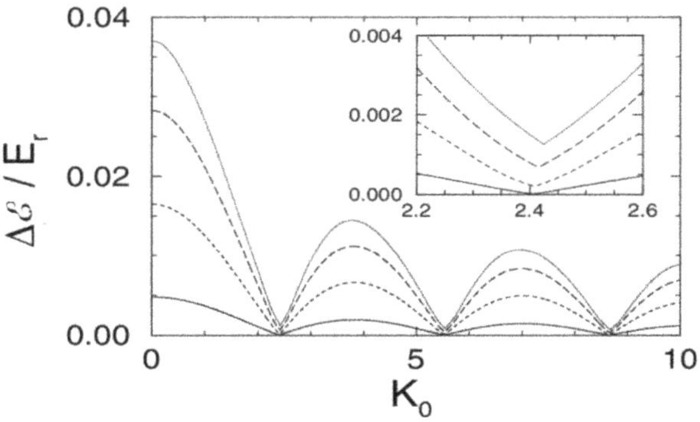


**Figure 10.2.** We show the variation in width of the lowest band $\Delta_{\text{sin}}$ scaled by recoil energy $E_r = \frac{\hbar^2 k^2}{2m}$ as a function of $K_0 = \frac{F_1 d}{\hbar \omega m}$ for various values of $J$ corresponding to $V_0/E_r = 2$ (dots), 3 (long dashes), 5 (short dashes), and 10 (full line). The inset quantifies the extent of the band narrowing at the first zero, 2.405, of the Bessel function $J_0$. Reprinted figure with permission from [7], Copyright (2009) by the American Physical Society.

*Quantum supremacy over classical chaos.* In its classical evolution an ensemble of particles displays diffusion as it evolves in a system experiencing external periodic forcing in time. Dynamical localization, however, restricts the diffusive behavior in the quantum domain. In contrast to the classical counterpart, the corresponding quantum system follows classical evolution only up to a certain time, after which it displays suppression of the diffusion. The time in which the quantum dynamics mimics the corresponding classical dynamics is named as the quantum break time of the dynamical system. Before the quantum break time, the quantum system follows classical evolution.

## 10.8 Dynamical de-localization

'What happens to the energy spectrum in general when an external force is applied to a system?' is a non-trivial question. The answer to the question is based on rigorous mathematical research work that explains the evolution of a time-dependent system based on the time-independent spectrum of the system. For this purpose, we divide all the time-independent systems available in nature into two general classes: strong binding systems, which have a time-independent spectrum such that the *level spacing always increases* as we go up in energy, and therefore we need to provide higher energy by means of external forcing to go to higher states; and weak binding systems, which have a time-independent spectrum such that the *level spacing always decreases* as we go up in energy. The following two theorems reveal the dynamics of the time-dependent energy spectrum as a function of the time-independent spectrum of the system.

*Howland's theorem.* If the time-independent spectrum of energy, $E_n$, has energy levels such that the level spacing always increases going up in energy, that is,

$$E_{n+1} - E_n = cn^s, \tag{10.47}$$

where $c > 0$, and $s > 0$, then for a sufficiently smooth external modulation, for which a sufficient number of derivatives exist, we find a pure-point spectrum for the Floquet operator [1–5]. The existence of the pure-point spectrum is evidence of dynamical localization.

*DSS theorem.* In their work, Delyon, Soulliard, and Seimen (DSS) calculated for Anderson-like systems, that if it is possible to reduce the probability amplitude equation of motion to a difference equation, that is,

$$C_{n+1} - C_{n-1} = \frac{C_n}{\sqrt{\epsilon}}, \tag{10.48}$$

then there exists a critical modulation strength, $\epsilon_c$, above which the spectrum of the system changes from a pure-point spectrum to a purely continuous spectrum [6]. De-Oliveira *et al* have shown that this theorem also works in periodically driven time-dependent systems if the equation of motion for the probability amplitudes can be expressed in the above form. These systems have a time-independent undriven energy spectrum such that the level spacing reduces as we go up in energy, for example as in a triangular well and in a hydrogen atom. The existence of a purely continuous spectrum or a dense spectrum with negligible level spacing is a hallmark of quantum de-localization in a modulated system above a critical modulation strength.

In a quantum de-localization regime, the quantum distributions display diffusion. For this reason, a quantum wave packet maintains its widths and displays localization only up to a critical value of modulation, $\epsilon_c$, whereas beyond this value it starts spreading similar to the classical case.

## 10.9 Ratchet effect: Brownian motors

Richard Feynman revived and popularized the work of Marian von Smoluchowski, in which he introduced the ratchet effect that occurs due to the violation of the thermal equilibrium so that the second law of thermodynamics does not work. The ratchet effect describes the fact that there exists no constraint on the steady transport even in the absence of constant force or gradients [8]. To enable the ratchet effect, the control parameters are used in such a way that all the symmetries are destroyed, which leads to a non-zero average current, $J$, for an atom in a modulated optical system with spatial and temporal periodicity.

We express an atom in an optical lattice using a generalization of the Hamiltonian (10.1) so that

$$H = \frac{p^2}{2m} + V(x, t), \tag{10.49}$$

where

$$V(x, t) = V(x + d, t) = V(x, t + T). \tag{10.50}$$

The symmetries involved in the system dynamics are translational and temporal, defined as

$$S_x[x, t]: [x, p, t] \longrightarrow [-x + d, -p, t + T],$$
$$S_t[x, t]: [x, p, t] \longrightarrow [x + d, -p, -t + T].$$

The symmetries imprint themselves on the Floquet operator and thus influence the transport properties of the Floquet states. The absence of any one of the two symmetries results in a ratchet effect. We can destroy both symmetries by choosing a bi-harmonic modulation such that

$$V(x, t) = \frac{V_0}{2}\cos[2k_{\mathrm{L}}\{x - f(t)\}], \tag{10.51}$$

where

$$f(t) = \Delta L\cos(\omega t) + \Delta L_1\cos(2\omega t + \theta), \tag{10.52}$$

for $\theta \neq 0, \pm\pi$. Hence, the system displays a non-vanishing transport current such that

$$J(t, t_0) = \frac{1}{M}\langle\psi(t, t_0)|\hat{p}|\psi(t, t_0)\rangle. \tag{10.53}$$

Here, $t_0$ is the initial time.

## Exercises

1. Apply a Kramers–Henneberger transformation on the time-dependent Schrödinger wave equation

$$i\hbar\frac{\partial|\Psi(t)\rangle}{\partial t} = \left\{\frac{p^2}{2m} + \frac{V_0}{2}\cos[2k_{\mathrm{L}}(x - \Delta L\sin(\omega_m t))]\right\}\Psi(t). \tag{10.54}$$

   and obtain the effective Hamiltonian in equation (10.3).
2. The Hamiltonian of a particle of mass $m$ in a modulated exponentially decaying field in the presence of gravity is described as

$$H = \frac{p^2}{2m} + mgx + V_0 e^{-\kappa(x+\epsilon\sin\omega t)}. \tag{10.55}$$

   Write the transformed Hamiltonian by applying the Kramers–Henneberger transformation.
3. Calculate the effective vector potential for the system defined in exercise 2.
4. Calculate the group velocity $v_g$ and effective mass $m^*$ for the system defined in exercise 2.
5. Calculate the transport current for a cloud of atoms, with initial quasi-momentum as zero, that can be understood as the initial cloud of atoms distributed over many lattice sites. The atoms are evolving in the bi-harmonic modulation defined in equation (10.51).

## References


[1] Howland J S 1979 Scattering theory for Hamiltonians periodic in time *Indiana Univ. Math. J.* **28** 471

[2] Howland J S 1987 Perturbation theory of dense point spectra *J. Funct. Anal.* **74** 52–80

[3] Howland J S 1989 Floquet operators with singular spectrum. I *Ann. Inst. Henri Poincaré* **49** 309–23

[4] Howland J S 1989 Floquet operators with singular spectrum. II *Ann. Inst. Henri Poincaré* **50** 325–34

[5] Joye A 1994 Absence of absolutely continuous spectrum of Floquet operators *J. Stat. Phys.* **75** 929–52

[6] Delyon F, Simon B and Souillard B 1985 From power pure point to continuous spectrum in disordered systems *Ann. Inst. Henry Poincaré* **42** 283

[7] Eckardt A, Holthaus M, Lignier H, Zenesini A, Ciampini D, Morsch O and Arimondo E 2009 Exploring dynamic localization with a Bose-Einstein condensate *Phys. Rev.* A **79** 013611

[8] Astumian R D and Hänggi P 2002 Brownian Motors *Phys. Today* **55** 33–9

[9] Holthaus M 2016 Floquet engineering with quasi-energy bands of periodically driven optical lattice *J. Phys. B: At. Mol. Opt. Phys.* **49** 013001

# Chapter 11

## Atoms in modulated optical lattices—II

*'The symmetries of the causes are to be found in the effects.'*

—Pierre Curie

In experiments on ultra-cold atoms, we also consider optical lattices in the presence of a quadratic confining potential. Experimentally, the quadratic potential can be provided for the atoms by means of a magneto-optic trap. This generates a linearly changing force on the confined quantum particles. The physical system can be described by the governing Hamiltonian, written as

$$\hat{H} = \frac{\hat{p}^2}{2m} + \hat{V}(x) + \hat{V}_{\mathrm{I}}(x). \tag{11.1}$$

The one-dimensional periodic optical lattice potential, expressed as $\hat{V}(x)$ and defined in equation (3.61), is

$$\hat{V}(x) = \frac{V_0}{2} \cos 2k_{\mathrm{L}}\hat{x},$$

while the quadratic confining potential is

$$\hat{V}_{\mathrm{I}}(x) = \frac{1}{2} m\, \omega_{\mathrm{H}}^2\, \hat{x}^2 = F_{\mathrm{I}}\hat{x}^2.$$

Here, $\omega_{\mathrm{H}}$ is the oscillation frequency of the ultra-cold atom in the quadratic confining potential as the optical lattice is switched off.

*Wannier basis.* Following the procedure developed in section 6.1 for optical lattices, we write the above Hamiltonian in the orthonormal Wannier functions

(WFs) basis, $|n\rangle$. In the tight-binding approximation, the Hamiltonian for a single band becomes

$$\hat{H} = -J \sum_{n=-\infty}^{\infty} (|n\rangle\langle n+1| + |n+1\rangle\langle n|) + e_{\rm h} \sum_{n=-\infty}^{\infty} n^2 |n\rangle\langle n|. \tag{11.2}$$

The parameter $e_{\rm h}$ is given as $e_{\rm h} = \frac{1}{2} m\omega_{\rm H}^2 d^2$.

*Quasi-momentum basis.* We use the WF description in terms of Bloch states and re-express the Hamiltonian given in equation (11.2) as

$$\hat{H}(k) = -J \cos \hat{k} d + F_{\rm l} \hat{x}^2. \tag{11.3}$$

Here, $J$ is defined in equation (6.7). We substitute the position operator in quasi-momentum form,

$$\hat{x} = i \frac{\partial}{\partial k},$$

which transforms equation (11.3) into a second-order differential operator in quasi-momentum space, that is,

$$H(k) = -J \cos kd - F_{\rm l} \frac{\partial^2}{\partial k^2}. \tag{11.4}$$

The quasi-momentum $k$ belongs to the Brillouin zone $[-k_{\rm L}, +k_{\rm L}]$, where $k_{\rm L} = \pi/d$.

## 11.1 Linear force on optical lattices

### 11.1.1 Phase space analysis

Equation (11.3) leads us to write the semi-classical equations of motion as

$$\dot{x}(t) = \frac{\partial H}{\hbar\, \partial k} = \frac{Jd}{\hbar} \sin kd, \tag{11.5}$$

which describes the group velocity of the material wave packet, and

$$\hbar \dot{k}(t) = -\frac{\partial H}{\partial x} = -2F_{\rm l} x, \tag{11.6}$$

expresses the linear dependence of acceleration on position. Equation (11.6) is also known as the Bloch acceleration theorem. A direct solution of this equation expresses the quasi-momentum as

$$k(t) = k(0) - 2\frac{F_{\rm l}}{\hbar} \int_0^t x(t)\, dt.$$

The two coupled equations, (11.5) and (11.6), are solved together to obtain the semi-classical solution of the system, that is,

$$\ddot{k}(t) = -\frac{2JdF_{\rm l}}{\hbar^2} \sin(kd). \tag{11.7}$$

In the limit of very small values of $k$, the above equation reduces to the harmonic oscillator equation, that is,

$$\ddot{k}(t) = -\omega_{\mathrm{D}}^2 k,$$

where

$$\omega_{\mathrm{D}}^2 = \frac{2Jd^2F_1}{\hbar^2} = \frac{mJd^2\omega^2}{\hbar^2}. \tag{11.8}$$

To solve equation (11.7), we may consider a time-dependent solution to be

$$k(t) = k(0)\cos(\eta t),$$

where $\eta$ is an unknown frequency. This reduces equation (11.7) as

$$-k(0)\eta^2\cos(\eta t) = -\frac{2JdF_1}{\hbar^2}\sin(k(0)\,d\cos(\eta t)) \tag{11.9}$$

$$= -\frac{2JdF_1}{\hbar^2}\sum_{j=0}^{\infty} J_{2j+1}(k(0)\,d)\sin\left[(2j+1)\left(\eta t + \frac{\pi}{2}\right)\right]. \tag{11.10}$$

Here, $k(0)d \ll 1$. On comparing the coefficients of $\cos(\eta t)$ on both sides, we obtain

$$\eta^2 = \frac{2JdF_1}{\hbar^2 k(0)} J_1(k(0)\,d), \tag{11.11}$$

where the frequency of the oscillations depends upon the initial value of the quasi-momentum $k(0)$. Thus, the solution to $k(t)$ and $x(t)$ is given as

$$k(t) = k\cos(\eta t), \tag{11.12}$$

whereas

$$x(t) = \frac{\hbar k(0)}{2F_1}\eta\sin(\eta t) = \sqrt{\frac{Jk(0)d}{m\omega^2}}\,[J_1(k(0)d)]^{1/2}\sin(\eta t). \tag{11.13}$$

We consider $k(0)d \ll 1$ and, therefore, ignore $j \neq 0$ terms. Hence, the particle evolves in time with an initial quasi-momentum $k = k(0)$, and follows oscillation dynamics in the neighborhood of $(x, k) = (0, 0)$. It moves with a frequency $\eta$, as shown in figure 11.1.

### 11.1.2 Bloch functions and energy bands

In the tight-binding approximation, equation (11.4) leads to a Schrödinger-like second-order differential equation in quasi-momentum space:

$$\left[-J\cos kd - F_1\frac{\partial^2}{\partial k^2}\right]\psi(k) = \epsilon\,\psi(k). \tag{11.14}$$

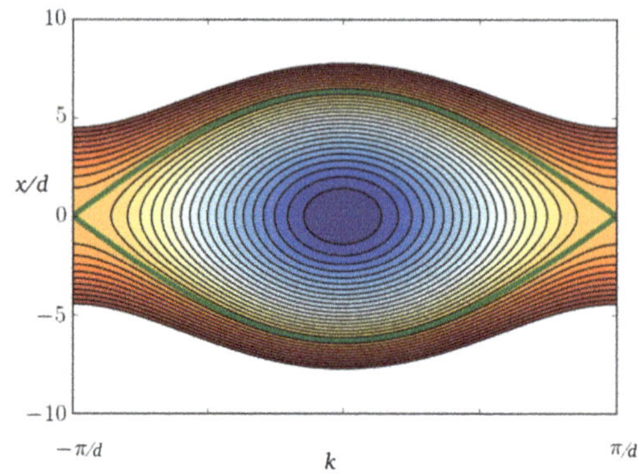


**Figure 11.1.** Phase space corresponding to the Hamiltonian (equation (11.3)). The green contours mark the separatrix, as the boundary at which the dynamics changes from oscillatory (within) to rotational (outside) of the separatrix. Adapted with permission of Christian Hill, from https://scipython.com/blog/the-separatrix-for-a-simple-pendulum/ (2016).

The substitution of $kd = 2\theta - \pi$, $q = \frac{2J}{F_1 d^2}$, and $\alpha = \frac{4\epsilon}{F_1 d^2}$ simplifies equation (11.14) to a Mathieu equation, that is,

$$\left[\frac{\partial^2}{\partial\theta^2} - 2q\cos 2\theta + \alpha\right]\psi(\theta) = 0. \tag{11.15}$$

Here, the Mathieu function, $\psi(\theta)$, and the Mathieu characteristic parameter, $\alpha(q)$, respectively, describe the eigenfunction and the eigenenergy. Equation (11.15) has $\pi$ periodicity, which will appear in the corresponding eigenfunction $\psi(\theta)$. A detailed discussion of the eigenfunctions and eigenenergies of the Mathieu equation is provided in chapter 5.

*Bloch functions.* Equation (11.15) states that for a fixed $q$ there occurs a countable infinite number of solutions, expressed by the index $\nu$. However, only for specific values of the parameter $\alpha_\nu$ are the solutions periodic. We write the periodic eigenfunction of the Mathieu equation by means of Bloch functions, that is,

$$\psi_\nu(\theta) = e^{i\nu\theta}u(\theta), \tag{11.16}$$

where $\nu$ is the characteristic exponent of the Mathieu function and $u(\theta)$ is a periodic function. If $\nu$ is an even integer, $u(\theta)$ is a $\pi$-periodic symmetric function, whereas if $\nu$ is an odd integer, $u(\theta)$ is a $2\pi$-periodic asymmetric function. The corresponding values of the Mathieu characteristic parameter are expressed as $a_\nu$ and $b_\nu$, respectively. In the quasi-momentum expression, we write

$$\psi_\nu(k) = e^{i\nu(kd+\pi)/2}u(kd). \tag{11.17}$$

In the WFs basis, we may express the solution as

$$\psi_\nu(k) = e^{i\nu(kd+\pi)/2}\langle k|u\rangle = e^{i\nu(kd+\pi)/2}\sum_n \langle k|n\rangle\langle n|u\rangle \tag{11.18}$$

$$=\mathrm{e}^{i\nu(kd+\pi)/2}\sum_n C_n(\nu,\, q)\mathrm{e}^{inkd}, \tag{11.19}$$

$$=\mathrm{e}^{i\nu\pi/2}\sum_n C_n(\nu,\, q)\mathrm{e}^{i(\nu+2n)kd/2}, \tag{11.20}$$

where $C_n(\nu, q) = \langle n|u\rangle$. We substitute equation (11.20) into the Mathieu equation, which reveals the following difference equation:

$$[\alpha_\nu(q) - (\nu + 2m)^2 d^2/4]C_m(\nu,\, q) + 2q[C_{m-1}(\nu,\, q) + C_{m+1}(\nu,\, q)] = 0. \tag{11.21}$$

The Mathieu characteristic parameters $\alpha_\nu(q)$ may be found by locating the zeros of Hill's determinant, namely, the roots of the characteristic equation associated with the above recurrence equation. Once the roots of interest are found, the expression for $C_m(\nu, q)$ is obtained recursively from the difference equation.

*Energy bands.* We develop an analytical understanding of the energy expression $\epsilon_\nu(q)$ as we consider the corresponding polynomial expressions for $q < 1$. Corresponding to $\pi$-periodic symmetric solutions, we have

$$\epsilon_\nu(q) = \frac{F_1 d^2}{4}\left\{\nu^2 + \frac{1}{2(\nu^2-1)}q^2 + \frac{5\nu^2+7}{32(\nu^2-1)^3(\nu^2-4)}q^4 + \cdots\right\}. \tag{11.22}$$

It is interesting to note that, for $q \to 0$, the energy spectrum displays quadratic dependence on $\nu$. In the other limiting case, when $q \gg 1$, we find

$$a_\nu \approx b_{\nu+1} = -2q + 2s\sqrt{q} - \frac{s^2+1}{2^3} - \frac{s^3+3s}{2^7\sqrt{q}} - \cdots, \tag{11.23}$$

where $s = 2\nu + 1$, which makes the energy expression

$$\begin{aligned}\epsilon_\nu(q) = \frac{F_1 d^2}{4}\Bigg\{&-2q + 4\sqrt{q}\left(\nu + \frac{1}{2}\right) - \frac{(2\nu+1)^2+1}{8}\\ &+ \frac{[(2\nu+1)^3 + 3(2\nu+1)]}{128\sqrt{q}} + \cdots\Bigg\}.\end{aligned} \tag{11.24}$$

Here, in the deep optical lattice limit, that is, $q \gg 1$, the width of the $n$th energy band is defined as

$$\delta\alpha = b_{\nu+1} - a_\nu \simeq \frac{2^{4\nu+5}\sqrt{\dfrac{2}{\pi}}\, q^{\frac{\nu}{2}+\frac{3}{4}}\exp(-4\sqrt{q})}{\nu!}, \tag{11.25}$$

which shows that in the deep optical lattice ($q \gg 1$), energy bands are realized as degenerate energy levels as the band width is negligible [1]. In the band structure of a parabolic optical lattice for large $q$, thin bands are seen near the bottom of the lattice. The band width increases and the band gap decreases for excited bands with increasing band index. From equation (11.14), the hopping matrix element, $J$, is

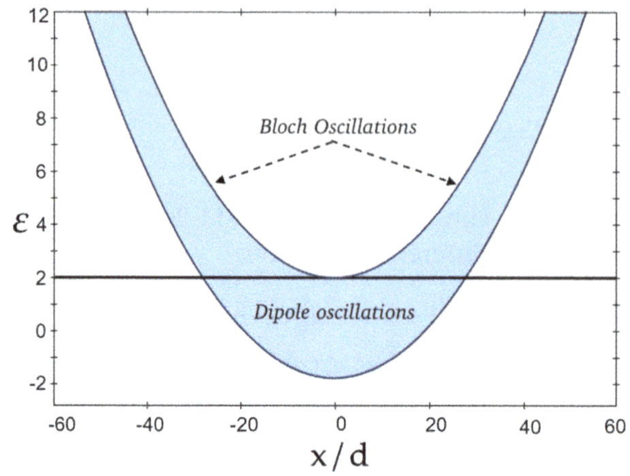


**Figure 11.2.** The lowest band structure is shown in a parabolic optical lattice. The thick horizontal line corresponds to a separatrix in phase space, shown also in figure 11.1. The region below the separatrix line supports dipole oscillations, whereas above the line we find Bloch oscillations. Adapted figure with permission from [9], Copyright (2014) by the American Physical Society.

related to the width of the band, that is, $\delta\alpha = 2J$. This explains the tunneling between adjacent sites for a deep optical lattice.

*Dispersion relation.* In the tight-binding limit, we obtain equation (11.3), which describes the energy of a particle in an optical lattice in the presence of a harmonic confinement as

$$\epsilon_\nu = -J\cos kd + F_1 x^2. \tag{11.26}$$

In the limiting case of the Brillouin zone, $k = 0$ and $k = \pi/d$, we obtain two distinct values for the lowest $\nu$, that is,

$$\epsilon_- = -J + F_1 x^2, \tag{11.27}$$

$$\epsilon_+ = +J + F_1 x^2. \tag{11.28}$$

This introduces a hyperbolic-shaped lowest band in energy space, as shown in figure 11.2. The particle follows harmonic oscillator dynamics with low-lying states, as shown in figure 11.2. The thick horizontal line corresponds to a separatrix in phase space, shown in figure 11.1. The region below the separatrix line supports dipole oscillations, where we find vibration dynamics, whereas above the line we find Bloch oscillations, indicating rotation dynamics [1, 2].

### 11.1.3 Dipole oscillations and Bloch oscillations

*Dipole oscillations.* Following equation (11.24), we find that for $q \gg 1$ and restricting ourselves to the linear case, the energies of the system appear in the lowest order as

$$\epsilon_\nu(q) \approx \frac{F_1 d^2}{4}\left[-J + 4\sqrt{q}\left(\nu + \frac{1}{2}\right)\right], \tag{11.29}$$

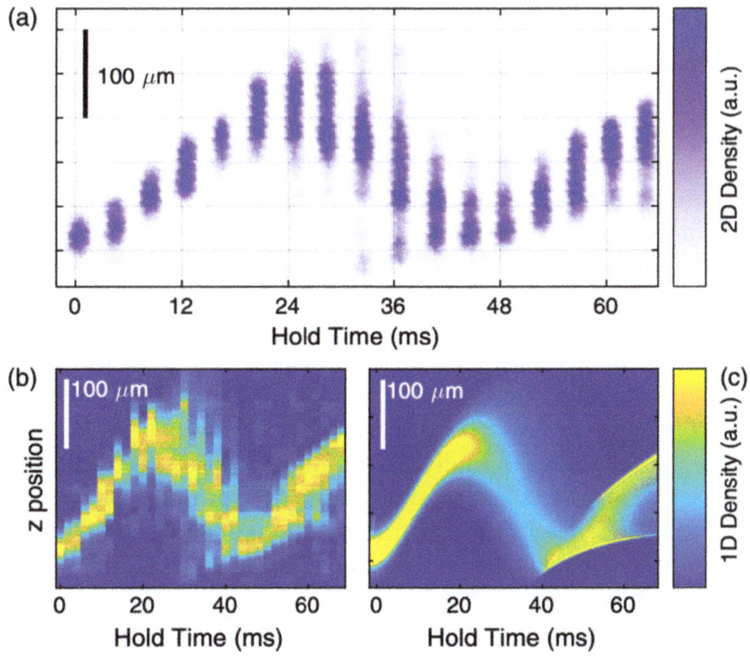


**Figure 11.3.** Position-space Bloch oscillations. (a) A time sequence of in situ absorption images of atoms in a deep optical lattice, subjected to a force corresponding to a Bloch frequency of 21.1 Hz. Each image corresponds to an individual experimental run with the indicated hold time. (b) Measured evolution of integrated density distribution as a function of axial position during a Bloch oscillation. (c) Corresponding numerical prediction for (b). Asymmetrical width variation is due to force inhomogeneity. Reproduced with permission from [2], copyright (2018) by the American Physical Society.

This corresponds to an equally spaced energy spectrum of a harmonic oscillator. Hence, the evolution of the particle follows harmonic oscillations with a frequency

$$\omega_{\rm D} = \frac{F_1 d^2}{4\hbar} 4\sqrt{q} = \frac{F_1 d^2}{\hbar}\sqrt{\frac{2J}{F_1 d^2}} = \sqrt{\frac{2F_1 d^2 J}{\hbar^2}}. \tag{11.30}$$

The dipole oscillation frequency is the same as that obtained in the semi-classical domain for $k$ approaching zero, and given in equation (11.8). Thus, the wave packet transverses both sides of the parabolic optical lattice in the high-$q$ regime. It displays oscillatory dynamics with frequency $\omega_{\rm D}$. The dipole oscillation dynamics for a narrowly peaked wave packet that initially resides within an optical lattice site and for a broad wave packet that spreads over a few optical sites are shown in figure 11.3.

*Bloch oscillations.* Following the relation obtained in equation (11.27), we find two turning points:

$$x_{c1} = +\sqrt{\frac{\epsilon + J}{F_1}}, \quad \text{and} \quad x_{c2} = -\sqrt{\frac{\epsilon + J}{F_1}}, \tag{11.31}$$

which define classical turning points. Equation (11.28) provides two more turning points:

$$x_{b1} = +\sqrt{\frac{\epsilon - J}{F_1}}, \quad \text{and} \quad x_{b2} = -\sqrt{\frac{\epsilon - J}{F_1}}, \tag{11.32}$$

which are inward points on the lowest band, as shown in figure 11.2, and referred to as Bragg turning points. The region between them is Bragg forbidden. In phase space they correspond to the rotation motion. Keeping in view figure 11.2, the Bloch oscillation in the left arm and right arm of the band have clockwise and anti-clockwise rotations, respectively.

Bloch oscillation of a broad atomic wave packet is shown in figure 11.3. The initial atomic wave packet spreads over many lattice sites. It displays minimum dispersion and revives at the Bloch period, $T_B$. However, a narrowly peaked Gaussian wave packet resides initially in an optical lattice site. It displays coherent dynamics and spreads as a coherent superposition over many sites of the periodic optical lattice. It also displays a reconstruction at a Bloch period, $T_B$. However, for an initial atomic wave packet that has a spread over many lattice sites, the wave packet displays minimum dispersion.

The contrast in the time evolution has its roots in the corresponding momentum space. The Fourier transform of the coordinate space Gaussian distribution is also a Gaussian with a momentum spread. For a narrowly peaked wave packet in the coordinate space the momentum distribution is broad, thus implying that the wave packet is spread over a broader region. By contrast, a broad coordinate space Gaussian distribution has a narrow spread in momentum space, which leads to plane-wave-like evolution with a dominant average momentum.

A weakly interacting Bose–Einstein condensate in the presence of periodic modulation near Bloch frequency exhibits super Bloch oscillations as giant center-of-mass oscillations in position space [10].

## 11.2 Modulated optical lattice

A periodic optical potential in the presence of a quadratic potential simulates a solid-state crystal in the presence of a quadrupole potential. The presence of amplitude modulation adds another degree of freedom that leads to complexity in the inherent atomic dynamics. Cold atoms in amplitude-modulated optical lattices have shown inter-band transitions, dynamical localization, dynamical tunneling, and led to characterization of the Mott insulator regime.

A so-called twinkling matter–field system is expressed by the governing Hamiltonian, $H(x, t)$, that is,

$$H(x, t) = \frac{p^2}{2m} + V_1(x) + V(t)\sin^2 kx, \tag{11.33}$$

where the external periodic in time potential is represented by $V(t)$. For the sake of simplicity, we write the physical system along one dimension, considered the $x$-axis. Here, the first term in the Hamiltonian describes the center-of-mass momentum of a cold atom of mass $m$. The second term in equation (11.33) expresses the confining harmonic potential for the atom, such that

$$V_l(x) = \frac{1}{2}m\omega_H^2 x^2 = F_l x^2. \tag{11.34}$$

Here, $\omega_H$ is the frequency associated with the atomic dynamics in the quadratic potential. The third term introduces a one-dimensional optical lattice with periodic in time modulation in amplitude, $V(t)$, such that

$$V(t + T) = V(t). \tag{11.35}$$

The amplitude modulation is eventually related to the intensity modulation of the optical lattice. It has static and dynamic parts, defined as

$$V(t) = V_0(1 + \epsilon \cos \omega t). \tag{11.36}$$

The time-independent or static part of the optical potential, $V_0$, is proportional to the unmodulated intensity. The time-dependent or dynamic parts, $\epsilon$ and $\omega$, are the amplitude and frequency of external modulation, respectively. Hence, the modulated optical potential, $V_{opt}(x, t)$, becomes

$$\begin{aligned} V_{opt}(x, t) &= V(t)\sin^2 k_L x, \\ &= \frac{V_0}{2}(1 + \epsilon \cos \omega t)(1 - \cos 2k_L x), \end{aligned} \tag{11.37}$$

such that

$$V_{opt}(x + d) = V_{opt}(x). \tag{11.38}$$

Here, $d = \frac{\pi}{k_L}$ describes the lattice constant.

In the absence of the confining potential, $V_c$, the system is periodic in space and time, whereas in the presence of a quadratic confining potential the system is symmetric in space and periodically changing in the time domain.

## 11.3 Mechanical action of light on atoms

### 11.3.1 Band formation

From equation (11.33), we note that in the absence of a harmonic confinement, that is, $F_l = 0$, the system has periodicity in space and in time. Furthermore, in the limit of a vanishing lattice depth ($V_0$ approaching zero), the band structure can be constructed by the superposition of parabolic energy spectra centered around all reciprocal points $E_n(k) = \hbar^2(k - nk_L)^2/2m$. Here, $n$ is an integer and $k_L = \pi/d$. For instance, bands 2 and 3 are constructed from the parabolas centered at $\pm \hbar k_L$, as shown in figure 11.4.

Within the limit of a small amplitude modulation, the modulation drives inter-band transitions that keep the quasi-momentum, $\hbar k$, unchanged. The inter-band transition frequencies that promote an atom from parabola $n$ to $n'$ are given by

$$\pm\omega_{n\to n'} = (E_n - E_{n'})/\hbar. \tag{11.39}$$

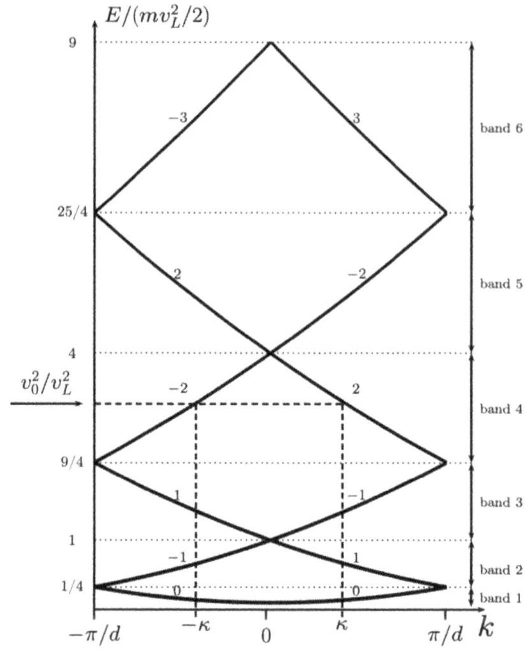


**Figure 11.4.** A band diagram is constructed by the superposition of parabolic energy spectra centered around all reciprocal points $E_n(k) = 2\hbar^2(k - nk_L)^2/2m$. For each band, we have indicated the index $n$ of the parabola from which the corresponding branches have been constructed. Reprinted figure with permission from [4], Copyright (2013) by the American Physical Society.

The sign + (−) corresponds to the transition to a lower (upper) band. By energy conservation, we have

$$\frac{1}{2}mv_0^2 = E_n(k).$$

Combining the above two equations, we obtain

$$\pm\omega_{n\to n'} = \frac{(E_n - E_{n'})}{\hbar} = (n - n')\frac{2\pi v_0}{d} - (n - n')^2\omega_L \tag{11.40}$$

In figure 11.5, we have represented schematically the experimentally observed depletion lines in velocity space. As a first example, let us consider the green dotted depletion line (see figure 11.5). The incident velocity domain $3v_L/2 < v_0 < 2v_L$ corresponds to a resonant transition between bands 4 and 3. According to figure 11.4, we have *a priori* two possibilities: either a transition from parabola $n = 2$ to parabola $n' = -1$ or a transition from parabola $n = -2$ to parabola $n' = 1$ occur at $+\kappa$ and $-\kappa$, respectively. The slope of the branch $n = -2$ is positive, which thus corresponds to a propagation from left to right (indicating a positive group velocity). The states of the incoming packet with positive velocity in the range of

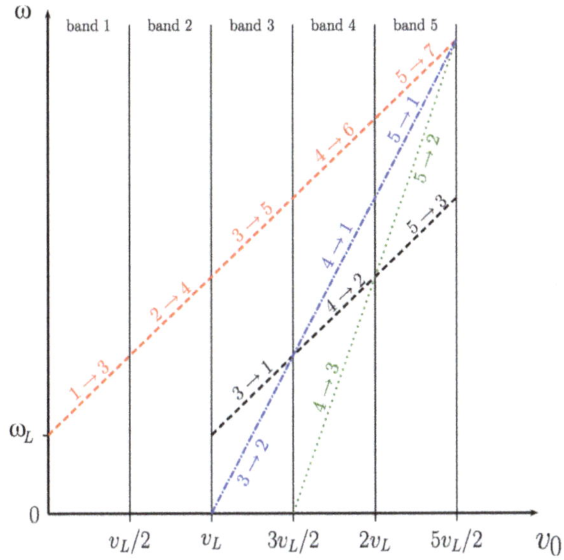


**Figure 11.5.** Schematic of the different depletion lines along with the numbers of the bands involved in the transition driven by the frequency of the amplitude modulation. With the experimental parameters $v_L \approx 7.1$ mm s$^{-1}$. Reprinted figure with permission from [4], Copyright (2013) by the American Physical Society.

energy that corresponds to band 4 will thus be projected onto the states that correspond to the branch $n = -2$. Knowing $n$ and $n'$, we deduce from equation (11.40) that the equation of the corresponding dotted line in figure 11.5 is $\omega = -9\omega_L + 6\pi v_0/d$ [4].

The lower (black) dashed line of figure 11.5 corresponds to the inter-band transition 3 → 1, 4 → 2, 5 → 3.... From equation (11.40), we deduce its equation as $\omega = -\omega_L + 2\pi v_0/d$. Similarly, we find for the upper (red) dashed line $\omega = \omega_L + 2\pi v_0/d$. We obtain for the (blue) dotted–dashed line the equation $\omega = -4\omega_L + 4\pi v_0/d$. Such an interpretation can be made for other inter-band transitions.

### 11.3.2 Dispersion relation for exact resonance

It is interesting to note that for the exact resonance case, the quasi-energy, $\mathcal{E}(k)$, is calculated as

$$\mathcal{E}(k) = \frac{\hbar k^2 \eta^2}{8F_1} + \frac{\epsilon_0^2 V_0^2(2\pi)}{128\pi F_1 d^2}, \tag{11.41}$$

$$\mathcal{E}(k) = \frac{1}{4}\left[Jkd \;\; J_1(kd) + \frac{\epsilon_0^2 V_0^2}{16F_1 d^2}\right]. \tag{11.42}$$

It is worth mentioning that the above expression reduces to a constant value as the quasi-momentum $k$ has values corresponding to the zeroes of the Bessel function, $J_1$. The constant values for quasi-energy lead to band narrowing as $E(k + 2\pi/d) - E(k) = 0$. The band narrowing leads to suppression of group velocity as it takes zero values for the corresponding quasi-momentum.

The group velocity is calculated as

$$v_g = \frac{\partial \mathcal{E}(k)}{\hbar \partial k} = \frac{1}{4\hbar}\left[ Jd \;\; J_1(kd) + Jkd\frac{\partial}{\partial k}(J_1(kd)) \right]. \tag{11.43}$$

Following the Bessel function identity,

$$\frac{\partial}{\partial z}(J_n(z)) = \frac{n}{z}J_n(z) - J_{n+1}(z), \tag{11.44}$$

we rewrite the group velocity as

$$v_g = \frac{Jd}{2\hbar}\left[ J_1(kd) - \frac{1}{2}kd \; J_2(kd) \right]. \tag{11.45}$$

The general expression for the quasi-energy $\mathcal{E}(k)$ is calculated in appendix C.

## 11.4 Chaos-assisted tunneling

Tunneling is a quantum mechanical characteristic associated with the wave nature of particles. It corresponds to the crossing of an energy barrier when a particle has less energy as compared with the barrier height. In this case, it is possible to solve analytically many simple one-dimensional systems, and the tunneling amplitudes are simple functions decreasing exponentially with the height and width of the barrier and the inverse of Planck's constant.

Generically, higher-dimensional coupled quantum systems exhibit more dynamical characteristics. For example, as discussed in chapter 8, two zones of a classical phase space separated by an invariant manifold of a dynamical system, also known as a Kolmogorov, Arnold, and Moser (KAM) surface, cannot be connected classically. However, in quantum mechanics it is possible to cross the invariant manifold through a tunneling process, called *dynamical tunneling.*

In general, multi-dimensional systems display various level of chaos. The most generic case corresponds to mixed systems, where islands of stability are embedded in a stochastic sea, existing together in the phase space. A classical particle in one of the islands of stability stays there, thus displaying a bounded motion. A quantum particle, however, tunnels from one island to the other. In this, it is preferable that it tunnels first to the chaotic sea and from there to another island of stability. In this case, the tunneling becomes a much more subtle process, called *chaos-assisted tunneling*.

Chaos-assisted tunneling corresponds to a much richer tunneling process, wherein the coupling between quantum states located in neighboring regular islands is mediated by other states spread over the chaotic sea. In the presence of complex classical dynamics associated with a mixed phase space, a quantum wave function can tunnel between two stable islands through a chaotic sea.

## 11.5 Inter-band transitions

The band dispersion of the Hamiltonian $H_B = -\partial^2/\partial x^2 + s\sin(x)$ is plotted in figure 11.6. Concerning the band indexes, we label the bands the ground, first, second, and so on. A classical Hamiltonian has two distinct dynamical regimes, namely, an oscillation mode and a rotation mode of a classical pendulum, respectively. With one notable difference that the roles of position and momentum appear interchanged in phase space separated by a separatrix, they correspond to the dipole mode and the Bloch mode. The dipole mode occurs around the origin of the harmonic trap, where $E_q^n$ is comparable to $\nu x^2$ so that the tunneling from one site to another is essential for characterizing the energy-band structure. Here, $q$ and $n$ are the quasi-momentum and the band index of $H_B$, respectively. However, in the region of the Bloch mode, the harmonic potential dominates so that $E_q^n$ may be ignored and thus the tunneling is suppressed. As a result, the eigenstates are spatially localized in this region.

Let us discuss how the preceding wave packet evolves in time in the presence of amplitude modulation. As shown in figure 11.7, if we tune the modulation frequency to the energy difference between the ground band and the top of the third band, the amplitude modulation produces two major couplings, namely, one between the ground (dipole) band and the third (dipole) band via a two-'photon' process, and the other between the first (Bloch) band and the third (dipole) band via a one-'photon' process. Here, the term 'photon' refers to an energy quantum emitted or absorbed due to the amplitude modulation. Figure 11.7 shows the band population dynamics [6]. Populations in the first and third bands both oscillate out

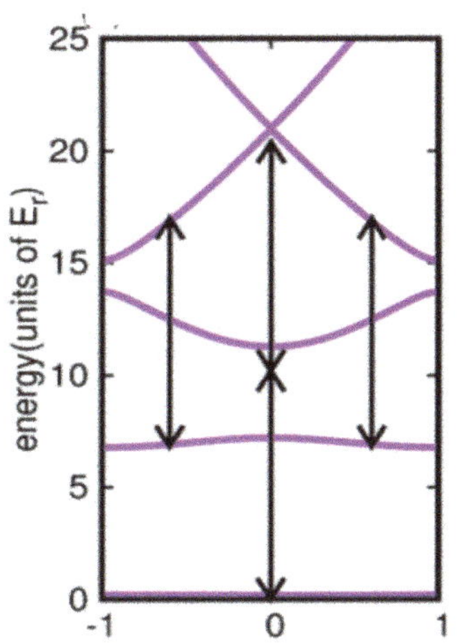


**Figure 11.6.** The band structure deriving from $H_B$. We note that the ground energies of $H_B$ are both set to zero. Double-headed arrows show where bands couple resonantly with $E_\omega = 10.2$. Reprinted figure with permission from [5], Copyright (2018) by the American Physical Society.

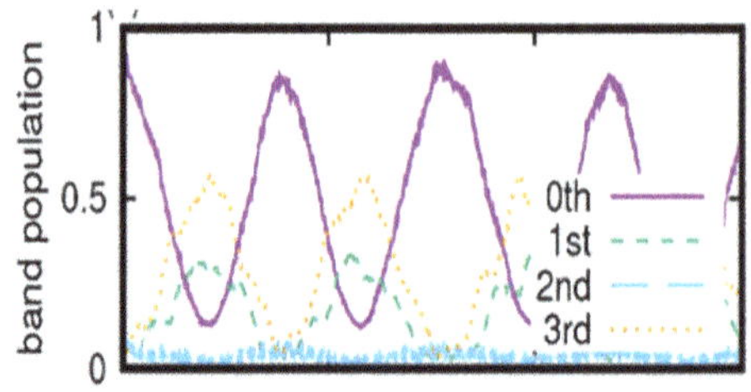


**Figure 11.7.** The band population. Purple solid line, green short dashed line, light blue long dashed line, and orange dots correspond to the ground (zeroth), first, second, and third bands, respectively. Reprinted figure with permission from [5, 6], Copyright (2018) by the American Physical Society.

of phase with the ground (zeroth) band, while the second band population always remains less than 0.1.

By introducing the method of a standing-wave pulse sequence, we can efficiently prepare ultra-cold atoms into a specific band in a one-dimensional optical lattice [7]. Further, a combination of one-dimensional bi-chromatic superlattices lead to preparing wave packets in excited bands [8]. Varying the lattice parameters leads to the so-called Dirac point where a pair of excited bands crosses.

## Exercise

1. Show that the incident velocity domain $3v_{\mathrm{L}}/2 < v_0 < 2v_{\mathrm{L}}$ corresponds to a resonant transition between bands 4 and 3.

## References

[1] Zheng M J, Chan Y S and Yu K W 2011 Photonic Bloch-dipole-Zener oscillations in binary parabolic optical waveguide arrays *J. Opt. Soc. Am.* B **28** 1339

[2] Geiger Z A *et al* 2018 Observation and uses of position-space Bloch oscillations in an ultracold gas *Phys. Rev. Lett.* **120** 213201

[3] Heinze J, Krauser J S, Fläschner N, Hundt B, Götze S, Itin A P, Mathey L, Sengstock K and Becker C 2013 Intrinsic photoconductivity of ultracold Fermions in optical lattices *Phys. Rev. Lett.* **110** 085302

[4] Cheiney P, Fabre C, Vermersch F, Gattobigio G L, Mathevet R, Lahaye T and Guery-Odelin D 2013 Matter-wave scattering on an amplitude-modulated optical lattice *Phys. Rev.* A **87** 013623

[5] Yamakoshi T, Saif F and Watanabe S 2018 Significantly stable mode of the ultracold atomic wave packet in amplitude-modulated parabolic optical lattices *Phys. Rev.* A **97** 023620

[6] Medhet S, Yamakoshi T, Ayub M, Saif F and Watanabe S 2020 Signatures of inter-band transitions on dynamical localization *Eur. Phys. J.* D **74** 175

[7] Zhai Y, Yue X, Wu Y, Chen X, Zhang P and Zhou X 2013 Effective preparation and collisional decay of atomic condensates in excited bands of an optical lattice *Phys. Rev.* A **87** 063638

[8] Yamakoshi T and Watanabe S 2019 Fast and selective interband transfer of ultracold atoms in bichromatic lattices permitting Dirac points *Phys. Rev.* A **99** 013621

[9] Kolovsky A R, Grusdt F and Fleischhauer M 2014 Quantum particle in a parabolic lattice in the presence of a gauge field *Phys. Rev. A* **89** 033607

[10] Haller E, Hart R, Mark M J, Danzl J G, Illner L R and Nägerl H C 2010 Inducing transport in a dissipation-free lattice with super Bloch oscillations *Phys. Rev. Lett.* **104** 200403

# Chapter 12

# Atoms in modulated evanescent wave fields

*'All things are made of atoms—little particles that move around in perpetual motion, attracting each other when they are a little distance apart, but repelling upon being squeezed to one another. In that one sentence ... there is an enormous amount of information about the world.'*

—Richard Feynman

Exponentially decaying or evanescent wave fields may obtain intensity modulation by means of an acousto-optic modulator. A periodic in time intensity modulation in an evanescent wave field exerts a periodic in time external force on interacting atoms. Mathematically speaking, the intensity modulation adds one more degree of freedom, and thus makes an explicitly time-dependent system. The corresponding classical system exhibits chaotic dynamics.

The dynamics of ultra-cold atoms interacting with an evanescent wave field is greatly modified as they experience periodic forcing in time. In a nutshell, the onset of modulation in an optical system is seen by the cold atoms as a transition of the classical dynamics from simple to complex. Whereas in the quantum domain, the atomic wave function displays non-trivial phenomena, such as quantum recurrences, dynamical tunneling, and dynamical localization, which is regarded as a signature of quantum chaos.

We consider ultra-cold atoms that, following a gravitational field, move toward an atomic mirror and exhibit normal incidence. The atomic mirror is modulated in time by an acousto-optic modulator joined to the forming evanescent wave field. The evanescent wave field undergoes periodic amplitude modulation, which modulates the intensity of the incident laser light field. Hence, the atoms bounce off the modulated atomic mirror and travel in the gravitational field, as shown in figure 12.1.

doi:10.1088/978-0-7503-2308-6ch12

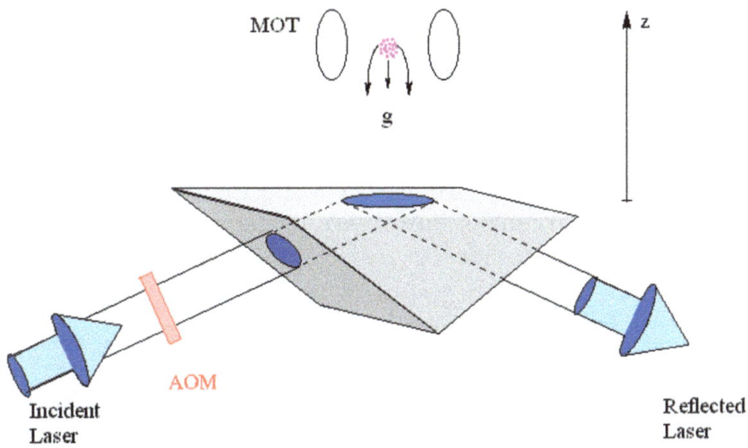


**Figure 12.1.** A cloud of atoms is trapped and cooled in a magneto-optic trap (MOT) to a few microkelvin. The MOT is placed at a certain height above a dielectric slab, where an evanescent wave is created by the total internal reflection of the incident laser beam from the surface of the dielectric. At the onset of the experiment, the MOT is switched off and the atoms move under gravity toward the exponentially decaying evanescent wave field. The evanescent wave field and gravitational field form a cavity for the atomic de Broglie waves. The atoms undergo bounded motion in this gravitational cavity. An acousto-optic modulator provides spatial modulation of the evanescent wave field. The model may serve as a realization of an atomic Fermi accelerator. Reprinted from [1], Copyright (2005), with permission from Elsevier.

The periodic modulation in the intensity of the evanescent wave field decaying along the $z$-axis is expressed as

$$I(z,\, t) = I_0(1 + \epsilon \sin(\omega t))\mathrm{e}^{-2\kappa z}. \tag{12.1}$$

Thus, the center-of-mass motion of the atom, experiencing the gravitational field in the $z$-direction, is bounded from above. This effectively follows the Hamiltonian

$$H = \frac{p^2}{2m} + mgz + \hbar\Omega_{\text{eff}}(1 + \epsilon \sin \omega t) \exp(-2\kappa z). \tag{12.2}$$

Here, $\Omega_{\text{eff}}$ denotes the coupling strength, and $\kappa$ indicates the inverse of the decay length. Moreover, $\epsilon$ and $\omega$ express the amplitude and frequency of the external modulation, respectively. In the absence of modulation, the effective Hamiltonian, given in equation (12.2), reduces to equation (7.8). The modulated system defined in equation (12.2) is referred to as a modulated gravitational cavity or Fermi accelerator.

## 12.1 Wave packet dynamics

The time evolution of a material wave packet in a periodically driven one-dimensional system is governed by the time-dependent Schrödinger wave equation. We rewrite the explicitly time-dependent Hamiltonian given in equation (12.2) as

$$H = H_0 + \epsilon_1\, u(z) \sin \omega t. \tag{12.3}$$

Here, $\epsilon_1 = \epsilon\hbar\Omega_{\mathrm{eff}}$, and $H_0$ is the Hamiltonian of the system in the absence of an external driving field. In order to understand the atomic dynamics in the absence of external periodic modulation, we follow the treatment given in chapter 7. The evolution of a material wave packet in the presence of external modulation is obtained by following section 9.4.2.

In order to simplify the calculations, we write the Hamiltonian expressed in equation (12.2) in dimensionless form. We introduce the scaling $\tilde{z}\equiv(\omega^2/g)z$, $\tilde{p}\equiv(\omega/mg)p$, and $\tau \equiv \omega t$, such that

$$[\tilde{z}, \tilde{p}] = \frac{\omega^2}{g}\frac{\omega}{mg}[z, p] = i\frac{\omega^2}{g}\frac{\omega}{mg}\frac{h}{2\pi} = i\bar{k}.$$

Here, $\bar{k}$ is the scaled Planck's constant for the driven system. Thus, the Hamiltonian becomes

$$\tilde{H} = \frac{\tilde{p}^2}{2} + \tilde{z} + V_0 \exp(-\kappa\tilde{z}) + \lambda \sin\tau \exp(-\kappa\tilde{z}), \tag{12.4}$$

$$= \tilde{H}_0 + \lambda\, u(\tilde{z}) \sin\tau. \tag{12.5}$$

Furthermore, we take $\tilde{H} = (\omega^2/mg^2)H$, $u(\tilde{z}) = \mathrm{e}^{-\kappa\tilde{z}}$, $V_0 = \hbar\Omega_{\mathrm{eff}}\omega^2/mg^2$, $\kappa = \kappa g/\omega^2$, and $\lambda = \epsilon\hbar\Omega_{\mathrm{eff}}\omega^2/mg^2$. For simplicity, we remove tilde sign (~) in the next sections of the chapter, and write $\tau = t$ as the scaled time.

Here, $\lambda$ is a dimensionless parameter that characterizes the Hamiltonian. In quantum mechanical solutions, another scaled parameter appears, that is, the scaled Planck's constant, $\bar{k}$. The parameter $\lambda$ controls the degree of non-integrability, while the scaled Planck's constant controls the scale at which its quantum mechanical counterpart can resolve phase-space structures.

## 12.2 Quantum recurrences

As discussed previously, in an explicitly time-dependent periodically driven system, the energy is no longer a constant of motion. Therefore, the corresponding time-dependent Schrödinger equation given in equation (12.5) is solved by using secular perturbation theory in the region of the resonances. The existence of resonances and their corresponding mathematical description is discussed in chapter 9. In the secular perturbation approach, faster frequencies are averaged out and the dynamical system is effectively reduced to one degree of freedom around a non-linear resonance [6]. The reduced Hamiltonian is integrable and its eigenenergies and eigenstates are considered as the quasi-energy and the quasi-energy eigenstates of the periodically driven system. The general solution of a periodically driven, time-dependent system is given in equations (9.61), (9.62), and (9.64).

The complete eigenstates are expressed as

$$|\psi_{j,m}\rangle = \mathrm{e}^{i\mathcal{E}_{\nu(j),m}t/\bar{k}}\, |u_{\nu(j),m}\rangle, \tag{12.6}$$

where

$$|u_{\nu(j),m}\rangle = \frac{1}{2\pi}\sum_n \int_0^{2\pi} g^{(n)}_{\nu(j),\,m} d\theta\;|n\rangle, \tag{12.7}$$

and

$$g^{(n)}_{\nu(j),\,m} = \mathrm{e}^{-i\frac{(n-\bar{n})(N\theta+t)}{N}} \exp\left\{i\left(\frac{j}{N} - \frac{N\omega_1 - 1}{N^2\xi\bar{k}}\right)\left(N\theta - \frac{\pi}{2}\right)\right\} P_{\nu(j),m}\left(\frac{N\theta - \pi/2}{2}\right). \tag{12.8}$$

The quasi-energy of the system becomes

$$\mathcal{E}_{\nu(j),m} = \left[\frac{N^2\bar{k}^2\zeta}{8}a_{\nu(j),m}(q) + \frac{N\omega_1 - 1}{N^2\zeta\bar{k}} + E_{\bar{n}} + \bar{k}\frac{j}{N}\right] \bmod \bar{k}. \tag{12.9}$$

Following equation (9.58) the Mathieu function $P$ is $\pi$ or $2\pi$ periodic which occurs for fixed value of $q$. More precisely, for a given (real) $q$ such periodic solutions exist for an infinite number of values of $a$, indexed as two separate sequences $a_m$ and $b_m$, for integer m. The corresponding functions are denoted as $cs_m(\theta, q)$ and $se_m(\Theta, q)$, respectively referred as *cosine-elliptic* and *sine-elliptic* Mathieu functions of the first kind. They satisfy the ortho-normality condition, that is,

$$\int_0^{2\pi} ce_m(\theta, q)ce_{m'}(\theta, q)d\theta = \delta_{m,m'}, \tag{12.10}$$

$$\int_0^{2\pi} se_m(\theta, q)se_{m'}(\theta, q)d\theta = \delta_{m,m'}. \tag{12.11}$$

The ortho-normality of the Floquet functions makes the quasi-energy eigenstates ortho-normal, therefore we can express an arbitrary wave packet, $|\Psi\rangle$, in the Hilbert space spanned by the quasi-energy eigenstates. Therefore, we write $|\Psi\rangle$ as a superposition of the quasi-energy eigenstates, that is,

$$|\Psi\rangle = \sum_{m,j} A_{m,j}|\psi_{j,m}\rangle, \tag{12.12}$$

$$|\Psi\rangle = \sum_{m,j} A_{m,j}\mathrm{e}^{-i\mathcal{E}_{\nu(j),m}t/\bar{k}}|u_{\nu(j),m}\rangle, \tag{12.13}$$

where $A_{m,j}$ is the probability amplitude.

In one degree-of-freedom systems the phenomenon of quantum revivals has been well studied both theoretically and experimentally. It has been established that quantum recurrences are generic to periodically driven one degree-of-freedom quantum systems, and to coupled two degrees-of-freedom quantum systems [4, 5]. The quantum nature of the recurrence phenomenon makes it suitable to use it as a probe to study quantum chaos [20].

An initial excitation produced at a mean quantum number, $r$, observes various time scales at which it recurs completely or partially during its quantum evolution.

In order to find these time scales at which the recurrences occur in a quantum mechanical driven system, we employ the quasi-energy, $\mathcal{E}_{\nu(j),m}$, of the driven system and write the auto-correlation function:

$$C(t) = \sum_{m,j} |A_{m,j}|^2 \exp\{-i(\omega_\lambda^{(0)} + (m-r)\omega_\lambda^{(1)} + (m-r)^2\omega_\lambda^{(2)} + \cdots)t\}. \quad (12.14)$$

There occurs a time scale, $T_\lambda^{(j)}$, inversely proportional to the frequencies, $\omega^{(j)}$, such that

$$T_\lambda^{(s)} = \frac{2\pi}{\omega_\lambda^{(s)}}, \quad (12.15)$$

where $s$ is an integer. We may define the frequency, $\omega_\lambda^{(s)}$, of the reappearance of an initial excitation in the dynamical system as

$$\omega_\lambda^{(s)} = \frac{1}{s!\hbar}\frac{\partial^{(s)}\mathcal{E}_{\nu(j),m}}{\partial m^{(s)}}, \quad (12.16)$$

calculated at $m = r$. The index $s$ describes the order of differentiation of the quasi-energy, $\mathcal{E}_{\nu(j),m}$, with respect to the quantum number, $m$. Equation (12.16) indicates that as the value of $s$ increases we get smaller frequencies, which lead to longer recurrence times.

We substitute the quasi-energy, $\mathcal{E}_{\nu(j),m}$, as given in equation (12.9), in the expression for the frequencies defined in equation (12.16). This leads to the classical period, $T_\lambda^{(1)} = T_\lambda^{(\mathrm{cl})}$, for the driven system as

$$T_\lambda^{(\mathrm{cl})} = [1 - M^{(\mathrm{cl})}]T_0^{(\mathrm{cl})}, \quad (12.17)$$

and the quantum revival time, $T_\lambda^{(2)} = T_\lambda^{(Q)}$, as

$$T_\lambda^{(Q)} = [1 - M^{(Q)}]T_0^{(Q)}. \quad (12.18)$$

Here, the time scales $T_0^{(\mathrm{cl})}$ and $T_0^{(Q)}$ express the classical period and the quantum revival time in the *absence* of external modulation, respectively, given in equations (7.30) and (7.31).

The time modification factors, $M^{(\mathrm{cl})}$ and $M^{(Q)}$, are given as

$$M^{(\mathrm{cl})} = -\frac{1}{2}\left(\frac{\lambda V\zeta}{\omega_1^2}\right)^2\frac{1}{(1-\mu^2)^2} \quad (12.19)$$

and

$$M^{(Q)} = \frac{1}{2}\left(\frac{\lambda V\zeta}{\omega_1^2}\right)^2\frac{3+\mu^2}{(1-\mu^2)^3}, \quad (12.20)$$

where

$$\mu = \frac{N^2\hbar\zeta}{2\omega_1}. \quad (12.21)$$

Equations (12.17) and (12.18) express the classical period and the quantum revival time in the periodically modulated system. These time scales are a function of the strength of the periodic modulation, $\lambda$, and the matrix element, $V$. Moreover, they also depend on the frequency, $\omega_1$, and the non-linearity, $\zeta$, associated with an *un*driven or *un*modulated system. As the modulation term approaches zero, that is, $\lambda \longrightarrow 0$, the modification terms $M^{(cl)}$ and $M^{(Q)}$ disappear.

### 12.2.1 Quantum recurrences in a Fermi accelerator

For the atomic dynamics in a Fermi accelerator, the classical frequency, $\omega_1$, and the non-linearity, $\zeta$, read as

$$\omega_1 = \left(\frac{\pi^2}{3r\bar{k}}\right)^{1/3} \tag{12.22}$$

and

$$\zeta = -\left(\frac{\pi}{9r^2\bar{k}^2}\right)^{2/3}, \tag{12.23}$$

respectively.

In an undriven system, we define the classical period as $T_0^{(cl)} = 2\pi/\omega_1$ and the quantum revival time as $T_0^{(Q)} = 2\pi/(\hbar\zeta)$. Hence, we find an increase in the classical period and in the quantum revival time of an undriven Fermi accelerator as the mean quantum number $r$ increases.

The calculation for the matrix element, $V$, for large $N$ yields the result

$$V \cong -\frac{2E_0}{N^2\pi^2}, \tag{12.24}$$

and the resonance, $N$, takes the value

$$N = \frac{\sqrt{2E_N}}{\pi}. \tag{12.25}$$

The time required to sweep one classical period and the time required for a quantum revival are modified in the driven system. In the classical domain and in the quantum domain, the modification terms become

$$M^{(cl)} = \frac{1}{8}\left(\frac{\lambda}{E_N}\right)^2 \frac{1}{(1-\mu^2)^2} \tag{12.26}$$

and

$$M^{(Q)} = \frac{1}{8}\left(\frac{\lambda}{E_N}\right)^2 \frac{3+\mu^2}{(1-\mu^2)^3}, \tag{12.27}$$

respectively, where

$$\mu = -\frac{\bar{k}}{4}\frac{1}{E_0}\sqrt{\frac{E_N}{E_0}}. \tag{12.28}$$

Therefore, the present approach is valid for smaller values of the modulation strength, $\lambda$, and larger values of the resonance, $N$.

A comparison between the quantum revival time, calculated with the help of equation (12.27), and those obtained numerically for an atomic wave packet bouncing in a Fermi accelerator shows a good agreement, and thus supports theoretical predictions [2, 3].

## 12.3 Quantum recurrences as a probe to study quantum chaos

The quantum recurrences change drastically at the onset of a periodic modulation. However, the change depends upon the initial location of the wave packet in the phase space. In order to emphasize this fascinating feature of quantum chaos, we note that for an initially propagated atomic wave packet in an *un*driven Fermi accelerator the quantum revivals exist. However, they disappear completely in the presence of modulation if the atom originates from the center of a non-linear resonance [20]. In the present situation the atomic wave packet reappears after a classical period only [4, 5], as shown in figure 12.2.

We may understand this interesting property as we note that, at a given resonance, the dynamics is effectively described by an oscillator Hamiltonian:

$$H = -\frac{\partial^2}{\partial\varphi^2} + V_0 \cos\varphi. \tag{12.29}$$

Therefore, when $\varphi \ll 1$, the Hamiltonian reduces to the Hamiltonion for a harmonic oscillator, given as

$$H \approx -\frac{\partial^2}{\partial\varphi^2} - \frac{V_0}{2}\varphi^2. \tag{12.30}$$

The Hamiltonian controls the evolution of an atomic wave packet placed at the center of a resonance. This analogy provides us with evidence that if an atomic wave packet is placed initially at the center of a resonance, it will always observe revivals after each classical period only, the same as in the case of a harmonic oscillator.

In addition, this analogy provides information about the level spacing and level dynamics at the center of a resonance in the Fermi accelerator as well. Since in case of a harmonic oscillator the spacing between successive levels is always equal, we conclude that the spacing between quasi-energy levels is equal at the center of the resonance in a periodically driven system.

For a detailed study of the effect of initial conditions on the existence of the quantum recurrences, the information about the underlying Floquet quasi-energies, and the use of quantum recurrences to understand different dynamical regimes, we refer to the reference [1].

## 12.4 Classical period and quantum revival time: interdependence

The non-linearity, $\zeta$, present in the energy spectrum of an *un*driven system contributes to the classical period and the quantum revival time, both in the

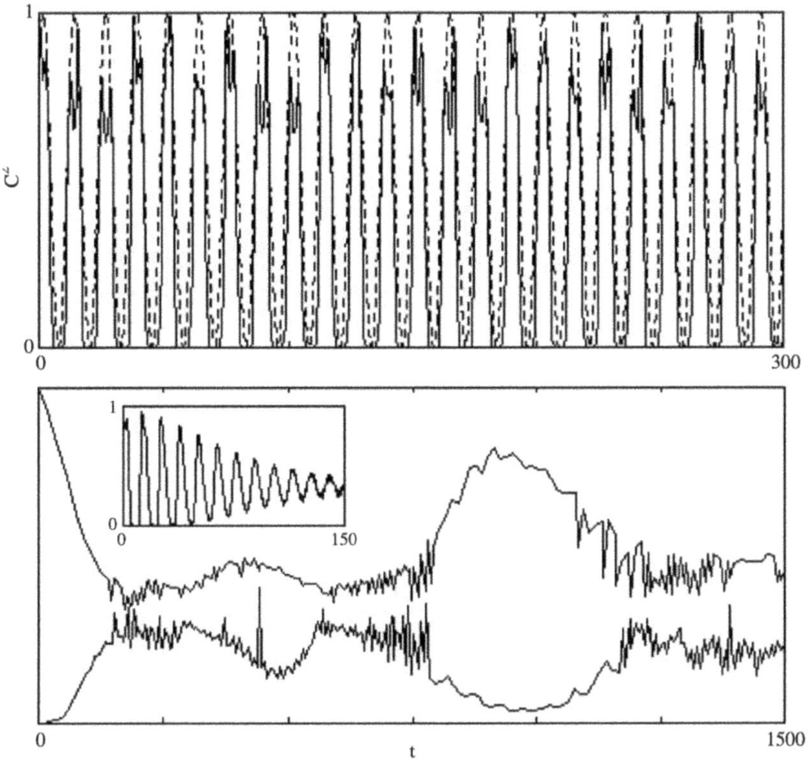


**Figure 12.2.** The change in revival phenomena for a wave packet originating from the center of a resonance. Top: The square of the auto-correlation function, $C^2 = |\langle\psi(0)|\psi(t)\rangle|^2$, is plotted as a function of time of evolution, $t$, for the atomic wave packet. The wave packet initially propagates from the center of a resonance in an atomic Fermi accelerator, and its mean position and mean momentum are $z_0 = 14.5$ and $p_0 = 1.45$, respectively. The thick line corresponds to the numerically obtained result for the Fermi accelerator, and the dashed line indicates the quantum revivals for a wave packet in a harmonic oscillator. The classical period is calculated to be $4\pi$ for $\lambda = 0.3$. Bottom: The square of the auto-correlation function as a function of time for $\lambda = 0$. As the modulation strength $\lambda$ vanishes, the evolution of the material wave packet changes completely. The wave packet experiences collapse after many classical periods and, later, displays quantum revivals at $T_0^{(Q)}$. The inset displays the short-time evolution of the wave packet comprising many classical periods in the absence of external modulation. Reprinted from [1], Copyright (2005), with permission from Elsevier.

presence and in the absence of an external modulation [7]. Below, we analyze different situations as the non-linearity varies.

### 12.4.1 Vanishing non-linearity

In the presence of a vanishingly small non-linearity in the energy spectrum of an undriven system, that is, for $\zeta \approx 0$, the time modification factors for the classical period $M^{(\mathrm{cl})}$ and the quantum revival time $M^{(Q)}$ vanish, as we find in equations (12.19) and (12.20). Hence, the modulated linear system displays recurrence only after a classical period, which is $T_\lambda^{(\mathrm{cl})} = T_0^{(\mathrm{cl})} = 2\pi/\omega_1$. The quantum revival takes place after an infinite time, that is, $T_\lambda^{(Q)} = T_0^{(Q)} = \infty$.

### 12.4.2 Weak non-linearity

For a weakly non-linear energy spectrum, $\zeta$, the classical period, $T_\lambda^{(\mathrm{cl})}$, and the quantum revival time, $T_\lambda^{(Q)}$, in the modulated system are related with the $T_0^{(\mathrm{cl})}$ and $T_0^{(Q)}$ of an *un*modulated system as

$$3T_\lambda^{(\mathrm{cl})}T_0^{(Q)} + T_0^{(\mathrm{cl})}T_\lambda^{(Q)} = 4T_0^{(Q)}T_0^{(\mathrm{cl})}. \tag{12.31}$$

As discussed earlier, the quantum revival time, $T_0^{(Q)}$, depends inversely on the non-linearity, $\zeta$, in the unmodulated system. As a result, $T_0^{(Q)}$ and $T_\lambda^{(Q)}$ are much larger than the classical periods, $T_0^{(\mathrm{cl})}$ and $T_\lambda^{(\mathrm{cl})}$.

The time modification factors, $M^{(\mathrm{cl})}$ and $M^{(Q)}$, are related as

$$M^{(Q)} = -3M^{(\mathrm{cl})} = 3\alpha, \tag{12.32}$$

where

$$\alpha = \frac{1}{2}\left(\frac{\lambda V \zeta}{\omega_1{}^2}\right)^2. \tag{12.33}$$

Hence, we conclude that the modification factors $M^{(Q)}$ and $M^{(\mathrm{cl})}$ are directly proportional to the *square* of the non-linearity, $\zeta^2$, in the system. From equations (12.31) and (12.32), we note that, as the quantum revival time $T_\lambda^{(Q)}$ reduces by $3\alpha T_0^{(Q)}$, $T_\lambda^{(\mathrm{cl})}$ increases by $\alpha T_0^{(\mathrm{cl})}$.

In the asymptotic limit, that is, for $\zeta$ approaching zero, the quantum revival time in the modulated and unmodulated systems are equal and infinite, that is, $T_\lambda^{(Q)} = T_0^{(Q)} = \infty$. Furthermore, the classical periods in the modulated and unmodulated cases are related as $T_\lambda^{(\mathrm{cl})} = T_0^{(\mathrm{cl})}$.

### 12.4.3 Strong non-linearity

For relatively strong non-linearity, the classical period, $T_\lambda^{(\mathrm{cl})}$, and the quantum revival time, $T_\lambda^{(Q)}$, in a modulated system are related with the $T_0^{(\mathrm{cl})}$ and $T_0^{(Q)}$, of an *un*modulated system as

$$T_\lambda^{(\mathrm{cl})}T_0^{(Q)} - T_0^{(\mathrm{cl})}T_\lambda^{(Q)} = 0. \tag{12.34}$$

The time modification factors, $M^{(Q)}$ and $M^{(\mathrm{cl})}$, follow the relation

$$M^{(Q)} = M^{(\mathrm{cl})} = -\beta, \tag{12.35}$$

where

$$\beta = \frac{1}{2}\left(\frac{4\lambda V}{N^2 \zeta \hbar^2}\right). \tag{12.36}$$

Hence, for a relatively strong non-linear case, $\beta$ may approach zero. Thus, the time modification factors vanish both in the classical and in the quantum domains. As a result, equations (12.17) and (12.18) reduce to $T_{\lambda}^{(\mathrm{cl})} = T_0^{(\mathrm{cl})}$ and $T_{\lambda}^{(Q)} = T_0^{(Q)}$, which proves the equality given in equation (12.34).

The quantity $\beta$, which determines the modification both in the classical period and in the quantum revival time, is inversely dependent on the fourth power of the scaled Planck's constant $\hbar$. Hence, for highly quantum mechanical cases, we find that the recurrence times remain unchanged.

## 12.5 Fermi acceleration modes

Within the limit of exponential potential as infinite potential, we obtain the Pustyl'nikov model on a Fermi accelerator. In his classical work, Leonid M Pustyl'nikov guaranteed the existence of a set of initial data, such that trajectories which originate from the set always speed-up to infinity. The existence of these accelerating modes requires the presence of certain windows of modulation strength, $\lambda$. Furthermore, the initial data appear always in circles of a finite radius that have a positive Lebesgue measure. The windows of modulation strength that support accelerated trajectories read as

$$s\pi \leqslant \lambda < \sqrt{1 + (s\pi)^2}. \tag{12.37}$$

Here, $s$ takes integer and half integer values for the periodic modulation of the atomic mirror.

The windows of modulation strength that support unbounded acceleration appear in the wave packet's dynamics as we calculate its dispersion in momentum space, $\Delta p \equiv \sqrt{\langle p^2 \rangle - \langle p \rangle^2}$, as a function of modulation strength. Here, $\langle p \rangle$ and $\langle p^2 \rangle$ are the first and the second moment of momentum, respectively. For very small modulation strengths, the widths remain small and almost constant, which indicates no diffusive dynamics, as shown in figure 12.3. We find that at the modulation strengths which correspond to windows on modulation strengths, given in equation (12.37), the diffusion of the wave packet is at its maximum. This behavior occurs as the parts of the initial distribution which correspond to the areas of phase space

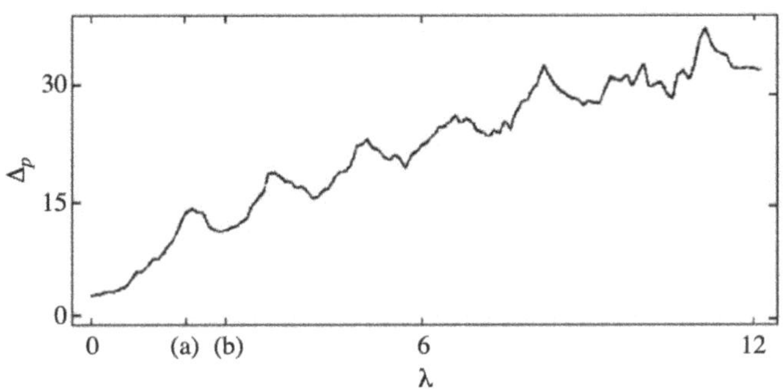


**Figure 12.3.** The width of the momentum distribution $\Delta p$ is plotted as a function of the modulation strength $\lambda$. An ensemble of particles, initially in a Gaussian distribution, is propagated for a time, $t = 500$. The numerical results depict the presence of accelerated dynamics for the modulation strengths expressed in equation (12.37). Reprinted from [1], Copyright (2005), with permission from Elsevier.

supporting accelerated dynamics undergo coherent acceleration, whereas the rest of the initial probability distribution displays maximum diffusion. For this reason, as the modulation strengths increase beyond these values, the dispersion reduces, as we find in figure 12.3. Equation (12.37) leads to the modulation strength, $\lambda_m$, for which maximum accelerated trajectories occur in the system. We find these values of $\lambda_m$ to be

$$\lambda_m = \frac{s\pi + \sqrt{1 + (s\pi)^2}}{2}. \tag{12.38}$$

The value is confirmed by the numerical results, as expressed in figure 12.3.

For a modulation strength, given in equation (12.37), around $\lambda_m$ and an initial ensemble originating from an area of phase space which supports accelerated dynamics, we find a sharply suppressed value of dispersion. This is a consequence of a coherent acceleration of the entire distribution [8].

## 12.6 Non-dispersive accelerated matter waves

A classical ensemble of particles displays a global diffusion in a Fermi accelerator beyond a critical modulation strength. The classical dynamics is described by a Chirikov map or the standard map. This occurs as the resonance overlap takes place above the critical modulation strength, $\lambda_l = 0.24$ [1]. In contrast, in the corresponding quantum mechanical domain there occurs another critical modulation strength, $\lambda_u$, which depends purely on quantum laws [9–11]. At the critical modulation strength, $\lambda_u$, a phase transition occurs and the quasi-energy spectrum of the Floquet operator changes from a point spectrum to a continuum spectrum [12–14]. We may define the critical modulation strength for the Fermi accelerator as $\lambda_u = \sqrt{k\!\!\!^{-}}/2$, when the exponential potential of the atomic mirror is considered as an infinite potential. The scaled Planck's constant is defined as $k\!\!\!^{-} = (\Omega/\Omega_0)^3$. The scaling frequency, $\Omega_0$, is defined as $\Omega_0 = (2\pi m g^2/h)^{1/3}$, where $h$ is the Planck's constant. Beyond this critical modulation strength, quantum diffusion takes place. Hence, the two conditions together make a window on the modulation strength. In this window a drastic difference takes place between classical dynamics and the corresponding quantum dynamics for a modulation strength which fulfils the condition

$$\lambda_l < \lambda < \lambda_u. \tag{12.39}$$

Classical diffusion applies within the window, whereas the corresponding quantum dynamics displays localization. For this reason, we call it the localization window.

Equations (12.39) and (12.37) explain that a bouncing quantum particle behaves differently by tuning the modulation strength, $\lambda$, from a localization window to an acceleration window. The quantum dynamics of a material wave packet modifies drastically for larger values of the scaled Planck's constant, $k\!\!\!^{-}$. It is important to note that as the scaled Planck's constant increases, the dynamical localization window broadens. Therefore, an overlap between the localization window and acceleration window, defined by equations (12.39) and (12.37), takes place for certain higher values of the scaled Planck's constant. Hence, the control over the two windows by means of the scaled Planck's constant, $k\!\!\!^{-}$, leads to the ability to generate accelerated and non-dispersive ultra-cold atoms that act as *atomic bullets*. Equations (12.39) and (12.37) combined together, therefore, define new windows on the modulation amplitude, $\lambda$, that is,

$$s\pi \leqslant \lambda \leqslant \frac{\sqrt{k}}{2}, \qquad (12.40)$$

which supports accelerated and dynamically localized evolution. Here, $\frac{\sqrt{k}}{2} \leqslant \sqrt{1 + (s\pi)^2}$. The lower bound defined by equation (12.40) indicates unbounded acceleration, whereas the upper bound corresponds to the absence of dynamical delocalization. As an example, if $k=12$ the localization window and the first acceleration window, respectively, are $0.24 \leqslant \lambda < 1.73$ and $1.57 \leqslant \lambda < 1.86$. An overlap between the two windows occurs for $1.57 \leqslant \lambda < 1.73$, which leads to an accelerated non-dispersive dynamics of the matter wave. Whereas, for $k = 14$ the first acceleration window defined by equation (12.37) comes within the localization window. For this reason, accelerated non-dispersive dynamics exists when $1.57 \leqslant \lambda < 1.86$ for a particle originating from the regions of phase space supporting accelerated dynamics.

In order to analyze wave packet evolution, we propagate an atomic wave packet, $\psi(z, 0)$, initially in a Gaussian distribution that fulfils the Heisenberg minimum uncertainty principle. Let us place it above an atomic mirror, with an initial average momentum $p_0 = 0$. The time evolution of the atomic wave packet is stored in the wave function $\psi(z, t)$. The probability distributions in position space, $|\psi(z, t)|^2$, and in momentum space, $|\psi(p, t)|^2$, are the marginal integration of the probability distribution function in phase space. These are shown for a fixed evolution time in figure 12.4. The sharp spikes appear when the modulation strength satisfies the acceleration condition, given in equation (12.37), and gradually disappear otherwise.

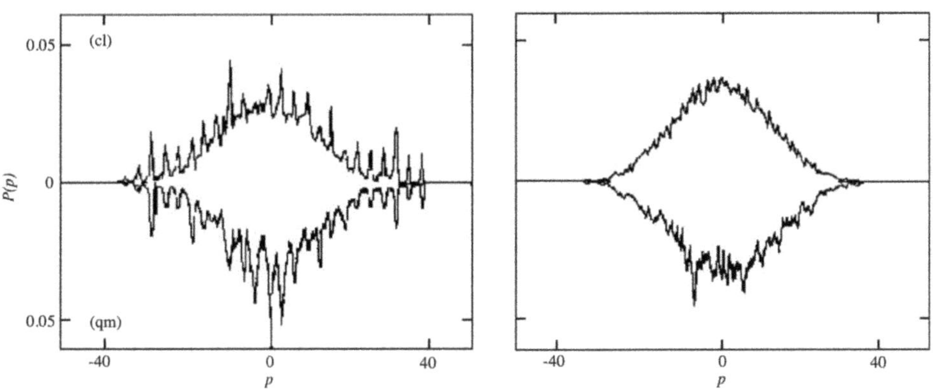


**Figure 12.4.** The classical and the quantum dynamics are compared by calculating the momentum distributions after the evolution time, $t = 500$, (left) for $\lambda = 1.7$ and (right) for $\lambda = 2.4$, marked as (a) and (b) in figure 12.3, respectively. We plot the momentum distributions, both in the classical and in the quantum space, as mirror images. We find spikes appearing in the momentum distributions for $\lambda = 1.7$, which are due to the presence of coherent accelerated dynamics. Reprinted from [1], Copyright (2005), with permission from Elsevier.

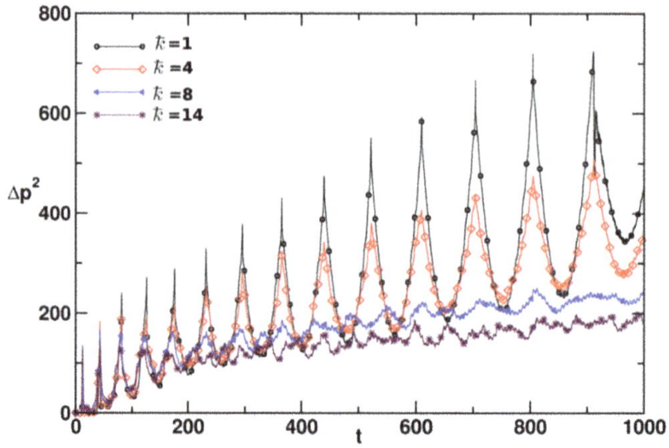


**Figure 12.5.** Variance in the momentum space as a function of time, for a fixed scaled modulation strength $\lambda = 1.7$. Here, we take different values of the scaled Planck's constant, $k$. We note that for higher values of $k$, the acceleration window comes within the localization window. As a result, the behavior of variance alters and it saturates with the evolution time [16]. Copyright © 2014, EDP Sciences, SIF, Springer-Verlag Berlin Heidelberg. With permission of Springer.

This spiky behavior in momentum distribution is therefore a hallmark of coherent accelerated dynamics. In contrast, the portions of the initial probability distribution originating from the regions of the phase space that do not support accelerated dynamics undergo diffusion and form a background. In figure 12.5, we show the variance in momentum space, $\Delta p^2 = \langle p^2 \rangle - \langle p \rangle^2$, as a function of evolution time for different values of $k$. In all these calculations the modulation strength, $\lambda$, is fixed to 1.7. For smaller $k$ values, we observe a linear growth with time due to dynamical delocalization, as discussed in [15]. The difference between two peaks in these plots corresponds to the time of flight of the accelerated wave packet between two impacts with the modulated atomic mirror. The periodicity in these plots indicates coherence in evolution. In addition, a gradual increase of the time interval between peaks corresponds to a larger time that the matter wave takes during its accelerated dynamics between two successive impacts with the modulated surface. Furthermore, drastic variation in $\Delta p^2$ during the time interval reveals different behavior in the wave packet during its evolution over the atomic mirror at the two turning points, one on the surface of the mirror and the other in the gravitational field. For larger values of $k$, however, dynamical localization restricts drastic variations, resulting in a transition to relatively smooth and saturation behavior in $\Delta p^2$ with time. Hence, for larger values of the scaled Planck's constant, there occurs localized dynamics for the accelerated matter wave, as shown in figure 12.5.

Following classical dynamics in the Fermi accelerator [17], the atomic wave packet follows classical trajectories and bounces off the moving surface, as in figure 12.6. Here, we find the contour of the marginal probability distribution $P(z, t)$ for a propagation time $t = 500$ for $\lambda = 1.7$ and $k = 14$. Each bounce takes longer than the previous one, and after every bounce the maximum average position (i.e. the turning point of the wave packet in the gravitational field) becomes higher,

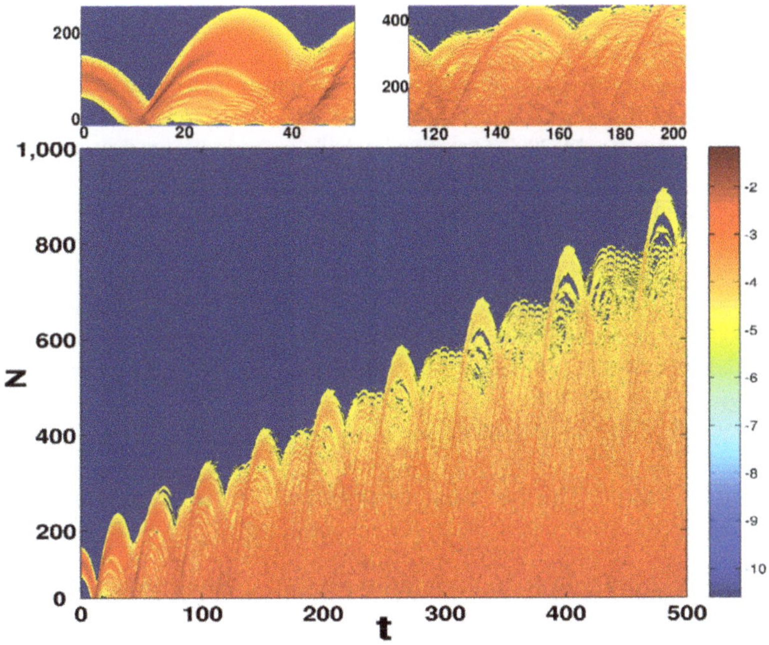


**Figure 12.6.** Contour plots of the quantum mechanical position distribution after a propagation time, $t = 500$, for $\lambda = 1.7$ for $k = 14$ [16]. Copyright © 2014, EDP Sciences, SIF, Springer-Verlag Berlin Heidelberg. With permission of Springer.

which corresponds in turn to a longer time of flight above the mirror. The two insets show the two turning points, one due to the atomic mirror (left) and the other in the gravitation field (right). It is notable that portions of the wave packet following acceleration modes are visible as jets of space-time distributions following classical trajectories [16]. However, the rest of the wave packet reaches those areas which do not support coherent acceleration and, as a consequence, a stochastic background profile appears, as shown in figure 12.6.

## Exercises

1. Obtain the quasi-energy and quasi-energy eigenstates for an ultra-cold atom in a Fermi accelerator using the semi-classical approach, developed in section 9.4.1 (see reference [18]).
2. Show that the Mathieu functions satisfy the orthogonality criteria (see reference [19]).

## References


[1] Saif F 2005 Classical and quantum chaos in atom optics *Phys. Rep.* **419** 207–58

[2] Saif F 1999 *Dynamical Localization and Quantum Revivals in Driven Systems* (Berlin: Lehmann Verlag)

[3] Saif F 2002 Dynamical revivals in Fermi accelerator *J. Phys. Soc. Japan* **69** L2363

[4] Saif F 2005 Quantum recurrences in periodically driven systems *J. Opt. B: Quantum Semiclass. Opt.* **7** S116
[5] Saif F 2006 Nature of quantum recurrences in coupled higher dimensional systems *Eur. Phys. J.* D **39** 87–91
[6] Saif F and Fortunato M 2002 Quantum revivals in periodically driven systems close to nonlinear resonances *Phys. Rev.* A **65** 013401
[7] Iqbal S, Qurat-ul-Ann and Saif F 2006 Quantum recurrences in driven power-law potentials *Phys. Lett.* A **356** 231
[8] Yannocopoulos A N and Rowlands G R 1993 A model for the coexistence of diffusion and accelerator modes in a chaotic area-preserving map *J. Phys. A: Math. Gen.* **26** 6231
[9] Saif F 2000 Dynamical localization and signatures of classical phase space *Phys. Lett.* **274** 98
[10] Saif F, Bialynicki-Birula I, Fortunato M and Schleich W P 1998 Fermi accelerator in atom optics *Phys. Rev.* A **58** 4779
[11] Chen W-Y and Milburn G J 1997 Quantum chaos in the atomic gravitational cavity *Phys. Rev.* E **56** 351
[12] Benvenuto F, Casati G, Guarneri I and Shepelyansky D L 1991 A quantum transition from localized to extended states in a classically chaotic system *Z. Phys. B: Condens. Matter* **84** 159
Benvenuto F, Casati G and Shepelyansky D L 1997 Dynamical localization: hydrogen atoms in magnetic and microwave fields *Phys. Rev.* A **55** 1732
[13] de Oliveira C R, Guarneri I and Casati G 1994 From power-localized to extended quasi-energy eigenstates in a quantum periodically driven system *Europhys. Lett.* **27** 187
[14] Brenner N and Fishman S 1996 Excitation of small quantum systems by high-frequency fields *Phys. Rev. Lett.* **77** 3763
[15] Saif F and Rehman I 2007 Coherent acceleration of material wave packets in modulated optical fields *Phys. Rev.* A **75** 043610
[16] Saif F, Naseer K and Ayub M 2014 Non-dispersive, accelerated matter-waves *Eur. Phys. J.* D **68** 95
[17] Akram J, Naseer K, Rehman I and Saif F 2009 Acceleration of material waves in Fermi accelerator *Math. Probl. Eng.* **2009** 246438
[18] Flatte M E and Holthaus M 1996 Classical and quantum dynamics of a periodically driven particle in a triangular well *Ann. Phys. NY* **245** 113–46
[19] Wu B and Diebold G J 2012 Mathieu function solutions for photo-acoustic waves in sinusoidal one-dimensional structures *Phys. Rev.* E **86** 016602
[20] Saif F 2000 Quantum recurrences: probe to study quantum chaos *Phys. Rev.* E **62** 6308

# Chapter 13

## Nano-opto-mechanics

*'Measure what can be measured, and make measurable what cannot be measured.'*.

—Galileo Galilei

The science of opto-mechanics covers macroscopic systems where radiation pressure force is sufficient to introduce microscopic modulation in optical cavities or resonators. Macroscopic quantum systems are vital for engineering devices and sensors based on quantum laws. The radiation pressure force has successfully been used to develop the kilometers-long Laser Interferometer Gravitational Waves Observatory as well as micro- and nano-sensors of ultra-high sensitivity.

Because gravitational waves are waves of gravitational tidal force, they apply a differential acceleration to a mass quadrupole (a pair of separated test masses). Assuming a wave amplitude $h = q/L$, we can treat $L$ as the effective spacing of the test masses, and $q$ is the induced small vibration motion. Thus, the detection of gravitational waves becomes the task of measuring very small differential motions $q/L$, which can be accurately measured. Therefore, in gravitational-wave detectors, opto-mechanics is used to create a new route to improved sensitivity, up to the order of 1 part in $10^{20}$.

The breakthrough came with the recognition that the same physics of gravitational-wave detectors can be realized with opto-mechanics. A millimeter-scale mirror resonator coupled to a light field and pumped with blue-detuned laser light creates a cavity with negative dispersion for a signal beam. However, this configuration is intrinsically unstable; we discuss here the key concepts to making it practical.

As a paradigm system, we consider a Fabry–Pérot optical cavity that consists of two highly reflecting mirrors separated by a distance $L$ with confined optical modes within these mirrors. These modes are classified in terms of the resonance

doi:10.1088/978-0-7503-2308-6ch13

frequencies. Maxwell's equations suggest that solutions with frequencies $\omega_j = j\pi c/L$ are available for the cavity field, which define the *modes*. Here, $c$ is the speed of light in vacuum and $j$ is a positive integer. The frequency difference between two adjacent modes, that is, the *free spectral range* (FSR), $\omega_{\rm FSR}$, is therefore obtained as

$$\omega_{\rm FSR} = \frac{\pi c}{L}.$$

The inverse dependence of $\omega_{\rm FSR}$ on the length of the cavity indicates that the difference between neighboring frequencies increases as the cavity length $L$ becomes smaller. This makes it possible to generate a single cavity mode of frequency $\omega_{\rm c} = \pi c/L$ in smaller cavities.

A finite *photon lifetime*, $\tau$, of the cavity mode is a result of cavity imperfections that exist in the cavity due to finite mirror transparencies. Thus, the photon escapes the cavity after this time. The reciprocal of the photon lifetime defines the *photon decay rate*, $\kappa$, that is, $\kappa = \frac{1}{\tau}$. More precisely, a good cavity is identified by high *optical finesse*, $\mathcal{F}$, which is the average number of round trips of a photon before it leaves the cavity, given as

$$\mathcal{F} \equiv \frac{\omega_{\rm FSR}}{\kappa} = \frac{\pi\tau c}{L}.$$

Hence, the optical finesse is directly proportional to the photon lifetime $\tau$ and inversely proportional to length $L$ of the cavity.

We may characterize the quality of a cavity by defining its *quality factor*, $Q_{\rm opt}$, proportional to the photon lifetime, given by $Q_{\rm opt} = \omega_{\rm c}\tau$. In fact, the decay rate $\kappa$ of the cavity is the sum of two decay rates, the internal decay rate, $\kappa_{\rm in}$, and the external losses, $\kappa_{\rm ex}$. The former is due to transmission absorption and scattering losses on the inner side of mirror walls, while the latter is because of input coupling between the source and cavity on the interface between them. Thus, we write the decay rate as

$$\kappa = \kappa_{\rm in} + \kappa_{\rm ex}.$$

## 13.1 Fabry–Pérot cavity with moving end-mirror

*Radiation pressure force* is applied by the confined optical mode on the end-mirrors. The force moves the delicate end-mirrors by an amount, $q$, and thus modifies the resonance frequency, such that

$$\omega_{\rm c}(q) = \frac{\pi c}{L + q}.$$

Considering the displacement $q$ to be very minute compared with the length of the cavity $L$, we use binomial expansion of the above expression:

$$\omega_{\rm c}(q) = \omega_{\rm c} - \omega_{\rm c}\frac{q}{L} + \omega_{\rm c}\frac{q^2}{L^2} - \cdots. \tag{13.1}$$

This leads to linearizing the above expression by considering the higher powers of $q/L$ as negligibly small. Hence, we find the cavity frequency:

$$\omega_c(q) \approx \omega_c - \frac{\omega_c}{L}q, \tag{13.2}$$

The *radiation pressure force* exerted by a photon on the end-mirror is calculated by noting that a photon carries a momentum of amount, $\hbar k$, where $k = 2\pi/\lambda$ describes the wave number associated with a field of wavelength $\lambda$. Following the law of the conservation of momentum, the photon interacts and imparts its momentum to the end-mirror. Hence, a change in the momentum of the mirror, $\Delta p$, takes place, which is

$$\Delta p = \frac{2E}{c}, \tag{13.3}$$

where $E$ is the energy of the photon. If the round trip time of a photon is $\Delta t = 2L/c$, the radiation pressure force, $F_{\text{rp}}$, exerted by $n$ photons on the mirror is

$$F_{\text{rp}} = n\frac{\Delta p}{\Delta t} = n\frac{E}{L}. \tag{13.4}$$

The corresponding potential, $V_{\text{rp}}(q)$, due to mirror displacement $q$ is

$$V_{\text{rp}}(q) = -n\frac{E}{L}q. \tag{13.5}$$

The quantum mechanical description of the potential is obtained by replacing the $c$-number representation by respective operators. We, therefore, introduce the position operator $\hat{q}$ and the number operator $\hat{n} = \hat{a}^\dagger\hat{a}$. As we consider the energy of the cavity photon to be $E = \hbar\omega_c$, the quantum mechanical expression of the potential experienced by the mirror becomes

$$\hat{V}_{\text{RP}} = -\hbar\frac{\omega_c}{L}\hat{a}^\dagger\hat{a}\hat{q}. \tag{13.6}$$

### 13.1.1 Interacting moving end-mirror with electromagnetic field

We consider that a laser source of frequency $\omega_0$ that drives a cavity through a fixed partially transparent mirror $M_1$, as shown in figure 13.1, develops a cavity field mode with frequency $\omega_c$. The radiation pressure force exerted by the photons of the cavity mode sets the moveable end-mirror $M_2$ in motion. The dynamics of the movable end-mirror is expressed as a single mechanical mode of frequency $\omega_m$. We develop the cavity field under the condition of a large FSR, that is, $\omega_{\text{FSR}} \gg \omega_m$. This leads to the absence of scattering of photons from one mode to another and ensures a single mode.

We write the Hamiltonian of the opto-mechanical system as

**Figure 13.1.** Paradigmatic model for an opto-mechanical system: an optical mode of frequency $\omega_c$ is confined between two mirrors: a left fixed mirror, $M_1$, and a right oscillating mirror, $M_2$. Reproduced from [1]. 

$$\hat{H} = \hat{H}_c + \hat{H}_m + \hat{H}_{cm} + \hat{H}_{cl}. \tag{13.7}$$

Here, $\hat{H}_c$ corresponds to the Hamiltonian of the single-mode cavity field, obtained as

$$\hat{H}_c = \hbar\omega_c\left(\hat{a}^\dagger\hat{a} + \frac{1}{2}\right). \tag{13.8}$$

The operator $\hat{a}(\hat{a}^\dagger)$ is the photon annihilation (creation) operator following the commutation relation, $[\hat{a}, \hat{a}^\dagger] = 1$. The Hamiltonian $\hat{H}_m$ describes the Hamiltonian of the end-mirror, written as

$$\hat{H}_m = \frac{\hat{p}^2}{2m} + \frac{1}{2}m\omega_m^2\hat{q}^2. \tag{13.9}$$

Here, $\hat{p}$ $(\hat{q})$ is the center-of-mass momentum (position) operator, and $m$ is the mass of the moving end-mirror. Moreover, these coordinates satisfy the commutation relation, $[\hat{q}, \hat{p}] = i\hbar$. Whereas, $\hat{H}_{cm}$ describes the interaction between the single-mode cavity field and the moving end-mirror, expressed as

$$\hat{H}_{cm} = V_{rp} = -\hbar\frac{\omega_c}{L}\hat{a}^\dagger\hat{a}\hat{q}. \tag{13.10}$$

The incident field is expressed by the Hamiltonian, $\hat{H}_{cl}$, given by

$$\hat{H}_{cl} = i\hbar\varepsilon_L(\hat{a}^\dagger e^{-i\omega_0 t} - \hat{a}e^{i\omega_0 t}). \tag{13.11}$$

Here, the amplitude, $\varepsilon_L$, is a complex number related to the incident laser power, $P_l$, following the mathematical relation, i.e. $|\varepsilon_L| = \sqrt{\frac{2P_l\kappa}{\hbar\omega_0}}$. Thus, we express the complete Hamiltonian governing the evolution of the opto-mechanical system as

$$\hat{H} = \frac{\hat{p}^2}{2m} + \frac{1}{2}m\omega_{\rm m}^2\hat{q}^2 + \hbar\omega_{\rm c}\hat{a}^\dagger\hat{a} - \hbar\frac{\omega_{\rm c}}{L}\hat{a}^\dagger\hat{a}\,\hat{q} + i\hbar\varepsilon_{\rm L}(\hat{a}^\dagger{\rm e}^{-i\omega_0 t} - \hat{a}{\rm e}^{i\omega_0 t}). \quad (13.12)$$

*Rotating frame transformation.* A time-independent description of the opto-mechanical Hamiltonian is obtained by transforming it into a frame rotating with laser field frequency, $\omega_0$. For this purpose, we define a unitary transformation, $\hat{U}(t)$, as

$$\hat{U}(t) = {\rm e}^{i\omega_0\,\hat{a}^\dagger\hat{a}\,t}. \quad (13.13)$$

The system Hamiltonian, $H$, therefore attains the shape of a time-independent transformed Hamiltonian, $\hat{H}_0$, that is,

$$\hat{U}(\hat{H} - i\hbar\partial_t)\hat{U}^\dagger = H_0 - i\hbar\partial_t, \quad (13.14)$$

Here, $\partial_t = \frac{\partial}{\partial t}$. In obtaining the expression, the Baker–Campbell–Hausdorff identity plays an important role, which has the expression

$${\rm e}^{\hat{A}}\hat{B}{\rm e}^{-\hat{A}} = \hat{B} + [\hat{A}, \hat{B}] + \frac{1}{2!}[\hat{A}, [\hat{A}, \hat{B},]]+\cdots, \quad (13.15)$$

Therefore, we find

$$\hat{U}\hat{a}\hat{U}^\dagger = \hat{a}{\rm e}^{-i\omega_0 t},$$
$$\hat{U}\hat{a}^\dagger\hat{U}^\dagger = \hat{a}^\dagger{\rm e}^{+i\omega_0 t},$$

and

$$i\hbar\hat{U}\partial_t\hat{U}^\dagger = \hbar\omega_0\hat{a}^\dagger\hat{a} + i\hbar\partial_t. \quad (13.16)$$

This leads us to write the transformed Hamiltonian of the opto-mechanical system as

$$\hat{H}_0 = \hbar\Delta_{\rm c}\hat{a}^\dagger\hat{a} + \frac{\hbar\omega_{\rm m}}{2}(\hat{q}^2 + \hat{p}^2) - \hbar\xi\hat{q}\hat{a}^\dagger\hat{a} + i\hbar\varepsilon_{\rm L}(\hat{a}^\dagger + \hat{a}). \quad (13.17)$$

Here, $\Delta_{\rm c}$ is the difference between the cavity frequency and laser frequency, that is,

$$\Delta_{\rm c} = \omega_{\rm c} - \omega_0, \quad (13.18)$$

and $\xi = \frac{\omega_{\rm c}}{L}$.

## 13.2 Time evolution in opto-mechanics

The evolution of the opto-mechanical system follows the quantum Langevin equation, that is,

$$\frac{d\hat{O}}{dt} = \frac{i}{\hbar}\left[\hat{H}_0, \hat{O}\right] - \frac{1}{2}\{\hat{\Gamma}, \hat{O}\} + \hat{N}. \quad (13.19)$$

The first term on the right-hand side provides the commutation relation between the total Hamiltonian of the system and a general operator, $\hat{O}$. The second term accounts for the decay associated with the observable related to the operator $\hat{O}$, and the last term is a noise operator corresponding to the noise in the system. The emergence of decay or dissipation in a system due to its interaction with the fluctuating environment is known as the fluctuation–dissipation theorem. The operator $\Gamma$ is the dissipation operator, whereas $\hat{F}$ describes the noise operator. For further details on the fluctuation–dissipation theorem, see [2].

The Heisenberg–Langevin equation for the optical field operator $\hat{a}$ becomes

$$\frac{d\hat{a}}{dt} = \dot{a} = (-i\Delta_c + i\xi\hat{q} - \kappa)\hat{a} + \varepsilon_L + \sqrt{2\kappa}\,\hat{a}_{in}. \tag{13.20}$$

Here, $\kappa$ describes the decay rate of the single-mode field inside the cavity, whereas $\sqrt{2\kappa}\,\hat{a}_{in}$ is the noise operator arising due to the interaction of the field with the environment modes.

We write equation (13.19) for the momentum operator for the moving mirror, $\hat{p}$, that is,

$$\frac{d\hat{p}}{dt} = \dot{p} = -m\omega_m^2\hat{q} - \hbar\xi\hat{a}^\dagger\hat{a} - \gamma\hat{p} + \hat{f}\,, \tag{13.21}$$

and for the position operator $\hat{q}$, that is,

$$\frac{d\hat{q}}{dt} = \dot{\hat{q}} = \frac{1}{m}\hat{p}, \tag{13.22}$$

Here, $\gamma$ is the mechanical damping associated with the mirror, and $\hat{f}$ is the corresponding fluctuation operator. On substituting equation (13.22) into equation (13.21), we obtain

$$\ddot{\hat{q}} + \gamma\dot{\hat{q}} + \omega_m^2\hat{q} = -\frac{\hbar\xi}{m}\hat{a}^\dagger\hat{a} + \frac{1}{m}\hat{f}\,. \tag{13.23}$$

In its classical counterpart, equation (13.23) describes a damped harmonic oscillator driven by an external force. The equation shows that the mirror's dynamics is modified due to its coupling with the field. The second term, $m\gamma\dot{\hat{q}} = \gamma\hat{p}$, describes the drag force or friction force, whereas, the first term on the right-hand side of equation (13.23) expresses the external force due to the optical field, and $f$ is the external force due to the environment.

Here, $\hat{a}_{in}$ describes the zero-mean input vacuum noise operator; therefore, we have $\langle\hat{a}_{in}\rangle = 0$. Furthermore, $\hat{a}_{in}$ satisfies the following correlation functions:

$$\langle\hat{a}_{in}^\dagger(t)\hat{a}_{in}(t')\rangle = 0, \tag{13.24}$$

$$\langle\hat{a}_{in}(t)\hat{a}_{in}^\dagger(t')\rangle = \delta(t - t'). \tag{13.25}$$

Moreover, $\hat{f}$ describes the Brownian stochastic random force that arises due to a random force being applied on the mirror by the collisions of environmental particles. We get $\langle f \rangle = 0$ and, taking the noise as a memory-less Markvian process, the corresponding correlation function:

$$\langle \hat{f}(t)\hat{f}(t') \rangle = \gamma(2n_{\mathrm{m}} + 1)\delta(t - t'). \tag{13.26}$$

Here, $n_{\mathrm{m}} = \left[\exp\left(\frac{\hbar\omega m}{K_{\mathrm{B}}T}\right) - 1\right]^{-1}$ is the mean thermal phonon number [3]. The symbol $T$ describes the mechanical mode environment temperature, and $K_{\mathrm{B}}$ is the Boltzmann constant.

### 13.2.1 Steady-state solution

We consider that no appreciable average change takes place in the time evolution of the cavity field, and moreover in the position and momentum of the mirror in a steady state. Thus, we substitute

$$\left\langle \frac{d\hat{c}}{dt} \right\rangle = 0,$$

$$\left\langle \frac{d\hat{q}}{dt} \right\rangle = 0,$$

$$\left\langle \frac{d\hat{p}}{dt} \right\rangle = 0.$$

Equations (13.20), (13.21), and (13.22) therefore provide the steady-state values, which are

$$\alpha_{\mathrm{s}} = \frac{\varepsilon_{\mathrm{L}}}{\kappa + i(\Delta_{\mathrm{c}} - \xi q_{\mathrm{s}})}, \tag{13.27}$$

$$q_{\mathrm{s}} = \frac{\hbar\xi}{m\omega_{\mathrm{m}}^{2}}|\alpha_{\mathrm{s}}|^2, \tag{13.28}$$

$$p_{\mathrm{s}} = 0, \tag{13.29}$$

Here, $\langle \hat{a} \rangle_{\mathrm{s}} = \alpha_{\mathrm{s}}$ is the steady-state value of the field, whereas $\langle \hat{q} \rangle_{\mathrm{s}} = q_{\mathrm{s}}$ and $\langle \hat{p} \rangle_{\mathrm{s}} = p_{\mathrm{s}}$ are the steady-state values of the position and momentum of the mirror, respectively. These steady-state values make it possible to write the steady-state photon number, $n_{\mathrm{s}}$, as

$$n_{\mathrm{s}} = \alpha^{*}\alpha = |\alpha_{\mathrm{s}}|^2 \tag{13.30}$$

$$= \frac{\varepsilon_{\mathrm{L}}^{2}}{\kappa^2 + (\Delta_{\mathrm{c}} - \xi q_{\mathrm{s}})^2}, \tag{13.31}$$

It is interesting to note that equation (13.31) reveals that the steady-state photon number $n_s$ itself depends upon the position of the mirror, thus introducing a Kerr-like non-linearity in the end-mirror dynamics.

### 13.2.2 Effective Hamiltonian for moving end-mirror

We have developed equations (13.20), (13.21), and (13.22) that govern the dynamics of a moving end-mirror. Taking the second derivative of equation (13.22), we obtain

$$\frac{d^2\hat{q}}{dt^2} = -\omega_{\mathrm{m}}^2\hat{q} - \frac{(\hbar\xi/m)\varepsilon_{\mathrm{L}}^2}{\kappa^2 + (\triangle_{\mathrm{c}} - \xi\hat{q})^2}. \tag{13.32}$$

Hence, in the absence of an electromagnetic field, that is, $\varepsilon_{\mathrm{L}} = 0$, the second term in equation (13.32) disappears, showing harmonic oscillator dynamics for the end-mirror. The presence of a field thus introduces a non-linearity as discussed above.

The effective Hamiltonian for the moving end-mirror is the sum of the kinetic energy and potential energy, that is, $H_{\mathrm{eff}} = K + V$. Keeping in view the classical description, we write the dimensionless kinetic energy as

$$K = \frac{p^2}{2},$$

whereas the potential energy $V$ is obtained from the integration of the equation of force, that is,

$$\dot{p} = -\frac{\partial V}{\partial q},$$

and we get

$$\begin{aligned} V &= -\int \frac{dp}{dt}dq = -m\int\left(\frac{d^2q}{dt^2}\right)dq, \\ &= \frac{1}{2}m\omega_{\mathrm{m}}^2q^2 + \int \frac{\hbar\xi\beta}{1 + [\mu - \mu_1 q]^2}\,dq, \end{aligned}$$

Here, we introduce new parameters: $\beta = \varepsilon_{\mathrm{L}}^2/\kappa^2$, $\mu = \Delta_{\mathrm{c}}/\kappa$, and $\mu_1 = \xi/\kappa$. Therefore, the effective Hamiltonian becomes

$$H_{\mathrm{eff}} = \frac{p^2}{2m} + \frac{1}{2}m\omega_{\mathrm{m}}^2q^2 + \int \frac{\hbar\xi\beta}{1 + [\mu - \mu_1 q]^2}\,dq, \tag{13.33}$$

Hence, we conclude that the mirror dynamics follows a harmonic oscillator in the absence of a field, whereas it is a non-linear oscillator with a non-linearity directly proportional to the coupling strength, $\xi$, and to the power of the incident laser field, since $\beta = \varepsilon_{\mathrm{L}}^2/\kappa^2$. Therefore, we obtain

$$\hat{H}_{\mathrm{eff}} = \frac{\hat{p}^2}{2m} + \frac{1}{2}m\omega_{\mathrm{m}}^2\hat{q}^2 + \hbar\Omega_{\mathrm{eff}}\arctan(\mu - \mu_1\hat{q}), \tag{13.34}$$

where $\Omega_{\text{eff}} = \frac{\xi\beta}{\mu_1} = \frac{2P_1}{\hbar\omega_0}$.

## 13.3 Bistability in opto-mechanics

The interaction between the cavity field and the moving end-mirror leads to bistable behavior in the opto-mechanics. We substitute the value of $q_s$ from equation (13.28) into equation (13.31), and obtain

$$
\begin{aligned}
n_s &= \frac{\varepsilon_L^2}{\kappa^2 + (\Delta_c - \xi q_s)^2}, \\
&= \frac{\varepsilon_L^2}{\kappa^2 + \left(\Delta_c - \xi\left(\frac{\xi n_s}{\omega_m}\right)\right)^2}, \\
&= \frac{\varepsilon_L^2}{\kappa^2 + (\Delta_c - \Omega_m n_s)^2},
\end{aligned}
\tag{13.35}
$$

where $\Omega_m$ is a function of opto-mechanical coupling, $\xi$, such that $\Omega_m = \frac{\hbar\xi^2}{m\omega_m^2}$. By rearranging the terms in equation (13.35), we obtain a third-order polynomial of the steady-state photon number, $n_s$, that is,

$$
n_s^3 + an_s^2 + bn_s + c = 0, \tag{13.36}
$$

where the coefficients $a$, $b$, and $c$ are

$$
\begin{aligned}
a &= \frac{-2\Delta_c}{\Omega_m}, \\
b &= \frac{\kappa^2 + \Delta_c^2}{\Omega_m^2}, \\
c &= \frac{-\varepsilon_L^2}{\Omega_m^2}.
\end{aligned}
$$

It is evident from equation (13.36) that there exist three roots of the equation, two corresponding to a stable regime and a third corresponding to an unstable regime of the steady-state photon number, $n_s$. The bistability depends on the coupling parameters $\xi$ through $\Omega_m$. In figure 13.2, bistable behavior is plotted as a function of $\varepsilon_L$ for three different values of detuning, $\Delta_c$.

Similarly, on substituting equation (13.31) into equation (13.28), we obtain

$$
q_s = \frac{\hbar\xi}{m\omega_m^2}|\alpha_s|^2 = \frac{\varepsilon_L^2}{\kappa^2 + (\Delta_c - \xi q_s)^2}\left(\frac{\hbar\xi}{m\omega_m^2}\right), \tag{13.37}
$$

By rearranging the terms in the above equation, we obtain a third-order polynomial of the steady-state position of the mirror, $q_s$, that is,

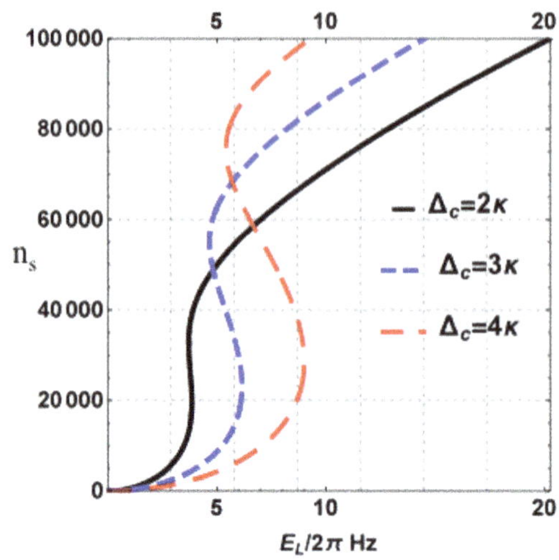


**Figure 13.2.** The bistable behavior of the intra-cavity photon number $n_S$, vs the driving field amplitude. The black, blue, and red dashed curves represent the bistable information for different values of $\Delta_c = 2\kappa$, $3\kappa$, and $4\kappa$.

$$q_s^3 + a_1 q_s^2 + b_1 q_s + c_1 = 0, \tag{13.38}$$

where the coefficients $a_1$, $b_1$, and $c_1$ are

$$a_1 = \frac{-2\Delta_c}{\xi},$$
$$b_1 = \frac{\kappa^2 + \Delta_c^2}{\xi^2},$$
$$c_1 = \frac{-\hbar\varepsilon_L^2}{m\omega_m^2\xi}.$$

## 13.4 Linearized equations of motion

Opto-mechanics in linear regimes opens up a number of fascinating domains which are of immense importance, thanks to their experimental feasibility. We write the dynamics around a stable point by writing the operators as the sum of the unit operator with c-number steady-state values and an operator for incremental change, that is,

$$\hat{O} = O_s\hat{I} + \delta\hat{O}, \tag{13.39}$$

where $\hat{I}$ is the unit operator, $O_s$ is the steady-state value of the observable, and $\delta$ indicates a small step away from the steady-state value. Therefore, we present the field and mirror operators as

$$\begin{aligned}\hat{q} &= q_s + \hat{\delta q},\\ \hat{p} &= p_s + \hat{\delta p},\\ \hat{a} &= \alpha_s + \hat{\delta a}.\end{aligned} \tag{13.40}$$

For a strong laser field $\alpha_s = \sqrt{n_s}$, where $n_s$ describes the steady-state field intensity in the cavity. Similarly, $q_s$ and $p_s$ express the steady-state position and momentum of the mirror.

We make these substitutions in equations (13.21) and (13.22). This provides the dynamical equations for fluctuating operators, which are

$$\delta\dot{q} = \frac{1}{m}\delta p, \tag{13.41}$$

$$\delta\dot{p} = m\omega_m^2\delta q - \gamma\delta p + i\hbar\xi(\alpha_s^*\delta\hat{a} + \alpha_s\delta\hat{a}^\dagger) + f. \tag{13.42}$$

Here, we further simplify the expression

$$\hat{a}^\dagger\hat{a} = \alpha_s^2 + (\alpha_s^*\delta\hat{a} + \alpha_s\delta\hat{a}^\dagger) + \delta\hat{a}^\dagger\delta\hat{a}.$$

Here, the last term depends on $\delta^2$, therefore we may discard it. Moreover, equation (13.20) provides

$$\delta\dot{\hat{a}} = -(\kappa + i\Delta)\delta\hat{a} + i\xi\alpha_s\delta\hat{q} + \sqrt{2\kappa}\,\hat{a}_{in}, \tag{13.43}$$

where

$$\Delta = \Delta_c - \xi q_s = \Delta_c - \frac{\hbar\xi^2|\alpha_s|^2}{m\omega_m^2}.$$

Following the discussion developed in section 2.2, we write the operators

$$\begin{aligned}\delta\hat{X} &= \frac{\delta\hat{a} + \delta\hat{a}^\dagger}{\sqrt{2}},\\ \delta\hat{Y} &= \frac{\delta\hat{a} - \delta\hat{a}^\dagger}{\sqrt{2}i}.\end{aligned}$$

We use these quadrature operators and include equations (13.41), (13.42), and (13.43) as a set of four coupled equations, which are

$$\delta\dot{\hat{q}} = \frac{1}{m}\delta\hat{p}, \tag{13.44}$$

$$\delta\dot{\hat{p}} = -m\omega_m^2\delta\hat{q} - \gamma\delta\hat{p} + \chi_1\delta\hat{X} + \chi_2\delta\hat{Y} + \hat{f}, \tag{13.45}$$

$$\delta\dot{\hat{X}} = -\kappa\delta\hat{X} + \Delta\delta\hat{Y} + \sqrt{2\kappa}\delta\hat{X}_{in}, \tag{13.46}$$

$$\delta\dot{\hat{Y}} = -\Delta\delta\hat{X} - \kappa\delta\hat{Y} + \chi_{mc}\delta\hat{q} + \sqrt{2\kappa}\delta\hat{Y}_{in}, \tag{13.47}$$

Here, $\delta\hat{X}_{\text{in}}$ and $\delta\hat{Y}_{\text{in}}$ describe the quadrature noise operators in following similar expressions as $\delta X$ and $\delta Y$. Moreover,

$$\chi_{\text{mc}} = \sqrt{2}\,\xi\alpha_{\text{s}},$$
$$\chi_1 = \hbar\xi\frac{\alpha_{\text{s}} + \alpha_{\text{s}}^*}{\sqrt{2}},$$
$$\chi_2 = \hbar\xi\frac{\alpha_{\text{s}} - \alpha_{\text{s}}^*}{\sqrt{2}}.$$

We write equations (13.44), (13.45), (13.46), and (13.47) in a matrix form as

$$\dot{\mathbf{R}}(t) = \mathbf{M}\mathbf{R}(t) + \mathbf{F}(t). \tag{13.48}$$

Here, $\mathbf{R} = \begin{pmatrix}\delta\hat{q} & \delta\hat{p} & \delta\hat{X} & \delta\hat{Y}\end{pmatrix}^T$ is the column matrix. The superscript $T$ expresses the transpose of the matrix. In addition, $\mathbf{M}$ is the kernel matrix of the system, which is

$$\mathbf{M} = \begin{pmatrix} 0 & \dfrac{1}{m} & 0 & 0 \\ -m\omega_{\text{m}}^2 & -\gamma & \chi_1 & \chi_2 \\ 0 & 0 & -\kappa & \Delta \\ \chi_{\text{mc}} & 0 & -\Delta & -\kappa \end{pmatrix}, \tag{13.49}$$

and $\mathbf{F} = \mathbf{N}_{\text{mf}} = \begin{pmatrix}0 & \hat{f} & \sqrt{2\kappa}\,\hat{X}_{\text{in}} & \sqrt{2\kappa}\,\hat{Y}_{\text{in}}\end{pmatrix}^T$ is the noise column matrix.

### 13.4.1 Stability criteria

Equation (13.48) has a solution that maps each realization of the noise column matrix to a stochastic realization of the output quadrature column matrix. Therefore, the solution becomes

$$\mathbf{R}(t) = \mathbf{S}(t)\mathbf{R}(0) + \int_0^t d\tau\mathbf{S}(\tau)\mathbf{F}(t - \tau), \tag{13.50}$$

where

$$\mathbf{S}(t) = \mathrm{e}^{\mathbf{M}t}. \tag{13.51}$$

Equation (13.51) is important as it determines the stability of the solution, $\mathbf{R}(t)$. According to Routh–Hurwitz criteria, the solution $\mathbf{R}(t)$ is stable asymptotically if all the eigenvalues of the kernel matrix $\mathbf{M}$ have negative real parts. This implies $\mathbf{S}(t \to \infty) = 0$ [4], necessary for a convergent solution. In order to obtain the Routh–Hurwitz criteria, we calculate the eigenvalues of the matrix $\mathbf{M}$. Therefore, we calculate

$$\det[\mathbf{M} - \lambda\mathbf{I}] = 0.$$

This yields the fourth-order characteristic equation:

$$\lambda^4 + c_3\lambda^3 + c_2\lambda^2 + c_1\lambda + c_0 = 0, \tag{13.52}$$

where

$$c_3 = 2\kappa + \gamma,$$
$$c_2 = \kappa(\kappa + 2\gamma) + \Delta^2 + \omega_{\mathrm{m}}^2,$$
$$c_1 = \kappa^2\gamma + \Delta^2\gamma + 2\omega_{\mathrm{m}}^2\kappa - \frac{\chi_1\chi_2}{m},$$
$$c_0 = \kappa^2\omega_{\mathrm{m}}^2 + \Delta^2\omega_{\mathrm{m}}^2 - \frac{1}{m}\chi_{\mathrm{mc}}\chi_1\Delta.$$

The solution of the deterministic equation (13.52) are the eigenvalues of the matrix **M**. In general, an analytical solution of a polynomial equation is very complicated, especially when the degree of the polynomial has higher integer values. For the reason, we are interested in the character of the real part of the eigenvalues, which implies whether they are positive, negative, or zero. For the fourth-order equation (13.52), the criteria provide constraints on the parameters of an opto-mechanical system that must be satisfied for a stable output. These are

$$s_1 = (c_3c_2 - c_1)c_1 - c_3^2c_0 > 0, \tag{13.53}$$

which provides

$$\begin{aligned} s_1 = &\left(\Delta^2\gamma + \kappa^2\gamma + 2\kappa\ \omega_{\mathrm{m}}^2 - \frac{\chi_1\chi_2}{m}\right)\left(2\Delta^2\kappa + 2\kappa^3 + 4\kappa^2\gamma + 2\kappa\gamma^2 + \gamma\omega_{\mathrm{m}}^2 + \frac{\chi_1\chi_2}{m}\right) \\ &- (2\kappa + \gamma)^2\left(-\frac{1}{m}\Delta\chi_{\mathrm{mc}}\chi_1 + \Delta^2\omega_{\mathrm{m}}^2 + \kappa^2\omega_{\mathrm{m}}^2\right) > 0, \end{aligned} \tag{13.54}$$

and the other self-evident condition from the above equation (13.52) becomes

$$s_2 = c_0 = -\frac{1}{m}\chi_{\mathrm{mc}}\chi_1 + \Delta^2\omega_{\mathrm{m}}^2 + \kappa^2\omega_{\mathrm{m}}^2 > 0, \tag{13.55}$$

From these conditions, we see that for $\Delta > 0$ the first condition, $s_1$, is always satisfied and we need to check only $s_2$, whereas for $\Delta < 0$ the second condition is always satisfied and we need to check $s_1$ only.

### 13.4.2 Coherence in the mirror and optical field

A column vector, $\mathbf{R} = (R_1\ \ R_2\ \ R_3\ \ R_4)^T$, of the random variables has a corresponding covariance matrix, $\boldsymbol{\sigma}$, such that the diagonal elements describe the variance of each random variable, whereas the off-diagonal elements express covariance. In the opto-mechanical discussion $R_1$, $R_2$, $R_3$, and $R_4$ correspond to $\delta q$, $\delta p$, $\delta X$, and $\delta Y$, respectively. We write the covariance matrix as

$$\boldsymbol{\sigma} = \begin{pmatrix} \sigma_{11} & \sigma_{12} & \sigma_{13} & \sigma_{14} \\ \sigma_{21} & \sigma_{22} & \sigma_{23} & \sigma_{24} \\ \sigma_{31} & \sigma_{32} & \sigma_{33} & \sigma_{34} \\ \sigma_{41} & \sigma_{42} & \sigma_{43} & \sigma_{44} \end{pmatrix}. \tag{13.56}$$

Here, we express the variance associated with the random variable as

$$\sigma_{ii} = \langle (R_i - \mu_i)^2 \rangle, \tag{13.57}$$

where $\mu_i$ is the average value of the $i$th random variable, that is, $\mu_i = \langle R_i \rangle$. Moreover, the covariance between two random variables is

$$\sigma_{ij} = \langle (R_i - \mu_i)(R_j - \mu_j) \rangle = \langle R_i R_j \rangle - \langle R_i \rangle \langle R_j \rangle. \tag{13.58}$$

In case the random variable follows a Gaussian-nature random process, we write the general expression for the variance as

$$\sigma_{ij} = \frac{\langle \{R_i, R_j\} \rangle}{2} = \frac{\langle R_i R_j + R_j R_i \rangle}{2}. \tag{13.59}$$

In quantum mechanics a state is of a Gaussian nature if its quasi-density function or characteristic function is Gaussian with zero average value.

*Dynamical coherence.* The dynamics of the coherence can be obtained from the rate equation for the covariance matrix:

$$\dot{\sigma}_{ij} = \frac{\langle \{\dot{R}_i, R_j\} + \{R_i, \dot{R}_j\} \rangle}{2}. \tag{13.60}$$

Equation (13.60) can be simplified by applying equation (13.48) for the matrix elements, that is,

$$\dot{R}_i(t) = \sum_l M_{il} R_l(t) + F_i. \tag{13.61}$$

On substituting equation (13.61) in equation (13.60), we obtain

$$\dot{\sigma}_{ij} = \sum_l M_{il} \sigma_{lj} + \sum_l \sigma_{il} M_{lj} + D_{ij}, \tag{13.62}$$

where

$$D_{ij} = \frac{\langle \{R_i, F_j\} + \{F_i, R_j\} \rangle}{2}. \tag{13.63}$$

In matrix form, equation (13.62) appears as

$$\dot{\boldsymbol{\sigma}} = \mathbf{M}\boldsymbol{\sigma}(t) + \boldsymbol{\sigma}(t)\mathbf{M}^T + \mathbf{D}, \tag{13.64}$$

Equation (13.64) provides the solution of the covariance matrix, $\boldsymbol{\sigma}(t)$.

*Steady-state coherence.* Following Routh–Hurwitz criteria, the eigenstates of $\mathbf{M}$ have negative real values; this implies $\mathbf{S}(t \to \infty) = 0$. Therefore, equation (13.50), becomes

$$R_i(\infty) = \int_0^\infty d\tau \sum_j S_{ij}(\tau) F_j(t - \tau). \tag{13.65}$$

The steady-state covariance matrix, $\boldsymbol{\sigma}(\infty)$, is obtained as

$$\boldsymbol{\sigma}(\infty) = \sum_{ij} \frac{\langle R_i(\infty)R_j(\infty) + R_j(\infty)R_i(\infty)\rangle}{2}, \tag{13.66}$$

where $R_i(\infty)$ is the steady-state value of $R_i$. The covariance matrix fulfils the Lyapunov equation, which we obtain as we consider that the time variation in the covariance matrix disappears at the steady state; therefore, we rewrite equation (13.64) as

$$-\mathbf{D} = \mathbf{M}\boldsymbol{\sigma}(\infty) + \boldsymbol{\sigma}(\infty)\mathbf{M}^T, \tag{13.67}$$

where the matrix $\mathbf{M}$ is given in equation (13.49). Moreover, $\mathbf{D}$ is the diffusion matrix defined in equation (13.63), and is obtained as $\mathbf{D} = \mathrm{diag}\big(0 \quad \gamma(2n_{\mathrm{m}} + 1) \quad \kappa \quad \kappa\big)$.

The general representation of the covariance matrix, $\sigma$, is

$$\boldsymbol{\sigma} = \begin{pmatrix} A & C \\ C^T & B \end{pmatrix}. \tag{13.68}$$

The covariance matrix is written in $2 \times 2$ block form. Here, $A$ and $B$ describe a $2 \times 2$ reduced covariance matrix for the mirror and field, respectively, whereas $C$ describes a $2 \times 2$ matrix showing in-between correlation. The covariance matrix is real-valued and symmetric in nature.

### 13.4.3 Entanglement

Entanglement is an important resource for the implementation of quantum information processing devices, and in performing communication and computations tasks more efficiently up to limits not achievable classically. Therefore, it sets a boundary between the quantum and classical worlds. Correlation between two distinct parties or the non-separability of their states defines entanglement; hence, the quantum states of the two are described with respect to each other.

In continuous variable opto-mechanical systems, we measure the entanglement between a mechanical mirror and an optical field by calculating the logarithmic negativity, $E_{\mathrm{mc}}$, defined as

$$E_{\mathrm{mc}} = \max[0, -\ln 2\nu], \tag{13.69}$$

where

$$\nu \equiv 2^{-1/2}\left\{\sum(\sigma_{\mathrm{mc}}) - \left[\sum(\sigma_{\mathrm{mc}})^2 - 4\det\sigma\right]^{1/2}\right\}^{1/2}, \tag{13.70}$$

with

$$\sum(\sigma_{\mathrm{mc}}) \equiv \det A + \det B - 2\det C. \tag{13.71}$$

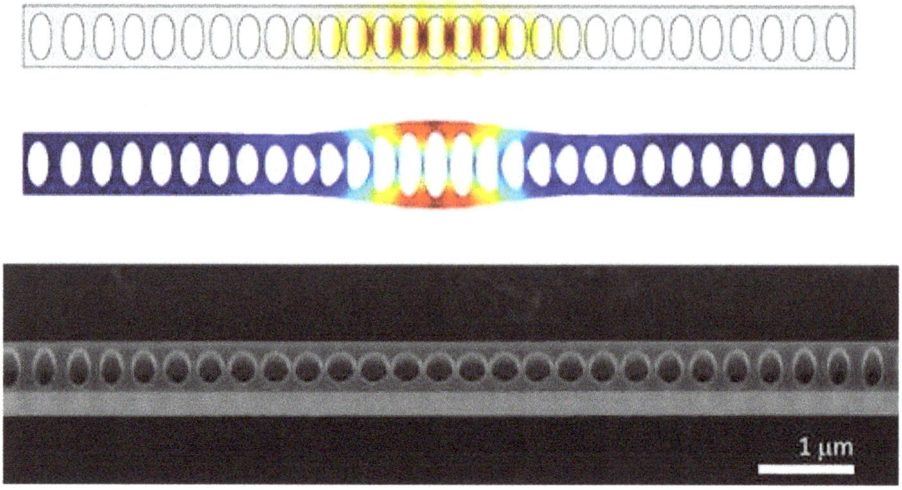


**Figure 13.3.** Top: Schematic diagram of crystal geometry, overlaid with a simulation of an optical resonance. Middle: Schematic diagram showing a mechanical breathing mode. Bottom: Scanning electron micrograph of silicon opto-mechanical crystal. Reprinted from [5], with the permission of AIP Publishing.

The Gaussian state becomes entangled only if $\nu < 1/2$, which is Simon's necessary-and-sufficient entanglement non-positive partial transpose criterion of the Gaussian states.

## 13.5 Opto-mechanical crystals

As a direct extension of a single-cavity opto-mechanical system, opto-mechanical crystals are being developed (see figure 13.3). These are intended to study their mechanical vibrations and to manipulate and modify optical signals [5]. Thereby, opto-mechanical crystals are setting the stage to produce integrated devices. The major focus is on engineering the radiation-pressure-based interaction between light and mechanical vibrations in these systems. In addition to studying the basic quantum interactions between light and matter, this interaction will lead to the development of devices enabling applications such as RF memorizers, accelerometers, torque sensors, synchronization of mechanical oscillators, and wavelength converters.

## Exercises

1. Calculate the average radiation pressure force applied by an optical field of an average photon number $\bar{n} = 1$ and frequency $\omega = 2\pi \times 10$ KHz in a cavity of length 1 km.
2. Calculate the average radiation pressure force eta applied by an optical field of an average photon number $\bar{n} = 1$ and frequency $\omega = 2\pi \times 10$ KHz in a cavity of length 1 $\mu$m.

3. Plot the bistable behavior of the steady-state position of a mirror, $q_s$, as a function of a driving field amplitude, $\varepsilon_L$, for different values of detuning $\Delta_c = 2\kappa, 3\kappa, 4\kappa$. Show that the two roots of equation (13.38) are stable solutions whereas the third root expresses an unstable solution of $q_s$.
4. Calculate the Ruth–Hurwitz stability criteria for the resonance case, that is, $\Delta = 0$.
5. Show that the time evolution of a covariance matrix is expressed as

$$\dot{\boldsymbol{\sigma}} = \mathbf{M}\boldsymbol{\sigma}(t) + \boldsymbol{\sigma}(t)\mathbf{M}^T + \mathbf{D},$$

   where $T$ is the transpose of a matrix and $\mathbf{D}$ expresses the diffusion matrix.
6. With the help of the Lyapunov equation,

$$-\mathbf{D} = \mathbf{M}\boldsymbol{\sigma}(\infty) + \boldsymbol{\sigma}(\infty)\mathbf{M}^T,$$

   obtain the $\boldsymbol{\sigma}(\infty)$ for the opto-mechanical system defined in equation (13.48).

## References

[1] Liu Y-C, Hu Y-W, Wong C W and Xiao Y-F 2013 Review of cavity optomechanical cooling *Chin. Phys. B* **22** 114213

[2] Kubo R 1966 The fluctuation–dissipation theorem *Rep. Prog. Phys.* **29** 255

[3] Gardiner C W and Zoller P 2000 *Quantum Noise* (Berlin: Springer)

[4] DeJesus E X and Kaufman C 1987 Routh–Hurwitz criterion in the examination of eigenvalues of a system of nonlinear ordinary differential equations *Phys. Rev.* A **35** 5288

[5] Davanço M, Ates S, Liu Y and Srinivasan K 2014 *Appl. Phys. Lett.* **104** 041101

# Chapter 14

## Hybrid opto-mechanics

*'Provide ships or sails adapted to the heavenly breezes, and there will be some who will brave even that void.'*

—Johannes Kepler

In this chapter, we discuss a hybrid opto-mechanical system obtained as ultra-cold atoms interact with an opto-mechanical system. We can create such hybrid opto-mechanical systems by considering the interaction of two-level atoms, multi-level atoms, nano-particles, nano-membranes, etc, with the single-mode or multi-mode fields of opto-mechanical systems in resonant or far-off resonant domains.

### 14.1 Ultra-cold atoms in an opto-mechanical cavity

As discussed in chapter 3, a two-level atom interacts with a single-mode field via dipole interaction. We follow the treatment developed in section 3.1 and write the atom–field interaction Hamiltonian, $H_{ac}$, in the dipole approximation and in the rotating-wave approximation as

$$\hat{H}_{ac} = \frac{\hat{p}^2}{2M} + \hbar g(x)(\hat{\sigma}_+\hat{c} + \hat{c}^\dagger\hat{\sigma}_-). \tag{14.1}$$

The first term expresses the kinetic energy of an atom of mass $M$ along the cavity field axis. Following the discussion in section 3.1, here we have introduced atomic lowering and raising operators, $\hat{\sigma}_-$ and $\hat{\sigma}_+$, respectively, and ladder operators, $\hat{c}$ and $\hat{c}^\dagger$, for the electromagnetic field. The atom–field coupling, $g(x) = g_0 u(x)$, is position dependent and real-valued.

We consider a far-off resonant regime where a large detuning takes place between the field frequency and the atomic transition frequency. Therefore, we adiabatically eliminate the excited-state dynamics of the atom in space and time, for the reason

doi:10.1088/978-0-7503-2308-6ch14

that we neglect the spontaneous emission. Hence, the atom–field interaction is expressed as

$$\hat{H}_{ac} = \frac{\hat{p}^2}{2m} + \hbar\frac{g_o^2}{\Delta_a}\,\hat{c}^\dagger\hat{c}\,\cos^2(kx). \tag{14.2}$$

Here, $g_o$ is the atom–field coupling strength, where $g_o$ is the atom–field coupling and $\Delta_a = \omega_c - \omega_{eg}$ is the detuning between the field frequency, $\omega_c$, and the atomic transition frequency, $\omega_{eg}$.

Let us consider a Bose–Einstein condensate (BEC) interacting with an opto-mechanical cavity under the condition of a low atomic density (figure 14.1). Therefore, we take the atom–atom interaction, discussed in section 6.4, as negligibly small. The atom–field Hamiltonian is written as

$$\hat{H}_{ac} = \int \hat{\psi}^\dagger(x)\left(-\frac{\hbar^2}{2m}\frac{d^2}{dx^2} + \hbar U_0\,\hat{c}^\dagger\hat{c}\,\cos^2 kx\right)\hat{\psi}(x)\,dx. \tag{14.3}$$

The operator $\hat{\psi}^\dagger$ ($\hat{\psi}$) is an atomic creation (annihilation) operator, $U_0 = g_0^2/\Delta_a$ expresses the effective frequency, and $k$ is the wave number of the electromagnetic field.

Due to its interaction with the field, the atom experiences a recoil momentum of an integer multiple of $\pm 2\hbar k$. The atomic interaction with the intra-cavity field thus develops different momentum side modes of the condensate separated by an integer multiple of $2\hbar k$. In a weak-field approximation, we take into account a low photon number. Recent experimental observations show that when the cavity photon number is not large, the atomic momentum side modes that interact significantly with the cavity field are those with momenta 0 and $\pm 2\hbar k$. Thus, the coupling generates zeroth- and first-order modes, and for simplicity higher-order modes are ignored. This approach provides $\hat{\psi}(x)$ in terms of the side-mode bosonic operators $\hat{b}_0$ and $\hat{b}_1$ as

$$\hat{\psi}(x) = \frac{1}{\sqrt{L}}\left\{\hat{b}_0 + \sqrt{2}\,\cos(2kx)\hat{b}_1\right\}.$$

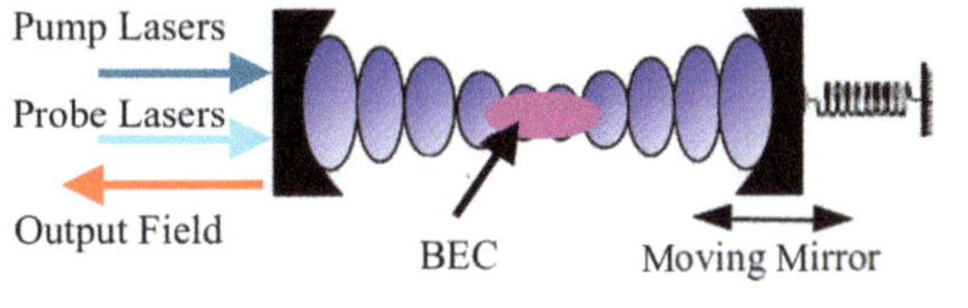


**Figure 14.1.** Paradigm model for a hybrid opto-mechanical system: an optical mode of frequency $\omega_c$ is confined between two mirrors, $M_1$ and $M_2$. Mirror $M_1$ is fixed, whereas mirror $M_2$ oscillates with a frequency $\omega_m$ around its equilibrium position under the effect of radiation pressure force. Both optical and mechanical modes are coupled with their local environment with couplings $\kappa$ and $\gamma_m$, respectively. The displacement of $M_2$ due to the radiation pressure force is $q$. A cavity mode is formed via an input laser of frequency $\omega_o$ and power $P$. A Bose–Einstein condensate (BEC) is coupled with the optical field in the cavity. Adapted figure with permission from [7], Copyright (2015) by the American Physical Society.

Here, $\hat{b}_0$ and $\hat{b}_1$ are the annihilation operators for the zeroth mode and the first mode, and are related as

$$\hat{b}_0^{\dagger}\hat{b}_0 + \hat{b}_1^{\dagger}\hat{b}_1 = N, \tag{14.4}$$

that is, the number of bosonic particles in all momentum side modes is $N$. As the population in the zeroth mode is much larger than the population in the first-order side mode, we can write $\hat{b}_0^{\dagger}\hat{b}_0 \simeq N$ or $\hat{b}_0$, and $\hat{b}_0^{\dagger} \to \sqrt{N}$ as the coupling to higher side modes is weak enough and therefore they are not populated significantly. Therefore, equation (14.3) becomes

$$H_{\rm ac} = \frac{2\hbar^2 k^2}{M}\hat{b}_1^{\dagger}\hat{b}_1 + \frac{\hbar U_0}{4}\hat{c}^{\dagger}\hat{c}\left\{\left(\hat{b}_0^{\dagger}\hat{b}_1 + \hat{b}_1^{\dagger}\hat{b}_0\right) + 2\left(\hat{b}_0^{\dagger}\hat{b}_0 + \hat{b}_1^{\dagger}\hat{b}_1\right)\right\}, \tag{14.5}$$

We substitute the dimensionless position and momentum coordinates for momentum side modes, that is,

$$\hat{Q} = \frac{1}{\sqrt{2}}(\hat{b}_1 + \hat{b}_1^{\dagger}), \tag{14.6}$$

$$\hat{P} = \frac{1}{\sqrt{2}}(\hat{b}_1 - \hat{b}_1^{\dagger}), \tag{14.7}$$

and use equation (14.4) to re-express the Hamiltonian as

$$\hat{H}_{\rm ac} = \frac{\hbar U_0 N}{2}\hat{c}^{\dagger}\hat{c} + \frac{\hbar\Omega}{2}(\hat{P}^2 + \hat{Q}^2) + \hbar\xi_{\rm sm}\hat{c}^{\dagger}\hat{c}\hat{Q}, \tag{14.8}$$

where $\Omega = 2\hbar k^2/M$ is the recoil frequency of an atom. In equation (14.8), the first term describes the energy of the field due to the condensate. Moreover, $U_0 = g_0^2/\Delta_a$ is the optical lattice barrier depth per photon, and $N$ is the averaged number of atoms. The second term expresses the harmonic oscillator dynamics of the condensate atoms in the cavity. The third term describes the coupling between the field and condensate atoms with coupling strength $\xi_{\rm sm} = U_0\sqrt{N}/2$.

## 14.2 Quantum suppression of classical diffusion

An opto-mechanical system in the presence of a coupled BEC expresses a hybrid system that is effectively controlled by opto-mechanical coupling and BEC–field coupling. The general system presents a two degrees-of-freedom system, thereby satisfying the minimum criteria to exhibit complex dynamics. Following the treatment developed in subsection 13.2.2, a hybrid opto-mechanical system in the presence of ultra-cold atoms can be expressed by coupled differential equations as follows:

$$\frac{d^2q}{dt^2} = -\omega_{\rm m}^2 q - \frac{(\hbar\xi/m)\varepsilon_{\rm L}^2}{\kappa^2 + (\tilde{\Delta}+\xi q - \xi_{\rm sm}Q)^2}, \tag{14.9}$$

$$\frac{d^2Q}{dt^2} = -\Omega^2 Q - \frac{\Omega \xi_{\mathrm{sm}} \varepsilon_{\mathrm{L}}^2}{\kappa^2 + (\tilde{\Delta} + \xi q - \xi_{\mathrm{sm}} Q)^2}. \tag{14.10}$$

Here, $\tilde{\Delta} = \Delta_c - NU_0/2$.

### 14.2.1 Dynamical localization of moving end-mirror

Let us consider that the matter–field coupling $\xi_{\mathrm{sm}}$ is small as compared with the coupling between the moving end-mirror and the field, $\xi$, that is, $\xi_{\mathrm{sm}} \ll \xi$. Therefore, neglecting the second term on the right-hand side, equation (14.10) provides the BEC evolution as a harmonic oscillator with frequency $\Omega$, such that $Q = Q_0 \cos(\Omega t)$, where $Q_0$ is the maximum displacement of the BEC from mean position. Equation (14.9) becomes

$$\frac{d^2q}{dt^2} = -\omega_{\mathrm{m}}^2 q - \frac{(\hbar\xi/m)\varepsilon_{\mathrm{L}}^2}{\kappa^2 + (\tilde{\Delta} + \xi q - \xi_{\mathrm{sm}} Q_0 \cos(\Omega t))^2}, \tag{14.11}$$

$$= -\omega_{\mathrm{m}}^2 q - \frac{(\hbar\xi/m)\varepsilon_{\mathrm{L}}^2}{\kappa^2 + (\tilde{\Delta} + \xi\{q - g_{\mathrm{sm}} \cos(\Omega t)\})^2}, \tag{14.12}$$

where $g_{\mathrm{sm}} = \xi_{\mathrm{sm}} Q_0/\xi$ has units of length. We write the corresponding potential energy

$$V(q) = \frac{1}{2} m\omega_{\mathrm{m}}^2 q^2 + \hbar\xi\varepsilon_{\mathrm{L}}^2 \int \frac{1}{\kappa^2 + (\tilde{\Delta} + \xi\{q - g_{\mathrm{sm}} \cos(\Omega t)\})^2} dq. \tag{14.13}$$

Applying a Kramers–Henneberger transformation, as discussed in chapter 10, we consider $x = q - g_{\mathrm{sm}} \cos \Omega t$. The effective Hamiltonian is obtained as

$$H_{\mathrm{eff}} = \frac{p^2}{2m} + \frac{1}{2} m\omega_{\mathrm{m}}^2 x^2 - m\Omega^2 g_{\mathrm{sm}}\, x \sin \Omega t + \int \frac{\hbar\xi\varepsilon_{\mathrm{L}}^2}{\kappa^2 + (\tilde{\Delta} + \xi x)^2} dx, \tag{14.14}$$

which has the same structure as that of a driven system:

$$H_{\mathrm{eff}} = H_0 - \lambda V(x) \sin \Omega t, \tag{14.15}$$

where $\lambda = m\Omega^2 g_{\mathrm{sm}}$ and $V(x) = x$.

The classical dynamics of a moving end-mirror in an opto-mechanical system, when the BEC–field coupling is small as compared with the mirror–field coupling, displays a dominant stochastic behavior for increasing mirror–field coupling. As the time evolves, an initial probability distribution showing various possible initial values for the moving end-mirror shows dispersion in position space and in momentum space.

The quantum mechanical mirror is expressed as an initial wave function, $\Psi(0)$, that follows the minimum uncertainty relation initially. As a consequence of dynamical localization, however, the quantum evolution of the mirror is significantly different both in position space and momentum space. It resists its expansion as it evolves and almost freezes in both spaces. Figure 14.2 shows the end-mirror

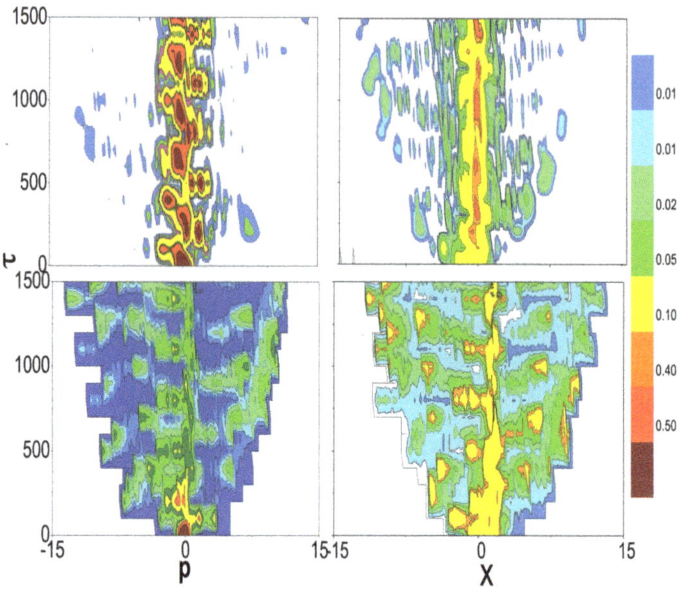


**Figure 14.2.** The upper two figures show the quantum mechanical dynamical localization of a moving end-mirror in momentum space (left) and position space (right) as a function of time. The two lower figures display the corresponding classical dispersive behaviour for the same parameters; see reference [1]. Taylor & Francis Ltd. http://tandfonline.com.

probability distributions in position space ($|\Psi(x, t)|^2$) and in momentum space ($|\Psi(p, t)|^2$) as a function of time. The space-time carpets, thus generated, exhibit a contrast in classical and corresponding quantum dynamics. We find that the classical dispersive dynamics of the end-mirror is restricted in the quantum domain due to dynamical localization in both the position and momentum spaces. The quantum mechanical wave packet follows classical dispersive dynamics up to a short time limit, named the quantum break time, $t^*$. Beyond that time limit, it exhibits an exponential localization in the probability distributions with maximum probability confined within a space attained during the quantum break time, both in position as well as in momentum spaces.

### 14.2.2 Dynamical localization of ultra-cold atoms

The classical dynamics of a BEC in an opto-mechanical cavity, when the mirror–field coupling is small as compared with the BEC–field coupling, displays a dominant stochastic behaviour for increasing BEC–field coupling. As the time evolves, an initially classical probability distribution showing the distribution of atoms in the BEC undergoes classical dispersive dynamics and, as a result, the variance increases linearly with time, both in position space and in momentum space. By contrast, as a consequence of dynamical localization, the quantum evolution in space and time differs significantly. The quantum mechanical dynamics of the initial BEC resists its expansion and almost freezes in both spaces, that is, in the position space and in the momentum space.

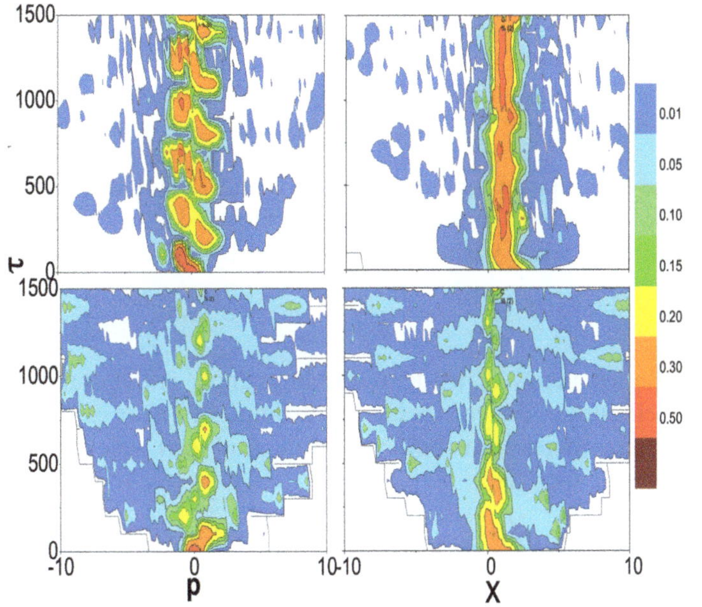


**Figure 14.3.** The upper two figures show the quantum mechanical dynamical localization of ultra-cold atoms in momentum space (left) and position space (right) as a function of time. The two lower figures display the corresponding classical dispersive behaviour for the same parameters; see reference [2]. Reproduced from [2]. 

Figure 14.3 shows the BEC evolution in position space and momentum space as a function of time. The generated space-time carpets exhibit a contrast in the classical and corresponding quantum dynamics. The quantum mechanical wave packet follows classical dispersive dynamics up to a quantum break time. Beyond the time limit, it displays dynamical localization with maximum probability confined within a space attained during the quantum break time, both in position as well as in momentum space. As a consequence, we find that the classical dispersive dynamics in both spaces is restricted due to dynamical localization.

## 14.3 Input–output formalism

In order to develop an electromagnetic field inside an opto-mechanical cavity, we couple an external laser light to the opto-mechanical cavity by using a partially reflecting mirror. Following the fluctuation–dissipation theorem, the partially reflecting mirror also couples the mode inside the cavity with the environment, comprising infinite external vacuum modes. Hence, the external laser light and the environment constitute the input field launched inside of the opto-mechanical cavity, whereas the electromagnetic field coming from the cavity is considered as the output. Following reservoir theory, we model the outside environment modes as a bath of an infinite set of oscillators interacting with the intra-cavity mode. We may express the system Hamiltonian,

$$\hat{H} = \hat{H}_c + \hat{H}_b + \hat{H}_{int}, \tag{14.16}$$

where $\hat{H}_c$ is the cavity Hamiltonian, such that $\hat{H}_c = \hbar\omega_c\hat{c}^\dagger\hat{c}$. Here, $\hat{H}_b$ describes the bath Hamiltonian, such that

$$\hat{H}_b = \int_0^\infty d\omega\ \hbar\omega\hat{b}^\dagger(\omega)\hat{b}(\omega). \tag{14.17}$$

Here, $\hat{b}(\omega)$ is the annihilation operator for the bath mode. Moreover, $\hat{H}_{int}$ describes the interaction of the cavity mode and the infinite modes of the bath. Considering the continuum in frequency variable, the bath Hamiltonian is given by

$$\hat{H}_{int} = i\hbar\int_0^\infty d\omega\ g(\omega)[\hat{c}\hat{b}^\dagger(\omega) - \hat{c}^\dagger\hat{b}(\omega)], \tag{14.18}$$

where $g(\omega)$ is the frequency-dependent coupling, and $\hat{c}$ is the annihilation operator for the cavity field mode. Except the mode of frequency near resonance, i.e. $\omega_c$, the modes of other frequencies contribute very little, and thus we extend the lower limit of integration to $-\infty$ such as

$$\hat{H}_{int} = i\hbar\int_{-\infty}^\infty d\omega\ g(\omega)[\hat{c}\hat{b}^\dagger(\omega) - \hat{c}^\dagger\hat{b}(\omega)]. \tag{14.19}$$

The corresponding commutation relations are

$$[\hat{c}, \hat{c}^\dagger] = 1, \tag{14.20}$$

$$\left[\hat{b}(\omega), \hat{b}^\dagger(\omega')\right] = \delta(\omega - \omega'). \tag{14.21}$$

We write the Heisenberg equation of motion for the operators $\hat{b}$ and $\hat{c}$, which are

$$\dot{\hat{b}} = \frac{i}{\hbar}[\hat{H}, \hat{b}] = -i\omega\hat{b}(\omega) + g(\omega)\hat{c}, \tag{14.22}$$

$$\dot{\hat{c}} = \frac{i}{\hbar}[\hat{H}, \hat{c}] = \frac{i}{\hbar}[\hat{H}_s, \hat{c}] - \int_{-\infty}^\infty d\omega\ g(\omega)\hat{b}(\omega). \tag{14.23}$$

The integral solution to equation (14.22) for $t > t_0$ is expressed as

$$\hat{b}(\omega) = e^{-i\omega(t-t_0)}\hat{b}_0(\omega) + g(\omega)\int_{t_0}^t dt' e^{-i(t-t')}\hat{c}(t'), \tag{14.24}$$

where $\hat{b}_0$ expresses the initial bath operator. In terms of the final time, i.e. $t < t_1$, the integral solution for equation (14.22) $t < t_1$ is expressed as

$$\hat{b}(\omega) = e^{-i\omega(t-t_1)}\hat{b}_1(\omega) - g(\omega)\int_t^{t_1} dt' e^{-i(t-t')}\hat{c}(t'). \tag{14.25}$$

Here, $\hat{b}_1$ expresses the bath operator after a time of interaction, $t_1$.

Inserting $\hat{b}(\omega)$ from equation (14.24) in equation (14.23), we obtain

$$\dot{\hat{c}} = \frac{i}{\hbar}[\hat{H}_c, \hat{c}] - \int_{-\infty}^\infty d\omega\ g(\omega)e^{-i\omega(t-t_0)}\hat{b}_0(\omega) - \int_{-\infty}^\infty d\omega\ g^2(\omega)\int_{t_0}^t dt' e^{-i\omega(t-t')}\hat{c}(t'). \tag{14.26}$$

In the Markov approximation, we consider that the coupling to environment has no memory, that the coupling becomes frequency independent and is expressed as $g(\omega) = \sqrt{\kappa/2\pi}$. Therefore, the resulting equation appears as

$$\dot{\hat{c}} = \frac{i}{\hbar}[\hat{H}_c, \hat{c}] - \sqrt{\frac{\kappa}{\pi}} \int_{-\infty}^{\infty} d\omega\ e^{-i\omega(t-t_0)}\hat{b}_0(\omega) - \kappa\hat{c}(t), \tag{14.27}$$

$$\dot{\hat{c}} = \frac{i}{\hbar}[\hat{H}_c, \hat{c}] + \sqrt{2\kappa}\,\hat{c}_{in} - \kappa\hat{c}(t). \tag{14.28}$$

Here, we have used the delta function identities,

$$\frac{1}{2\pi} \int_{-\infty}^{\infty} d\omega\ e^{-i\omega(t-t')} = \delta(t - t'), \tag{14.29}$$

$$\int_{-\infty}^{\infty} dy\delta(y - b)f(y) = f(b), \tag{14.30}$$

and introduced the input field operator $\hat{c}_{in}(t)$ as the continuous sum over frequency variable of the bath's initial and final values, respectively, of all oscillators as

$$\hat{c}_{in}(t) = \frac{-1}{\sqrt{2\pi}} \int_{-\infty}^{\infty} d\omega\ e^{-i\omega(t-t_0)}\hat{b}_0(\omega). \tag{14.31}$$

Following the same procedure and inserting $\hat{b}(\omega)$ from equation (14.25) in equation (14.23), we obtain

$$\dot{\hat{c}} = \frac{i}{\hbar}[\hat{H}_c, \hat{c}] + \sqrt{2\kappa}\,\hat{c}_{out} + \kappa\hat{c}(t). \tag{14.32}$$

Here, we have defined the output field operator as

$$\hat{c}_{out}(t) = \frac{-1}{\sqrt{2\pi}} \int_{-\infty}^{\infty} d\omega e^{-i\omega(t-t_0)}\hat{b}_1(\omega). \tag{14.33}$$

Subtracting equation (14.28) from equation (14.32), we obtain

$$\hat{c}_{out} = \hat{c}_{in} - \sqrt{2\kappa}\,\hat{c}. \tag{14.34}$$

## 14.4 Induced transparency and four-wave mixing

In order to investigate the optical property of the output probe field, we recall the input and output standard relation of the cavity field, discussed in section 14.3, that is,

$$\hat{c}_{out} = \hat{c}_{in}(t) - \sqrt{2\kappa}\ \hat{c}(t), \tag{14.35}$$

where $\hat{c}_{in}(t)$ and $\hat{c}_{out}(t)$ are the input and output operators, respectively. By using the above input–output relation, we obtain the expectation value of the output field in the steady state as

$$\langle c_{out}(t)\rangle = (E_l - \sqrt{2\kappa}\ c_s) + (E_p - \sqrt{2\kappa}\ c_-)\ e^{-i\delta t} - \sqrt{2\kappa}c_+e^{i\delta t}. \tag{14.36}$$

Here, we have considered

$$\langle \hat{c}_{\text{in}} \rangle = E_{\text{l}} + E_{\text{p}} e^{-i\delta t},$$

and the ansatz

$$\langle \hat{c} \rangle = c_{\text{s}} + c_{-}\, e^{-i\delta t} + c_{+}\, e^{+i\delta t}.$$

Equation (14.36) shows that the output field consists of three terms, which when multiplied by $e^{-i\omega_{\text{l}}t}$ indicates that the first term,

$$(E_{\text{l}} - \sqrt{2\kappa}\, c_{\text{s}})\, e^{-i\omega_{\text{l}}t},$$

corresponds to the output field at driving field frequency $\omega_{\text{l}}$. The second term,

$$(E_{\text{p}} - \sqrt{2\kappa}\, c_{-})\, e^{-i\omega_{\text{p}}t},$$

corresponds to the probe field with frequency $\omega_{\text{p}}$, while the third term,

$$\sqrt{2\kappa}\, c_{+}\, e^{-i(2\omega_{\text{l}}-\omega_{\text{p}})t},$$

corresponds to the output field with frequency $2\omega_{\text{l}} - \omega_{\text{p}}$, which is related to the Stoke field. The second term represents the mechanically induced transparency, while the third term shows the four-wave mixing process. In four-wave mixing two photons of a driving field interfere with a single photon of a probe field, each with frequencies $\omega_{\text{l}}$ and $\omega_{\text{p}}$, respectively, leading to the creation of a field of frequency, $2\omega_{\text{l}} - \omega_{\text{p}}$. This shows the non-linear behaviour of the system [3–5].

Mechanically induced transparency is shown in figure 14.4 by plotting transmission of the probe field as a function of normalized probe detuning $\delta/\omega_{\text{m}}$. The transmission coefficient of the probe field is defined as a ratio between the returned probe field from the coupled system and the sent probe field, that is,

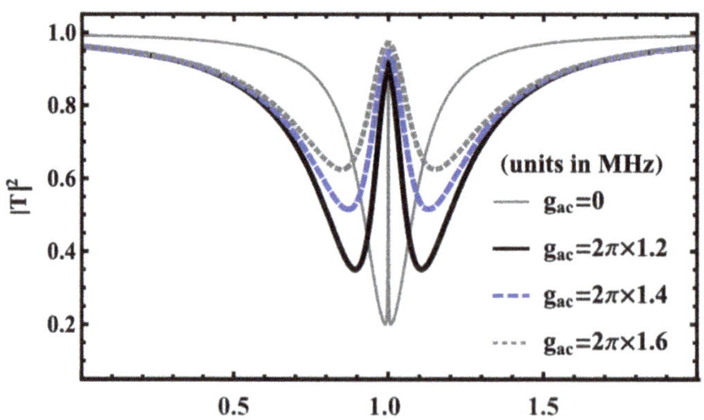


**Figure 14.4.** The transmission $|T|^2$ of the probe field as a function of normalized probe detuning $\delta/\omega_{\text{m}}$ shows induced transparency in a hybrid system. The atom–field coupling $g_{\text{ac}} = 0$ for the thin gray curve, and for the rest $g_{\text{ac}}/2\pi = 1.2$, 1.4, and 1.6 MHz. For a detailed description, see references [6] and [7]. Reprinted figure with permission from [7], Copyright (2015) by the American Physical Society.

$$T = \frac{E_{\rm p} - \sqrt{2\kappa}\, c_-}{E_{\rm p}} = 1 - \frac{\sqrt{2\kappa}\, c_-}{E_{\rm p}}. \tag{14.37}$$

## 14.5 Super-luminality and sub-luminality

In nano-electro-opto-mechanical systems the driven field frequency brings rapid changes in the phase of the transmission field. The transmitted output probe field is

$$TE_{\rm p}{\rm e}^{-i(\omega_{\rm p}t - \phi_{\rm t})}.$$

The phase, $\phi_{\rm t}(\omega_{\rm p})$, causes a transmission group delay, $\tau_{\rm g}$, which is obtained by expanding the phase up to first order around the average frequency, $\bar{\omega}$, that is,

$$\phi_{\rm t}(\omega_{\rm p}) \approx \phi_{\rm t}(\bar{\omega}) + (\omega_{\rm p} - \bar{\omega})\frac{\partial \phi_{\rm t}}{\partial \omega_{\rm p}}\bigg|_{\bar{\omega}},$$

Hence, we write

$$\tau_{\rm g} = \frac{\partial \phi_{\rm t}(\omega_{\rm p})}{\partial \omega_{\rm p}}\bigg|_{\bar{\omega}}, \tag{14.38}$$

and express the transmitted probe field as

$$TE_{\rm p}{\rm e}^{-i\omega_{\rm p}(t-\tau_{\rm g})},$$

together with a constant phase. This implies that the transmitted probe pulses peak at $t = \tau_{\rm g}$. The positive group delay $\tau_{\rm g} > 0$ corresponds to slow light and negative group delay $\tau_{\rm g} < 0$ corresponds to fast light, respectively (see figure 14.5). In addition

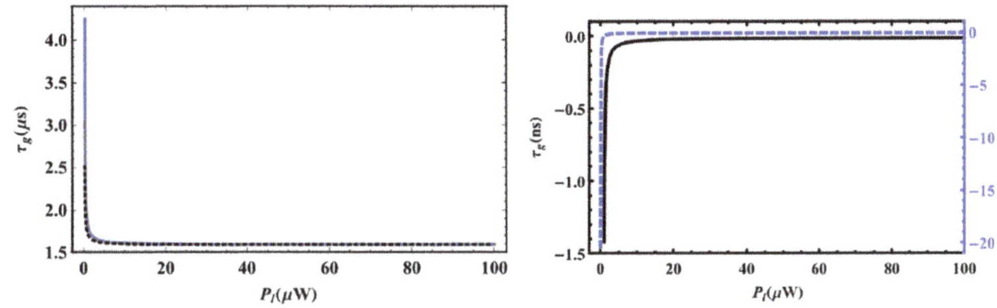


**Figure 14.5.** (a) Group delay ($\tau_{\rm g}$) as a function of the pump power ($P_{\rm l}$) for the case of $\Delta_a = -\omega_{\rm m}$, indicating the slow-light effect of the transmitted electromagnetic probe field for different values of the cavity decay rate $\kappa$, such that $\kappa/2\pi = 107.5$, 215, and 322.5 kHz, for blue solid, gray dashed, and black dotted curves, respectively. (b) Group delay ($\tau_{\rm g}$) as a function of the pump power ($P_{\rm l}$) for the case of $\Delta_a = +\omega_{\rm m}$, indicating the fast-light effect of the transmitted electromagnetic probe field for different values of the atom–field coupling $g_{\rm ac}$. The black solid and blue dashed curves correspond to the pulse advancement for $g_{\rm ac}/2\pi = 4$ and 8 MHz, respectively. For a detailed description, see reference [7]. Reprinted figure with permission from [7], Copyright (2015) by the American Physical Society.

to slow and fast light, new mechanically induced transparency-based applications include quantum memory efficiency enhancement, optical transistors [8], and photon counters and optical switches [9].

## Exercises

1. Show that an opto-mechanical system in the presence of ultra-cold atoms can be expressed by coupled differential equations, as follows:

$$\frac{d^2\hat{q}}{dt^2} = -\omega_{\rm m}^2\hat{q} - \frac{(\hbar\xi/m)\varepsilon_{\rm L}^2}{\kappa^2 + (\triangle + \xi\hat{q} - \xi_{\rm sm}\hat{Q})^2},$$

$$\frac{d^2\hat{Q}}{dt^2} = -\Omega^2\hat{Q} - \frac{\Omega\xi_{\rm sm}\varepsilon_{\rm L}^2}{\kappa^2 + (\triangle + \xi\hat{q} - \xi_{\rm sm}\hat{Q})^2}.$$

   Here, the mirror is expressed by $q$ and the position of the atoms is expressed by $Q$.
2. Following the treatment developed in section 13.3, show that the above system shows bistability in the presence of a BEC, as well.
3. Show that in the limiting case of a weak atom–field coupling, that is, $\xi_{\rm sm} \ll \xi$, we may express the hybrid opto-mechanical system as an explicitly time-dependent Hamiltonian, as given in equation (14.15).
4. Show that in the limiting case of a weak mirror–field coupling, that is, $\xi \ll \xi_{\rm sm}$, we may express the hybrid opto-mechanical system as an explicitly time-dependent Hamiltonian.

## References

[1] Yasir K A, Ayub M and Saif F 2014 Exponential Localization of moving-end mirror in optomechanics *J. Mod. Opt.* **61** 1318–23

[2] Ayub M, Yasir K A and Saif F 2014 Dynamical localization of matter waves in optomechanics *Int. J. Laser Phys.* **24** 115503

[3] Zhang H L, Saif F, Jiao Y and Jing H 2018 Loss induced transparency in opto-mechanics *Opt. Express* **26** 25199–210

[4] Lu T, Jiao Y, Zhang H, Saif F and Jing H 2019 Selective and switchable optical amplification with mechanical driven oscillators *Phys. Rev.* A **100** 013813

[5] Yamakoshi T and Watanabe S 2022 Dual-channel scattering problem in the cavity-like potential *Eur. J. Phys.* **43** 035401

[6] Akram M J, Ghafoor F and Saif F 2015 Electromagnetically induced transparency and tunable fano resonances in hybrid optomechanics *J. Phys. B: At. Mol. Opt. Phys.* **48** 065502

[7] Akram M J, Khan M M and Saif F 2015 Tunable fast and slow light in a hybrid optomechanical system *Phys. Rev.* A **92** 023846

[8] Kokab H, Siddiqui I A, A. Awan Z, Ghafoor F and Saif F 2023 Mechanically controlled quantum memory efficiency and optical transistor *Quantum Inf. Process.* **22** 88

[9] Fleischhauer M, Imamoglu A and Marangos J P 2005 Electromagnetically induced transparency: Optics in coherent media *Rev. Mod. Phys.* **77** 633–73

**IOP** Publishing

Optical Forces on Atoms

**Farhan Saif and Shinichi Watanabe**

# Epilogue

*'Whatever the final laws of nature may be, there is no reason to suppose that they are designed to make physicists happy.'*

—Steven Weinberg

***Optical Forces on Atoms*** explores the exciting world created by the latest inventions and innovations in quantum technology. The interactions of electromagnetic fields with atoms are essential in the implementation of cold-atom systems, including trapped ions, degenerate Fermi gases, Rydberg excitation blockade, ultracold molecules, among others. This book lays the foundations, therefore, of a better understanding of the concepts and tools that are emerging in quantum technologies.

**Quantum computing based on ultra-cold atoms** is an important area of research. Qubits and logic gates are being developed using atoms and optical fields. In quantum computing, the superposition of states ensures 'parallel processing'. Moreover, the unitary transformations at work imply reversibility. Thanks to its ability to perform parallel processing, quantum coherence may eventually produce quantum computers that are orders of magnitude faster than even the fastest supercomputers today. At the same time, it may also achieve much lower energy consumption. The phrases 'Quantum Advantages' and 'Quantum Supremacy' have been coined to describe these benefits. Quantum computers based on ultra-cold atoms may significantly increase efficiency in solving complicated contemporary issues such as drug development, investment strategies, dealing with encryption, and route scheduling for large vehicle fleets. The enormous computation power and low energy consumption make quantum computers the best candidates to address these complicated problems.

Ultra-cold atoms are a key component in the development of quantum devices such as **quantum clocks** (see Figure EP 1). Modern systems of time-keeping and global satellite navigation (such as GPS, BDS, Galileo, GLONASS, IRNSS, and QZSS) rely heavily on atomic clocks. Atomic clocks also enable deep-space surveys

doi:10.1088/978-0-7503-2308-6ch15  

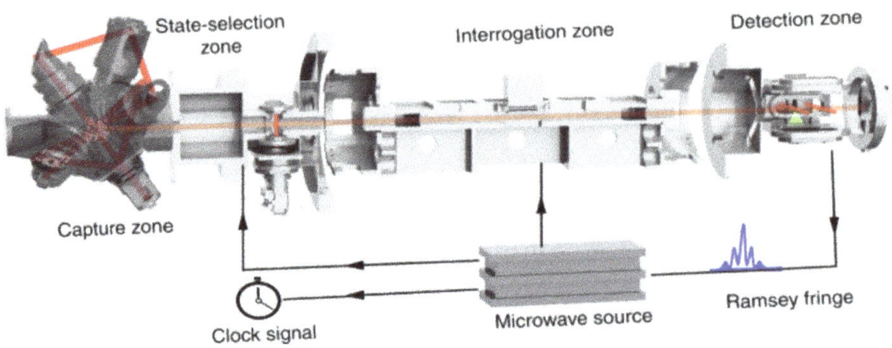


**Figure EP 1.** Principle and structure of a cold-atom clock. The capture zone is a magneto-optical trap (MOT) with a folded beam design. The ring interrogation cavity is used to enable the microwave field to study the cold atoms. The basis of the clock is the frequency standard. The atomic clock relies on the unperturbed atom's invariant frequency between a pair of states. The precision of the frequency measurement is enhanced by an ingenious use of interference fringes (Ramsey fringes) induced by a pair of cavities (whose actions are temporally separated). This is somewhat analogous to the interference fringes in Young's double-slit experiment, which gives a relationship between the fringe's spatial separation and the optical wavelength [1]. Copyright © 2018, The Author(s). With permission of Springer.

and more accurate tests of fundamental physics. However, traditional atomic clocks, which use hot atoms, have largely reached the limits of their long-term stability. Recently, a variety of cold-atom clocks, such as cesium atom fountain clocks, have been demonstrated both on the ground and in satellites. Ultra-cold-atom-based quantum gravimeters, gyroscopes, and interferometers are all helpful in achieving highly accurate and minutely sensitive measurement of parameters such as gravitational acceleration, angular velocity, and phase. This is laying the foundations of a new **quantum metrology**.

**Quantum gravimeters** use interferometry techniques based on cold atoms (see Figure EP 2). This involves rubidium atoms being first captured in a magneto-optical trap. They are then cooled to microkelvin temperatures before being released. During their free fall, they experience a sequence of three laser pulses that split and then recombine the atomic wave packets. Thus, we obtain a Mach–Zehnder interferometer in momentum space. Finally, the phase difference between the two arms of the interferometer is shown to be proportional to $g$. This is deduced from the measurement of the atomic state at the output of the interferometer (see figure EP 2).

**Position sensing at the nano scale** is performed by using optical microscopes to see, surface probe microscopes to sense, and electron microscopes to create images. Atom–field interaction can lead to images captured by optical microscopes being enhanced to a nano scale. For example, a stimulated emission depletion (STED) microscope can detect the difference in time between stimulated and spontaneous emissions of light from an excited atom on a probe surface and, from this, identify the surface structure. In this way, STED microscopes have attained a lateral resolution of 70–90 nm while examining a human protein (see Figure EP 3). Another study using a STED microscope measured the lengths of DNA fragments

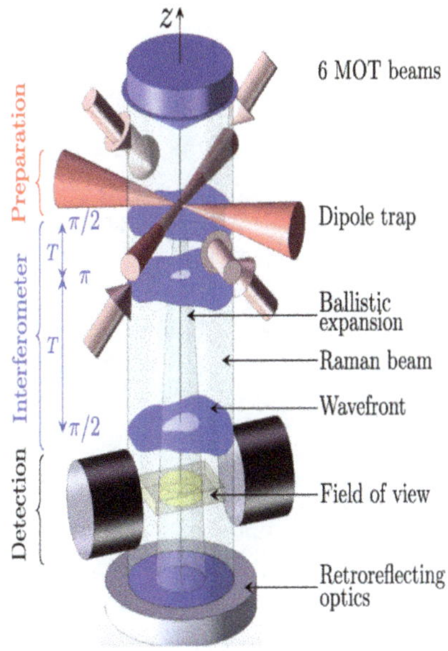


**Figure EP 2.** Principle and structure of an ultra-cold-atom-based gravimeter. Scheme of the experimental setup illustrates the effect of wavefront aberrations. Due to their ballistic expansion across the Raman beam, the ultra-cold atoms sample different parasitic phase shifts at the three $\pi/2$–$\pi$–$\pi/2$ Raman pulses due to wavefront distortions (displayed in blue as a distorted surface). This leads to a bias in the gravity measurement, resulting from the average of the effect over all atomic trajectories, filtered by finite-size effects, such as those related to the waist and clear aperture of the Raman beam and to the finite field of view at the detection [2, 4]. Reproduced from [4]. © IOP Publishing Ltd. CC BY 3.0.

far more accurately than conventional techniques using confocal microscopy. Surface scanning microscopes using ultra-cold atoms, such as recurrence tracking microscopes, are able to sense surface structures down to nanometer resolutions, overcoming difficulties encountered in the past with other techniques such as scanning tunneling and atomic force microscopes. These techniques are likely to have wide applications across a number of fields, ranging from biology to the material sciences.

**Atomtronics** promises to realize atomic circuit architectures using ultra-cold atoms. Atoms are manipulated in versatile micro-optical circuits generated by electromagnetic fields of different shapes and intensities or on micro-magnetic circuits, namely atom chips (see Figure EP 4). Atomtronics seeks to realize mesoscopic physics using cold atoms as constituents of matter flow in the same way that conventional electronics uses electrons as constituents of charge flow. Other fascinating areas of research in atomtronics include atom-based spintronics and quantum electronic structures like Josephson junction–based circuits.

The book will also be helpful in the study of ultra-cold-atom dynamics in **atom chips**, a versatile and compact tool used for generating magnetic fields for studying the quantum properties of cold atoms. An atom chip consists of one layer of mesoscopic copper wire structures and two layers of microscopic wire structures (see Figure EP 4). The cold atoms can be trapped close to the atom-chip surface at a

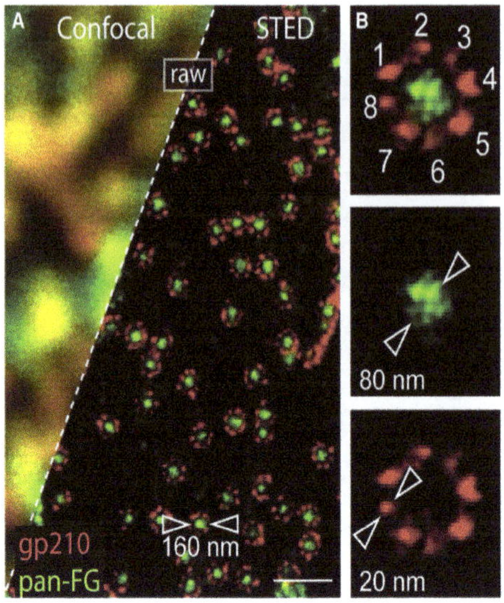


**Figure EP 3.** Comparison of confocal microscopy and STED microscopy, showing the improved resolution of a STED microscope compared with a traditional technique of a confocal microscope: this particular case, fluorescence nanoscopy of protein complexes with a compact near-infrared nanosecond-pulsed STED microscope. (A) STED reveals immunolabeled subunits in amphibian nuclear pore complex (NPC); raw data smoothed with a Gaussian filter extending over 14 nm in FWHM. The diameter of the octameric gp210 ring is established as 160 nm (scale bar, 500 nm). (B) Individual NPC image showing eight antibody-labeled gp210 homodimers as 20–40 nm-sized units and a 80 nm-sized localization of the subunits in the central channel. Reprinted from [5]. Copyright (2013), with permission from the Biophysical Society. Published by Elsevier Inc.

distance anywhere between a couple of hundred micrometers to a few millimeters. With this, it is possible to achieve high trap frequencies (up to 2 kHz) with just a few amperes of current. This makes the evaporative cooling, discussed in chapter 3, fast and power efficient. In addition, most structures on the chip are made from single wires, which results in much lower inductance than in a coil. In order to create the quadrupole fields required to operate a magneto-optic trap or Ioffe–Pritchard-type magnetic traps, the field created by the atom chip itself needs to be overlapped with homogeneous bias fields.

Richard Feynman introduced the idea of quantum emulators, based on the universality of quantum mechanics. His idea was to use one controllable quantum device to simulate the other system of interest and thereby avoid the difficulty of simulating a quantum system with classical computers. Ultra-cold degenerate atomic gases placed in periodic optical lattices can simulate ideal solid-state devices, without impurities and defects. An exciting prospect of new discoveries thus opens up as we explore the physics of **ideal solid-state many-body effects** with new optical laboratory experiments.

Ultra-cold atoms in optical lattices, in the presence of atom–atom interactions, lead to non-linear atom optics. Atom lasers based on Bose–Einstein condensation are an example of quantum coherence in which all the constituent atoms behave

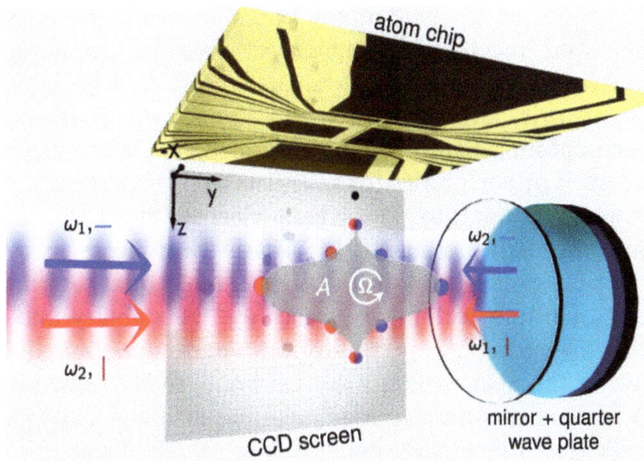


**Figure EP 4.** The atom-chip technology at the heart of an atomtronics experiment: the twin optical lattice is formed by retroreflecting light featuring two frequencies with linear orthogonal polarization. A quarter-wave plate in front of the retroreflector alters the polarization to generate two counter-propagating lattices (indicated in red and blue). After release from the atom chip and state preparation, the Bose–Einstein condensate is symmetrically split and recombined by the lattices driving double Bragg diffraction and Bloch oscillations. In this way, the interferometer arms form a Sagnac loop enclosing an area A (shaded in gray) for detecting rotations Ω. The interferometer output ports are detected on a CCD chip by absorption imaging [6]. Copyright © 2021, The Author(s). With permission of Springer.

identically. In this, a continuous cooling technique applies different cooling schemes separated in space and time. The atoms are cooled as they move toward the place where the atom laser is created. A continuously existing Bose–Einstein condensate thus generates a **continuous atom laser** as a coherent matter wave. This landmark achievement has kick-started a new era in many-body physics, with researchers now learning to control the strength of atomic interactions over wide ranges in these cold atomic gases.

As this book explains, the presence of periodic external forcing leads to greater degrees of freedom in a system as well as additional controlling parameters. Optical potentials, in the presence of external periodic modulations, have important applications in explaining the quantum dynamics of classically chaotic systems. Moreover, these controllable quantum chaotic systems play an essential role in the field of **quantum simulation**. For instance, in driven systems, it is possible to engineer effective magnetic fields enabling solid-state phenomena such as the Hall effect and fractional Hall effect to be studied. Effective magnetic fields are generated by using external modulation to alter the hopping potential. The hopping potential becomes complex-valued and thus acquires phase factors that correspond to Aharonov–Bohm phases. This technique is also known as **Floquet engineering**. In this way, generating effective magnetic fields makes it possible to study effective gauge fields. In addition, periodically driven optical lattices provide a control over the strength of tunneling in lattice sites that simulates Mott insulator–superfluid transitions.

This book also covers matter–field interaction in macroscopic systems that couple optical, electrical, and mechanical degrees of freedom in nano-scale devices. Macroscopic systems are at the focus of current research, with nano-devices being developed in laboratories around the world. **Nano-optical fibers**, as shown in Figure EP 5, **nano-opto-mechanical systems**, and **nano-electro-mechanical systems** all offer unprecedented opportunities for scientists to control the flow of light at high speed but low power consumption. These **nano-electro-opto-mechanical systems** also have the potential for highly efficient, low-noise transducers to operate between microwave and optical signals in both the classical and quantum domains.

These emerging quantum technologies may use reversibility and energy saving, which are generic properties of quantum computers, and link these to artificial intelligence. Thus, in the near future, what we know today as artificial intelligence may evolve into its next manifestation as **quantum artificial intelligence**. This shall enable us to handle much larger amounts of data than at present.

Ultra-cold atoms in optical lattices enable us to control not only the optically defined landscape of the potential energy surface in which the atoms reside, but also the strength of interaction between the atoms and the optical potential. In addition, even at temperatures in the millikelvin range, we are now able to control the quantum dynamics of single electrons on a quantum dot, a one-dimensional

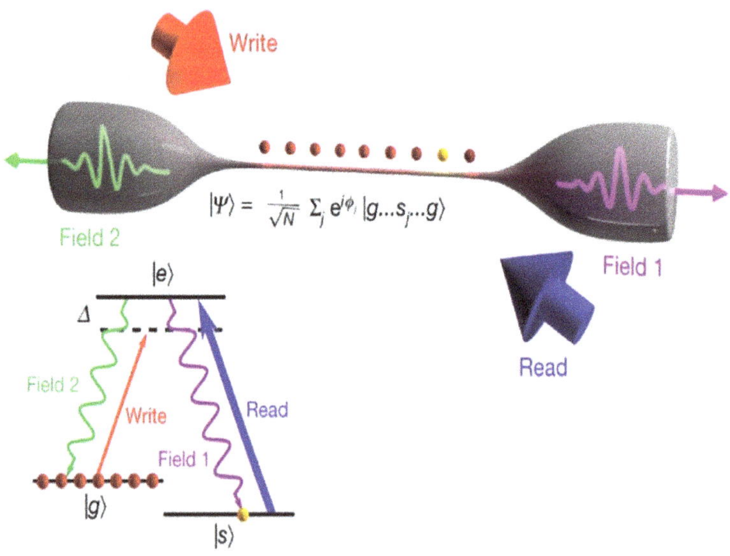


**Figure EP 5.** Using arrays of cold cesium atoms around a nano-fiber, the first wired entangled state of atoms is reported, with a capability to read this quantum state as a guided single photon. Via an external write pulse on the $|g\rangle \rightarrow |e\rangle$ transition, a collective entangled state with one 'spin-flip' excitation is created in a chain of $N$ individual cold atoms trapped in the evanescent field of a nano-scale waveguide. This process is heralded by the detection of a photon in the guided Field-1 mode. At a later time, a read pulse on the $|s\rangle \rightarrow |e\rangle$ transition converts the excitation into a single photon in the guided Field-2 mode. This scheme offers a fiber-addressable single collective excitation $|\Psi\rangle$ in an atomic register with on-demand readout [7]. Copyright © 2019, The Author(s), under exclusive licence to Springer Nature Limited. With permission of Springer.

quantum wire, or single Cooper pairs or flux quanta in the case of superconductors. Combining such solid-state devices and detectors with cold atoms could lead to the creation of novel architectures in the field of **quantum devices**.

Indeed, we may be on the verge of a *quantum leap* into a future of technologies enabling intelligent devices equipped with almost human-like decision-making capabilities. Roger Penrose explored these ideas in his book, *The Emperor's New Mind*, observing that humans may not have followed logically unique procedures in the way a Turing machine does [3]. However, there is no strong evidence—at least, not yet—that quantum mechanics can achieve human-like abilities.

The unprecedented level of control we have now achieved over macroscopic quantum systems and emerging quantum technologies puts us on the threshold of a second quantum revolution. The first such revolution saw us progress from lasers to medical imaging and computers, with impacts across our everyday lives. The second revolution will give us far greater control over macroscopic quantum systems based on ultra-cold atoms. As we have seen in this book, these quantum systems will have applications in atomtronics as well as in quantum simulators, information, computation, and metrology. We can look forward to a new era of miniaturized and portable devices which will be more intelligent, more energy efficient and more commercially useful than anything we have seen before. Companies such as Bosch, Siemens, Thales, Safran, ASML, Nokia, Airbus and Alcatel Lucent are already bringing quantum products to the market.

## References

[1] Liu L, Lü D S and Chen W B *et al* 2018 In-orbit operation of an atomic clock based on laser-cooled $^{87}$Rb atoms *Nat. Commun.* **9** 2760

[2] Gillot P, Cheng B, Imanaliev A, Merlet S and Pereira Dos Santos F 2016 The LNE-SYRTE cold atom gravimeter *2016 European Frequency and Time Forum (EFTF) (York, United Kingdom, 4–7 April 2016)* pp 1–3

[3] Penrose R 1989 The emperor's new mind: concerning computers *Minds and the Laws of Physics* (Oxford: Oxford University Press)

[4] Karcher R, Imanaliev A, Merlet S and Pereira Dos Santos F 2018 Improving the accuracy of atom interferometers with ultracold sources *New J. Phys.* **20** 113041

[5] Göttfert F, Wurm C A, Mueller V, Berning S, Cordes V C, Honigmann A and Hell S W 2013 Coaligned dual-channel STED Nanoscopy and molecular diffusion analysis at 20 nm resolution *Biophys. J.* **105** L01–3

[6] Gebbe M, Siemss J N and Gersemann M *et al* 2021 Twin-lattice atom interferometry *Nat. Commun.* **12** 2544

[7] Corzo N V, Raskop J and Chandra A *et al* 2019 Waveguide-coupled single collective excitation of atomic arrays *Nature* **566** 359–62

# Appendix A

## Unitary operators

The interaction of an atom with an electromagnetic field modifies the probability amplitudes, whereas the state vectors remain the same. This leads us to make an ansatz about the wave function of the atom–field system after an evolution time $t$, that is

$$|\psi(t)\rangle = \sum_{n=0}^{\infty} \{c_{\mathrm{e},n}(t)|e, n\rangle + c_{\mathrm{g},n}(t)|g, n\rangle\},$$

where $c_{\mathrm{e},n}(t)$ and $c_{\mathrm{g},n}(t)$ are unknown probability amplitudes. The time-evolved wave function $|\psi(t)\rangle$ has important information about the interacting system, which can be obtained by applying a time transformation operator, $\hat{U}$, on the initial state, $|\psi(t = 0)\rangle$, such that

$$|\psi(t)\rangle = \hat{U}|\psi(0)\rangle. \tag{A.1}$$

The unitary operator $\hat{U}$ is defined as

$$\hat{U} = \exp\left(\frac{-i\hat{V}t}{\hbar}\right).$$

It is straightforward to express the unitary operator in terms of real and imaginary parts, that is,

$$\hat{U} = \exp\left(\frac{-i\hat{V}t}{\hbar}\right) = \sum_{j=0}^{\infty} \frac{(-i\hat{V}t/\hbar)^j}{j!} = \hat{\alpha} + i\hat{\beta}. \tag{A.2}$$

As expressed in equation (3.40), the interaction Hamiltonian is described as

$$\hat{V} = \hbar\mu(\hat{\sigma}_+\hat{a}\mathrm{e}^{i\Delta t} + \hat{\sigma}_-\hat{a}^\dagger\mathrm{e}^{-i\Delta t}). \tag{A.3}$$

doi:10.1088/978-0-7503-2308-6ch16  

We consider the ground state $|g\rangle$ and excited state $|e\rangle$ in matrix form, that is,

$$|g\rangle = \begin{pmatrix} 0 \\ 1 \end{pmatrix},$$

and

$$|e\rangle = \begin{pmatrix} 1 \\ 0 \end{pmatrix},$$

respectively. This leads us to express $\sigma$ matrices as

$$\hat{\sigma}_- = |g\rangle\langle e| = \begin{pmatrix} 0 & 0 \\ 1 & 0 \end{pmatrix},$$

and

$$\hat{\sigma}_+ = |e\rangle\langle g| = \begin{pmatrix} 0 & 1 \\ 0 & 0 \end{pmatrix}.$$

Moreover, we obtain

$$\hat{\sigma}_-\hat{\sigma}_- = \hat{\sigma}_+\hat{\sigma}_+ = 0,$$

$$\hat{\sigma}_-\hat{\sigma}_+ = |g\rangle\langle g| = \begin{pmatrix} 0 & 0 \\ 0 & 1 \end{pmatrix},$$

$$\hat{\sigma}_+\hat{\sigma}_- = |e\rangle\langle e| = \begin{pmatrix} 1 & 0 \\ 0 & 0 \end{pmatrix}.$$

The identities help us to write

$$(\hat{\sigma}_-\hat{a}^\dagger \mathrm{e}^{-i\Delta t} + \hat{\sigma}_+\hat{a}\mathrm{e}^{i\Delta t})^2 = \hat{\sigma}_-\hat{\sigma}_+\hat{a}^\dagger\hat{a} + \hat{\sigma}_+\hat{\sigma}_-\hat{a}\hat{a}^\dagger, \tag{A.4}$$

$$= \begin{pmatrix} \hat{a}\hat{a}^\dagger & 0 \\ 0 & \hat{a}^\dagger\hat{a} \end{pmatrix}, \tag{A.5}$$

$$= \begin{pmatrix} \left(\sqrt{\hat{a}\hat{a}^\dagger}\right)^2 & 0 \\ 0 & \left(\sqrt{\hat{a}^\dagger\hat{a}}\right)^2 \end{pmatrix}. \tag{A.6}$$

Similarly,

$$(\hat{\sigma}_-\hat{a}^\dagger \mathrm{e}^{-i\Delta t} + \hat{\sigma}_+\hat{a}\mathrm{e}^{i\Delta t})^{2n} = \begin{pmatrix} \left(\sqrt{\hat{a}\hat{a}^\dagger}\right)^{2n} & 0 \\ 0 & \left(\sqrt{\hat{a}^\dagger\hat{a}}\right)^{2n} \end{pmatrix}. \tag{A.7}$$

Substituting equation (A.7) in the series expansion of the real part of the unitary operator, $\hat{\alpha}$, we obtain

$$\hat{\alpha} = \sum_{j=0}^{\infty} \frac{(-i\hat{V}t/\hbar)^{2j}}{(2j)!} = \begin{pmatrix} \cos(\mu t\sqrt{\hat{a}^\dagger\hat{a}+1}) & 0 \\ 0 & \cos(\mu t\sqrt{\hat{a}^\dagger\hat{a}}) \end{pmatrix}. \tag{A.8}$$

Here, we have applied the commutation relation between the operators $\hat{a}$ and $\hat{a}^\dagger$, that is, $[\hat{a}, \hat{a}^\dagger] = 1$.

Similarly, we calculate the imaginary part of the unitary operator, $\hat{\beta}$, that is,

$$\hat{\beta} = \sum_{j=0}^{\infty} \frac{(-i\hat{V}t/\hbar)^{2j+1}}{(2j+1)!}. \tag{A.9}$$

The operator identity in the series expansion is then simplified as we write

$$\begin{aligned}(\hat{\sigma}_-\hat{a}^\dagger e^{-i\Delta t} + \hat{\sigma}_+\hat{a}e^{i\Delta t})^{2n+1} &= (\hat{\sigma}_-\hat{a}^\dagger e^{-i\Delta t} + \hat{\sigma}_+\hat{a}e^{i\Delta t})^{2n}(\hat{\sigma}_-\hat{a}^\dagger e^{-i\Delta t} + \hat{\sigma}_+\hat{a}e^{i\Delta t}), \\ &= \begin{pmatrix} \left(\sqrt{\hat{a}\hat{a}^\dagger}\right)^{2n} & 0 \\ 0 & \left(\sqrt{\hat{a}^\dagger\hat{a}}\right)^{2n} \end{pmatrix} \begin{pmatrix} 0 & ae^{i\Delta t} \\ a^\dagger e^{-i\Delta t} & 0 \end{pmatrix},\end{aligned} \tag{A.10}$$

$$= \begin{pmatrix} 0 & \left(\sqrt{\hat{a}\hat{a}^\dagger}\right)^{2n} ae^{i\Delta t} \\ \left(\sqrt{\hat{a}^\dagger\hat{a}}\right)^{2n} a^\dagger e^{-i\Delta t} & 0 \end{pmatrix}, \tag{A.11}$$

$$= \begin{pmatrix} 0 & \dfrac{\left(\sqrt{\hat{a}\hat{a}^\dagger}\right)^{2n+1}}{\sqrt{\hat{a}\hat{a}^\dagger}}\hat{a}e^{i\Delta t} \\ \dfrac{\left(\sqrt{\hat{a}^\dagger\hat{a}}\right)^{2n+1}}{\sqrt{\hat{a}^\dagger\hat{a}}}\hat{a}^\dagger e^{-i\Delta t} & 0 \end{pmatrix}. \tag{A.12}$$

Hence, we obtain

$$\hat{\beta} = \begin{pmatrix} 0 & \dfrac{\sin(\mu t\sqrt{\hat{a}^\dagger\hat{a}+1})}{\sqrt{\hat{a}^\dagger\hat{a}+1}}\hat{a}e^{i\Delta t} \\ \dfrac{\sin(\mu t\sqrt{\hat{a}^\dagger\hat{a}})}{\sqrt{\hat{a}^\dagger\hat{a}}}\hat{a}^\dagger e^{-i\Delta t} & 0 \end{pmatrix}. \tag{A.13}$$

# Appendix B

## Representations and transformations

### Wannier–Stark states in quasi-momentum representation

With the help of equation (6.38), we express the Wannier–Stark states in the Hilbert space spanned by the quasi-momentum states as

$$\begin{aligned} |\Psi_m\rangle &= \int_{-\pi/a}^{+\pi/a} dk|k\rangle\langle k|\psi_m\rangle, \\ &= \sqrt{\frac{d}{2\pi}} \int_{-\pi/a}^{+\pi/a} dk \mathrm{e}^{-i(mkd+\gamma \sin kd)}|k\rangle. \end{aligned} \tag{B.1}$$

### Wannier–Stark states in Wannier state representation

The Wannier–Stark states can be expressed in the vector space spanned by the Wannier states as the basis vectors. Therefore, we write

$$\begin{aligned} |\Psi_m\rangle &= \sum_n |n\rangle\langle n|\Psi_m\rangle \\ &= \sum_n |n\rangle \int dk\langle n|k\rangle\langle k|\Psi_m\rangle \\ &= \sum_n |n\rangle \frac{d}{2\pi} \int dk \mathrm{e}^{-i[(n-m)kd-\gamma \sin kd]} \\ &= \sum_n J_{n-m}(\gamma)|n\rangle. \end{aligned} \tag{B.2}$$

In obtaining the expression, we have considered the Bessel function identity:

$$J_{n-m}(\mu) = \frac{1}{2\pi} \int du \ \mathrm{e}^{-i[(n-m)u-\mu \sin u]}. \tag{B.3}$$

doi:10.1088/978-0-7503-2308-6ch17  

## Time evolution in quasi-momentum space

The wave packet evolution in Wannier–Stark states is uniquely different. The propagator responsible for the evolution in the time domain,

$$\hat{U}(t) = \mathrm{e}^{-i\hat{H}t/\hbar},$$

is described in quasi-momentum space as

$$\begin{aligned}
\hat{U}(t) &= \int \int dk\, dk' |k'\rangle\langle k'| \mathrm{e}^{-i\hat{H}t/\hbar}|k\rangle\langle k|,\\
&= \sum_m \int \int dk\, dk' \langle k'| \mathrm{e}^{-i\hat{H}t/\hbar}|\Psi_m\rangle\langle\Psi_m|k\rangle|k'\rangle\langle k|,\\
&= \sum_m \int \int dk\, dk' \mathrm{e}^{-iE_m t/\hbar}\langle k'|\Psi_m\rangle\langle\Psi_m|k\rangle|k'\rangle\langle k|,\\
&= \sum_m \int \int dk\, dk' \mathrm{e}^{-i\gamma\{\sin k'd - \sin kd\}}\delta(k' - k + Ft/\hbar)|k'\rangle\langle k|.
\end{aligned} \tag{B.4}$$

Here, $\gamma = J/Fd$. This implies that the propagator takes the state $|k\rangle$ to $|k'\rangle$ following classical acceleration, as $\hbar k' = \hbar k - Ft$.

## Time evolution in Wannier space

In a similar way, we express the propagator using Wannier states as

$$\begin{aligned}
\hat{U}(t) &= \sum_n \sum_{n'} |n\rangle\langle n| \mathrm{e}^{-i\hat{H}t/\hbar}|n'\rangle\langle n'|,\\
&= \sum_m \sum_n \sum_{n'} \langle n| \mathrm{e}^{-i\hat{H}t/\hbar}|\Psi_m\rangle\langle\Psi_m|n'\rangle|n\rangle\langle n'|,\\
&= \sum_m \sum_n \sum_{n'} \mathrm{e}^{-i\hat{E}_m t/\hbar}\langle n|\Psi_m\rangle\langle\Psi_m|n'\rangle|n\rangle\langle n'|,\\
&= \sum_n \sum_{n'} U_{n,n'}(t)|n\rangle\langle n'|.
\end{aligned} \tag{B.5}$$

We simplify the expression using equation (B.2), and obtain

$$U_{n,n'}(t) = \sum_m \mathrm{e}^{-iE_m t/\hbar} J_{n-m}(\gamma) J^*_{n'-m}(\gamma). \tag{B.6}$$

Furthermore, substituting the integral expression presented in equation (B.3), and writing

$$\sum_m \mathrm{e}^{-im(u-u'+\omega_{\mathrm{B}}t)} = 2\pi\delta(u - u' + \omega_{\mathrm{B}}t),$$

we obtain

$$U_{n,n'}(t) = \frac{1}{2\pi} e^{-in\omega_B t} \int_{-\pi}^{+\pi} du \, e^{i(n-n')u} e^{-i\gamma(\sin(u-\omega_B t)-\sin u)}.$$

Here, $\omega_B = Fd/\hbar$ is the Bloch oscillation frequency. Applying a shift in the $u$-coordinate as $u + \omega_B t/2$ reveals

$$\begin{aligned} U_{n,n'}(t) &= \frac{1}{2\pi} e^{-in'\omega_B t} \int_{-\pi}^{+\pi} du e^{i(n-n')(u+\omega_B t/2)} \\ &\quad \exp\{-i\gamma(\sin(u-\omega_B t/2) - \sin(u+\omega_B t/2))\}, \\ &= \frac{1}{2\pi} e^{-in'\omega_B t} \int_{-\pi}^{+\pi} du e^{i(n-n')(u+\omega_B t/2)} \\ &\quad \exp\{-i\gamma(-2\cos u \sin \omega_B t/2)\}, \\ &= \frac{1}{2\pi} e^{-in'\omega_B t} \int_{-\pi}^{+\pi} du e^{i(n-n')(u+\pi/2+\omega_B t/2)} \\ &\quad \exp\{-i2\gamma \sin u \sin \omega_B t/2\}. \end{aligned} \tag{B.7}$$

In the last step, we introduced a constant shift $u = u + \pi/2$. Hence, we obtain

$$U_{n,n'}(t) = J_{n-n'}\left(2\gamma \sin \frac{\omega_B t}{2}\right) \exp\left\{i(n-n')\left(\frac{\pi}{2} + \frac{\omega_B t}{2}\right) - in'\omega_B t\right\}. \tag{B.8}$$

As a result, we write the evolution operator as

$$U(t) = \sum_n \sum_{n'} J_{n-n'}\left(2\gamma \sin \frac{\omega_B t}{2}\right) e^{i(n-n')\left(\frac{\pi}{2}-\frac{\omega_B t}{2}\right)-in'\omega_B t} |n\rangle\langle n'|. \tag{B.9}$$

**IOP** Publishing

Optical Forces on Atoms

**Farhan Saif and Shinichi Watanabe**

# Appendix C

## General solution of dispersion relation

In order to find a general solution for $\mathcal{E}(k)$, we consider

$$\mathcal{E}(k) = \frac{1}{T}\int_0^T [-J\cos(k(t')d) + F_1 x^2(t')]dt', \tag{C.1}$$

and solve the above equation by writing the integral expressions

$$I_1 = \frac{1}{T}\int_0^T (-J)\cos(k(t')d)dt', \tag{C.2}$$

$$I_2 = \frac{1}{T}\int_0^T F_1 x^2(t')dt'. \tag{C.3}$$

The general solution for the dispersion relation, $\mathcal{E}(k)$, is obtained by solving equations (C.2) and (C.3). Therefore, the first integral,

$$I_1 = -J\frac{1}{T}\int_0^T \cos\left[kd\cos(\eta_0\Omega t') - \frac{\epsilon V'(x)}{\hbar\Omega}\sin(\Omega t')\right]dt', \tag{C.4}$$

becomes

$$I_1 = -J\frac{1}{2\pi}\int_0^{2\pi} \cos\left[kd\cos(\eta_0\theta) - \frac{\epsilon V_0}{2\hbar\Omega}\sin(\theta)\cos\left(\frac{\hbar k^2\eta}{F_1}\sin(\eta_0\theta)\right)\right]d\theta. \tag{C.5}$$

The integral is further simplified by using the identities

$$\cos(A - B) = \cos A\cos B + \sin A\sin B, \tag{C.6}$$

 

and

$$\cos(\alpha \cos\theta) = \sum_m J_m(\alpha)\cos(m\theta + m\pi/2),$$
$$\sin(\alpha \cos\theta) = \sum_m J_m(\alpha)\sin(m\theta + m\pi/2),$$
$$\cos(\alpha \sin\theta) = \sum_{m'} J_{m'}(\alpha)\cos(m'\theta),$$
$$\sin(\alpha \sin\theta) = \sum_{m'} J_{m'}(\alpha)\sin(m'\theta).$$

Using these identities and the product-to-sum trigonometric relation, we obtain

$$I_1 = -J\frac{1}{2\pi}\int_0^{2\pi} d\theta \cos(kd\cos(\eta_0\theta))\cos\left[\epsilon'\sum_{m'} J_{m'}(\alpha)(\sin[(1+m'\eta_0)\theta] - \sin[(1-m'\eta_0)\theta])\right]$$
$$+ \sin(kd\cos(\eta_0\theta))\sin\left[\epsilon'\sum_{m'} J_{m'}(\alpha)(\sin[(1+m'\eta_0)\theta] - \sin[(1-m'\eta_0)\theta])\right],$$

where

$$\alpha = \frac{\hbar k^2 \eta}{F_1} \qquad \& \qquad \epsilon' = \frac{\epsilon V_0}{4\hbar\Omega}.$$

We introduce a mathematical simplicity by considering $\epsilon'$ as small; hence, we use the small-angle approximation,

$$\cos(\theta) \approx 1 \qquad \text{and} \qquad \sin(\theta) \approx \theta,$$

to obtain

$$I_1 = \frac{-J}{2\pi}\int_0^{2\pi} d\theta(\cos(kd\cos(\eta_0\theta))$$
$$+)\epsilon'\sum_{m'} J_{m'}(\alpha)\sin(kd\cos(\eta_0\theta))(\sin[(1+m'\eta_0)\theta] - \sin[(1-m'\eta_0)\theta])).$$

After a few mathematical steps, the detailed solution of the integral is obtained as

$$\begin{aligned} I_1 = \frac{-J}{2\pi}\Bigg[ & \sum_{m_1} J_{m_1}(kd)\frac{\sin(2\pi m_1\eta_0 + m_1\pi/2) - \sin(m_1\pi/2)}{m_1\eta_0} \\ & + \epsilon'\sum_{m'}\sum_{m_2} J_{m'}(\alpha)J_{m_2}(kd)\Bigg(\frac{\sin(2\pi((m_2 - m')\eta_0 - 1) + m_2\pi/2) - \sin(m_2\pi/2)}{(m_2 - m')\eta_0 - 1} \\ & - \frac{\sin(2\pi((m_2 + m')\eta_0 - 1) + m_2\pi/2) - \sin(m_2\pi/2)}{(m_2 + m')\eta_0 + 1} \\ & - \frac{\sin(2\pi((m_2 + m')\eta_0 - 1) + m_2\pi/2) - \sin(m_2\pi/2)}{(m_2 + m')\eta_0 - 1} \\ & + \frac{\sin(2\pi((m_2 - m')\eta_0 - 1) + m_2\pi/2) - \sin(m_2\pi/2)}{(m_2 - m')\eta_0 + 1}\Bigg)\Bigg]. \end{aligned} \tag{C.7}$$

The second integral, appearing in the expression for $\mathcal{E}(k)$, is expressed as

$$I_2 = \frac{F_1}{T}\int_0^T \left(\frac{\hbar k\eta}{2F_1}\sin(\eta_0\Omega t') + \frac{\epsilon V(x)}{2F_1 d}\cos(\Omega t')\right)^2 dt', \tag{C.8}$$

$$I_2 = \frac{F_1}{2\pi}\int_0^{2\pi} \left(\frac{\hbar k\eta}{2F_1}\sin(\eta_0\theta) + \frac{\epsilon V_0}{4F_1 d}\cos\left(\frac{\hbar k^2\eta}{F_1}\right)\sin(\eta_0\theta)\cos(\theta)\right)^2 d\theta, \tag{C.9}$$

$$\begin{aligned}
&= \frac{\hbar^2k^2\eta^2}{8F_1}\left(1 - \frac{\sin(4\pi\eta_0)}{4\pi\eta_0}\right) + \frac{\epsilon^2V_0^2}{128\pi F_1 d^2}\Bigg[2\pi + \frac{\sin(4\pi)}{2} \\
&+ \sum_{m'} J_{m'}\left(\frac{2\hbar k^2\eta}{F_1}\right)\left(\frac{\sin(2\pi m'\eta_0)}{m'\eta_0} + \frac{\sin(2\pi(m'\eta_0 + 2))}{(m'\eta_0 + 2)}\right. \\
&\left. + \frac{\sin(2\pi(m'\eta_0 - 2))}{(m'\eta_0 - 2)}\right) \\
&+ \frac{\hbar k\eta\epsilon V_0}{16\pi F_1 d}\sum_{m'} J_{m'}\left(\frac{\hbar k^2\eta_0}{F_1}\right)\left(\frac{\cos[(2\pi(m' + 1)\eta_0 + 1)] - 1}{(m' + 1)\eta_0 + 1}\right. \\
&\left. - \frac{\cos[(2\pi(m' - 1)\eta_0 - 1)] - 1}{(m' - 1)\eta_0 - 1}\right)\Bigg].
\end{aligned} \tag{C.10}$$

www.ingramcontent.com/pod-product-compliance
Ingram Content Group UK Ltd.
Pitfield, Milton Keynes, MK11 3LW, UK
UKHW051042120926
8556IPUK00030B/698

9 780750 323062